T0393748

GLOBAL ADVANCES IN SELENIUM RESEARCH FROM THEORY TO APPLICATION

PROCEEDINGS OF THE 4TH INTERNATIONAL CONFERENCE ON SELENIUM IN THE ENVIRONMENT AND HUMAN HEALTH, SÃO PAULO, BRAZIL, 18–21 OCTOBER 2015

Global Advances in Selenium Research from Theory to Application

Editors

Gary S. Bañuelos
US Department of Agriculture – Agricultural Research Service, Parlier, CA, USA

Zhi-Qing Lin
Southern Illinois University – Edwardsville, Edwardsville, IL, USA

Milton Ferreira de Moraes
Federal University of Mato Grosso, Barra do Garças, MT, Brazil

Luiz Roberto G. Guilherme
Federal University of Lavras, Lavras, MG, Brazil

André Rodrigues dos Reis
São Paulo State University, Ilha Solteira, SP, Brazil

CRC Press is an imprint of the
Taylor & Francis Group, an **informa** business

A BALKEMA BOOK

CRC Press/Balkema is an imprint of the Taylor & Francis Group, an informa business

© 2016 Taylor & Francis Group, London, UK

Typeset by MPS Limited, Chennai, India
Printed and bound in Great Britain by CPI Group (UK) Ltd, Croydon, CR0 4YY

All rights reserved. No part of this publication or the information contained herein may be reproduced, stored in a retrieval system, or transmitted in any form or by any means, electronic, mechanical, by photocopying, recording or otherwise, without written prior permission from the publishers.

Although all care is taken to ensure integrity and the quality of this publication and the information herein, no responsibility is assumed by the publishers nor the author for any damage to the property or persons as a result of operation or use of this publication and/or the information contained herein.

Published by: CRC Press/Balkema
P.O. Box 11320, 2301 EH Leiden, The Netherlands
e-mail: Pub.NL@taylorandfrancis.com
www.crcpress.com – www.taylorandfrancis.com

ISBN: 978-1-138-02731-2 (Hardback)
ISBN: 978-1-315-68756-8 (eBook PDF)

Global Advances in Selenium Research from Theory to Application – Bañuelos et al (Eds)
© 2016 Taylor & Francis Group, London, ISBN 978-1-138-02731-2

Table of contents

Preface   XI
List of contributors   XIII

Biogeochemistry of selenium

The global biogeochemical cycle of selenium: Sources, fluxes and the influence of climate   3
L.H.E. Winkel

Topsoil selenium distribution in relation to geochemical factors in main agricultural areas of China   5
T. Yu, Z.F. Yang, Q.Y. Hou, Y.Y. Lv, X.X. Xi & M. Li

Examining continental and marine sources of selenium in rainfall   7
T. Blazina, L.H.E. Winkel, A. Läderach, H. Wernli & J. Kirchner

The role of phytoplankton in marine selenium cycling   9
K.E. Luxem, B. Vriens, R. Behra & L.H.E. Winkel

Biogenic volatilization of nanoscale selenium particles in the soil-*Stanleya pinnata* system   11
J. Wang, R. Mahajan, L. Jones, Z.-Q. Lin & Y.H. Xie

Global predictions of selenium distributions in soils   13
G.D. Jones & L.H.E. Winkel

Selenium in agroecosystems in tropical areas: A focus in Brazil   15
L.R.G. Guilherme, G.S. Carvalho, E.C. Silva Júnior, L.B. Abreu, G.A. Souza & J.J. Marques

Selenium soil mapping under native Brazil nut forests in Brazilian Amazon   17
*K.D. Batista, K.E. da Silva, G.C. Martins, L.H.O. Wadt, L.M. da Silva, N.J.M. Júnior,
M.C. Guedes, R.C. de Oliveira Júnior, C.A.S. Magalhães & A.B.B. Tardini*

Fate of selenium in soil and engineered suboxic and anoxic environments   19
L. Charlet, B. Ma, A. Fernandez Martinez, R.M. Couture, M.R. Broadley & A.D.C. Chilimba

Soil selenium contents, spatial distribution and their influencing factors in Heilongjiang, China   21
Feng-qin Chi, En-jun Kuang, Jiu-ming Zhang, Qing-rui Su, Dan Wei & Qiang Xu

Selenium sorption in tropical agroecosystems   23
G. Lopes, L.R.G. Guilherme, A.M. Araujo & J.H.L. Lessa

Ionic strength effects upon selenate adsorption in cultivated and uncultivated Brazilian soils   25
A.M. Araujo, J.H.L. Lessa, G.A. Souza, L.R.G. Guilherme & G. Lopes

Soil cultivation affects selenate adsorption in Cerrado soils in Brazil   27
J.H.L. Lessa, A.M. Araujo, L.R.G. Guilherme & G. Lopes

The effect of calcination on selenium speciation in selenium-rich rock   29
H.Y. Zhang, M. Lui, Y. Xiao, Z.Y. Bao, C.H. Wei & X.L. Chen

The fraction of selenium in cumulated irrigated soil in Gansu Province, China: Effect of aging   31
J. Li, S.Y. Qin, P.Y. Feng, N. Man & D.L. Liang

Comparative study on the extraction methods and the bioavailability of soil available selenium   33
Z. Wang, S. Tu & D. Han

Partitioning of SeNPs in the water soluble and the exchangeable fractions and effects of soil
organic matter and incubation time   35
M.M. Rashid, Z.-Q. Lin & F. Kaniz

A rapid analytical method for selenium species by high performance liquid chromatography (HPLC) coupled with inductively coupled plasma mass spectrometry (ICP-MS)   37
H. Tian, Z.Y. Bao, X.L. Chen, C.H. Wei & Z.Y. Tang

Determination of selenium species in Se-enriched food supplement tablets by anion-exchange liquid chromatography-hydride generation-atomic fluorescence spectrometry   39
N. Zhang, L. Liu, W. Ren & S. Chen

Selenium speciation in plants by HPLC-ultraviolet treatment-hydride generation atomic fluorescence spectrometry using various mobile phases   41
D. Han, S. Xiong, S. Tu, Z. Xie, H. Li, M. Imtiaz, J. Zhou & D. Xing

Synchrotron studies of selenium interactions with heavy elements   43
I.J. Pickering, G.N. George, T.C. MacDonald, P.H. Krone & M. Korbas

Cellular and molecular functions of selenium

Selenium atom-specific modifications (SAM) of nucleic acids for human health   47
R. Abdur, W. Zhang, H. Sun & Z. Huang

Comparative cytotoxicity and antioxidant evaluation of biologically active fatty acid conjugates of water soluble selenolanes in cells   49
A. Kunwar, P. Verma, K.I. Priyadarsini, K. Arai & M. Iwaoka

Diselenodipropionic acid as novel selenium compound for lung radiotherapy   51
K.I. Priyadarsini, A. Kunwar, V.K. Jain, V. Gota & J. Goda

Association of selenoprotein and selenium pathway genetic variations with colorectal cancer risk and interaction with selenium status   53
D.J. Hughes, V. Fedirko, J.S. Jones, L. Schomburg, S. Hybsier, C. Méplan, J.E. Hesketh, E. Riboli & M. Jenab

Selenoenzymes iodothyronine deiodinases: 1. Effects of their activities in various rat tissues by administered antidepressant drug Fluoxetine   55
S. Pavelka

Selenoenzymes iodothyronine deiodinases: 2. Novel radiometric enzyme assays for extremely sensitive determination of their activities   57
S. Pavelka

The role of thioredoxin reductase 1 in cancer   59
E.S.J. Arnér

Selenoprotein T: A new promising target for the treatment of myocardial infarction and heart failure   61
I. Boukhalfa, N. Harouki, L. Nicol, A. Dumesnil, I. Rémy-Jouet, J.-P. Henry, C. Thuillez, A. Ouvrard-Pascaud, V. Richard, P. Mulder & Y. Anouar

Selenium status in humans: Measures, methods, and modifiers   63
L. Schomburg

Selenoneine is the major Se compound in the blood of Inuit consuming of traditional marine foods in Nunavik, Northern Canada   65
M. Lemire, A. Achouba, P.Y. Dumas, N. Ouellet, P. Ayotte, M. Martinez, L. Chan, B. Laird & M. Kwan

Selenium as a regulator of immune and inflammatory responses   67
P.R. Hoffmann

Effects of selenium exposure on neuronal differentiation of embryonic and induced pluripotent stem cells   69
M. Li & Z. Qiu

Functional deletion of brain selenoenzymes by methylmercury   71
N.V.C. Ralston & L.J. Raymond

The "SOS" mechanisms of methylmercury toxicity   73
N.V.C. Ralston & L.J. Raymond

Effects of selenium on animal, human and plant health

Role of the selenium in articular cartilage metabolism, growth, and maturation 77
C. Bissardon, L. Charlet, S. Bohic & I. Khan

Intestinal bioaccessibility and bioavailability of selenium in elemental selenium nano-/microparticles 79
G. Du Laing, R.V.S. Lavu, B. Hosseinkhani, V.L. Pratti, F.M.G. Tack & T. Van de Wiele

Autoimmunity against selenium transport in human sera 81
W.B. Minich, A. Schuette, C. Schwiebert, T. Welsink, K. Renko & L. Schomburg

Natural small molecules in protection of environment and health: Sulfur-selenium deficiency
and risk factor for man, animal and environment 83
R.C. Gupta

Incubation and crude protein separation from selenium-enriched earthworms 85
Y. Qiao, X. Sun, S. Yue, Z. Sun & S. Li

Health impact of dietary selenium nanoparticles on mahseer fish 87
K.U. Khan, A. Zuberi, Z. Jamil, H. Sarwar, S. Nazir & J.B.K. Fernandes

Pathways of human selenium exposure and poisoning in Enshi, China 89
J.-M. Zhu, H.-B. Qin, Z.-Q. Lin, T.M. Johnson & B.-S. Zheng

Selenium status in Iran: A soil and human health point of view 91
B. Atarodi & A. Fotovat

Impact of high selenium exposure on organ function & biochemical profile of the rural
population living in seleniferous soils in Punjab, India 93
R. Chawla, R. Loomba, R.J. Chaudhary, S. Singh & K.S. Dhillon

Keshan disease and Kaschin-Beck disease in China: Is there still selenium deficiency? 95
D. Wang, Y. Liu & D. Liu

Influence of canola oil, vitamin E and selenium on cattle meat quality and its effects on nutrition
and health of humans 97
M.A. Zanetti, L.B. Correa, A. Saran Netto, J.A. Cunha, R.S.S. Santana & S.M.F. Cozzolino

Selenium content of food and estimation of dietary intake in Xi'an, China 99
Z.W. Cui, R. Wang, J. Huang, D.L. Liang & Z.H. Wang

Assessment of selenium intake, status and influencing factors in Kenya 101
P. Biu Ngigi, G. Du Laing, C. Lachat & P. Wafula Masinde

Selenium improves the biocontrol activity of *Cryptococcus laurentii* against *Penicillium expansum*
in tomato fruit 103
J. Wu, Z. Wu, M. Li, G.S. Bañuelos & Z.-Q. Lin

Antifungal activity of selenium on two plant pathogens *Sclerotinia sclerotiorum* and *Colletotrichum
gloeosporioides* 105
Y. Zhang, H. Jiang, H. Zang, G. Tan, M. Li, L. Yuan & X. Yin

Exogenous selenium application influences lettuce on bolting and tipburn 107
X.X. Wang, Y.Y. Han & S.X. Fan

Biological uptake and accumulation of selenium

The genetics of selenium accumulation by plants 111
P.J. White

The genetic loci associated with selenium accumulation in wheat grains under soil surface
drenching and foliar spray fertilization methods 113
T. Li, A. Wang, G. Bai, L. Yuan & X. Yin

Molecular mechanisms of selenium hyperaccumulation in *Stanleya pinnata*: Potential key
genes *SpSultr1;2* and *SpAPS2* 115
M. Pilon, A.F. El Mehdawi, J.J. Cappa, J. Wang & E.A.H. Pilon-Smits

Effect of soil pH on accumulation of native selenium by Maize (*Zea mays var. L*) grains grown in Uasin Gishu, Trans-Nzoia Kakamega and Kisii counties in Kenya — 117
S.B. Otieno, T.S. Jayne & M. Muyanga

Quantification, speciation and bioaccessibility of selenium from Se-rich cereals cultivated in seleniferous soils of India — 119
N. Tejo Prakash

Wautersiella enshiensis sp. nov. – selenite-reducing bacterium isolated from a selenium-mining area in Enshi, China — 121
Z. Qu, L. Yuan, X. Yin & F. Peng

Selenium in Osborne fractions of Se-rich cereals and its bioaccessibility — 123
N.I. Dhanjal, S. Sharma & N. Tejo Prakash

New insights into the multifaceted ecological and evolutionary aspects of plant selenium hyperaccumulation — 125
E.A.H. Pilon-Smits, A.F. El Mehdawi, J.J. Cappa, J. Wang, A.T. Cochran, R.J.B. Reynolds & M. Sura-de Jong

Comparative effects of selenite and selenate on growth and selenium uptake in hydroponically grown pakchoi (*Brassica chinensis* L.) — 127
Q. Peng, Z. Li & D.L. Liang

Does selenium hyperaccumulation affect the plant microbiome? — 129
A.T. Cochran, J. Bauer, R.J.B. Reynolds, E.A.H. Pilon-Smits, M. Sura-de Jong, K. Richterova & L. Musilova

Effects of sulfur and selenium interaction on pakchoi growth and selenium accumulation — 131
S.Y. Qin, W.L. Zhao, Z. Li & D.L. Liang

Are all Brazil nuts selenium-rich? — 133
E.C. da Silva Júnior, L.R.G. Guilherme, G. Lopes, G.A. de Souza, K.E. da Silva, R.M.B. de Lima, M.C. Guedes, L.H.O. Wadt & A.R. dos Reis

Accumulation of mercury and selenium by *Oryza sativa* from the vicinity of secondary copper smelters in Fuyang, Zhejiang, China — 135
X. Yin, J. Song, Z. Li, W. Qian, C. Yao, Y. Luo & L. Yuan

Selenium and mycorrhiza on grass yield and selenium content — 137
S.M. Bamberg, M.A.C. Carneiro, S.J. Ramos & J.O. Siqueira

High selenium content reduces cadmium uptake in *Cardamine hupingshanesis* (Brassicaceae) — 139
Z.Y. Bao, H. Tian, Y.H. You & C.H. Wei

Selenium accumulation and its effects on heavy metal elements in garlic — 141
A.Q. Gu, Y.Y. Luo, H. Tian, Z.Y. Bao, C.H. Wei & X.L. Chen

Effect of selenium on cadmium uptake and translocation by rice seedlings — 143
Y.N. Wan, S.L. Yuan, Z. Luo & H.F. Li

Selenium biofortification

Environmental pathways and dietary intake of selenium in a selenium rich rural community in China: A natural biofortification case study — 147
G.S. Bañuelos, J. Tang, Y. Hou, X. Yin & L. Yuan

Effects of agronomic biofortification of maize and legumes with selenium on selenium concentration and selenium recovery in two cropping systems in Malawi — 149
A.D.C. Chilimba, S.D. Young & E.M. Joy

Potential roles of underutilized crops/trees in selenium nutrition in Malawi — 151
D.B. Kumssa, E.J.M. Joy, S.D. Young, M.R. Broadley, E.L. Ander, M.J. Watts & S. Walker

Necessity of biofortification with selenium of plants used as fodder and food in Romania — 153
R. Lăcătuşu, A.-R. Lăcătuşu, M.M. Stanciu-Burileanu & M. Lungu

The selenium content of organically produced foods in Finland *P. Ekholm, G. Alfthan & M. Eurola*	155
Effects of soil selenium ore powder application on rice growth and selenium accumulation *W. He, B. Du, Y. Luo, H. Chen, H. Liu, S. Xiao, D. Xing & J. Xu*	157
Agronomic biofortification of Brachiaria with selenium along with urea *L.A. Faria, M.C. Machado, A.L. Abdalla, P.P. Righeto, L.L. Campos, F.H.S. Karp & M.Y. Kamogawa*	159
Effects of different foliar selenium-enriched fertilizers on selenium accumulation in rice (*Oryza sativa*) *Q. Wang, X.F. Wang, J.X. Li, Y.B. Guo & H.F. Li*	161
Using agronomic biofortification to reduce micronutrient deficiency in food crops on loess soil in China *H. Mao, G.H. Lyons & Z.H. Wang*	163
Effect of selenium treatment on biomass production and mineral content in common bean varieties *M.A. de Figueiredo, D.P. Oliveira, M.J.B. de Andrade, L.R.G. Guilherme & L. Li*	165
Biofortification of irrigated wheat with Se fertilizer: Timing, rate, method and type of wheat *I. Ortiz-Monasterio, M.E. Cárdenas & G.H. Lyons*	167
Egg and poultry meat enrichment of selenium *A.G. Bertechini, V.A. Silva, F.M. Figueiredo & T.F.B. Oliveira*	169
Strategies for selenium supplementation in cattle: Se-yeast or agronomic biofortification *J.A. Hall & G. Bobe*	171
Selenoneine content of traditional marine foods consumed by the Inuit in Nunavik, Northern Canada *P. Ayotte, A. Achouba, P. Dumas, N. Ouellet, M. Lemire, L. Gautrin, L. Chan, B. Laird & M. Kwan*	173
Genotypic variation and agronomic biofortication of upland rice with selenium *H.P.G. Reis, J.P.Q. Barcelos, A.R. dos Reis & M.F. Moraes*	175
Effect of selenium fertilization on nitrogen assimilation enzymes in rice plants *J.P.Q. Barcelos, H.P.G. Reis, A.R. dos Reis & M.F. Moraes*	177
Selenium status in Brazilian soils and crops: Agronomic biofortification as a strategy to improve food quality *A.R. dos Reis*	179
Effects of selenium-enrichment on fruit ripening and senescence in mulberry trees *J. Wu, Z. Wu, M. Li, Y. Deng, G.S. Bañuelos & Z.-Q. Lin*	181
Wheat biofortification: Genotypic variation and selenium fertilization in Brazil *C.R.S. Domingues, J.A.L. Pascoalino, M.F. Moraes, C.L.R. Santos, A.R. dos Reis, F.A. Franco, A. Evangelista & P.L. Scheeren*	183
Improving selenium nutritional value of major crops *L. Li, F.W. Avila, G.A. Souza, P.F. Boldrin, M.A. de Figueiredo, V. Faquin, M.J.B. Andrade, L.R.G. Guilherme & S.J. Ramos*	185
The changing selenium content in vegetables and nutritional status of Chinese residents *W. Shi, S. Li, J. Min, L. Wu & G.S. Bañuelos*	187
Selenium and nano-selenium biofortified sprouts using micro-farm systems *H. El-Ramady, T. Alshaal, N. Abdalla, J. Prokisch, A. Sztrik, M. Fári & É. Domokos-Szabolcsy*	189
The standardization of selenium biofortification in China *X. Yin & F. Li*	191
The chemical form of selenium in dietary supplements *G.N. George, S.I. Yang & I.J. Pickering*	193

Selenium pollution control

Phytoremediation of selenium-contaminated soil and water *N. Terry*	197
Microbe-assisted selenium phytoremediation and phytomanagement of natural seleniferous areas *M. Yasin, M. Faisal, M. Yasin, A.F. El Mehdawi & E.A.H. Pilon-Smits*	199

Phytoremediation of selenium contaminated soils: strategies and limitations 201
K.S. Dhillon & S.K. Dhillon

Enhanced selenium removal and lipid production from wastewater by microalgae with low-energy ion implantation 203
Z. Wu, M. Li, M. Zhu & L. Qiu

Simultaneous production of biofuels and treatment of Se-laden wastewater using duckweed 205
Z. Wu, M. Li, L. Qiu & M. Zhu

Author index 207

Preface

Selenium (Se) is one of the most influential natural-occurring trace elements for humans and animals. The multi-faceted connections between the environment, crops, human and animal health have become focal topics in Se research for several decades. In particular, important selenoproteins and their functions have been better characterized and understood in recent years. More in depth research has also been conducted to quantify soil Se concentrations and selenium's bioavailability for crop uptake and accumulation. Generally, Se deficiencies are more widespread than Se toxicities in humans and animals worldwide. Thus, there is an essential need to improve both the bioavailability of Se in soil and in food and feed crops to ultimately preserve the biological function of Se for protecting human and ecosystem health, including interactions between Se and mercury or arsenic accumulation in organisms.

The 4th International Conference on Selenium in the Environment and Human Health will be held on 18–21 October 2015 in São Paulo, Brazil. This conference provides an effective scientific communication platform for researchers in different disciplines worldwide to elucidate and better understand those complex roles of Se as both essential nutrient and environmental contaminant. These proceedings include 97 peer-reviewed extended abstracts prepared by Se researchers from 27 countries. These most recent Se research studies address inter-relationships between the geological and atmospheric environment, agricultural food and selenium-biofortified food crops, human and animal health impacts, and genetic, biochemical, and cell and molecular processes. This book presents a unique myriad of Se research to further our international understanding of the role of Se in the context of biochemistry, food chain transfer, and health-related issues.

The success of this Se conference and publication of this book are attributed largely to the financial support from Brazilian funding agencies, including, but not limited to, the Brazilian National Council for Scientific and Technological Development (CNPq), CAPES Foundation – Ministry of Education of Brazil, Universidade Federal de Lavras (Soil Science Graduate Program, PACCSS grant AUXPE 2618-2012 from CAPES and FAPEMIG), Federal University of Mato Grosso, Federal University of Lavras, São Paulo State University, as well as University of Science and Technology of China. We are indebted to the international participants representing a multitude of disciplines from academia and industry, and to governments for sharing their knowledge to the worldwide Se research community. In particular, we would like to acknowledge the support from members of the International Society for Selenium Research.

Gary S. Bañuelos
Zhi-Qing Lin
Milton Ferreira de Moraes
Luiz Roberto G. Guilherme
André Rodrigues dos Reis

List of contributors

(*corresponding author)

Abdalla, A.L.: CENA/USP, Piracicaba/SP, Brazil
Abdalla, Neama: Plant Biotechnology Dept., Genetic Engineering Division, National Research Center, Egypt
Abdur, Rob*: School of Arts and Sciences, Georgia State University, Atlanta, USA; arob1@gsu.edu
Abreu, L.B.: Department of Soil Science, Federal University of Lavras (UFLA), Lavras (MG), Brazil
Achouba, A.: Axe santé des populations et pratiques optimales en santé, Centre de recherche du CHU de Québec, Québec, Canada, G1V 2M2; Institut national de santé publique du Québec, Québec, G1V 5B3, Canada
Alfthan, G.: National Institute of Health and Welfare, Helsinki, Finland
Alshaal, Tarek: Soil and Water Dept., Faculty of Agriculture, Kafrelsheikh University, Egypt
Ander, E. Louise: Centre for Environmental Geochemistry, British Geological Survey, Keyworth, Nottingham, NG12 5GG, UK
Andrade, M.J.B.: Federal University of Lavras, Lavras, Minas Gerais, Brazil
Anouar, Youssef: Inserm U982, Mont Saint Aignan, France; University of Rouen, Institute for Research and Innovation in Biomedicine, Rouen, France
Arai, K.: Department of Chemistry, School of Science, Tokai University, Kitakanaem, Hiratsuka-shi, Kanagawa 259-1292, Japan
Araujo, A.M.: Soil Science Department, Federal University of Lavras (UFLA), Lavras (MG), Brazil
Arnér, E.S.J.*: Division of Biochemistry, Department of Medical Biochemistry and Biophysics, Karolinska Institutet, SE-171 77 Stockholm, Sweden; Elias.Arner@ki.se
Atarodi, B.*: Department of Soil Science, Ferdowsi University of Mahhad, Mashhad, Iran; basir.atarodi@gmail.com
Avila, F.W.: Federal University of Lavras, Lavras, Minas Gerais, Brazil
Ayotte, P.*: Axe santé des populations et pratiques optimales en santé, Centre de recherche du CHU de Québec, Québec, Canada, G1V 2M2; Institut national de santé publique du Québec, Québec, G1V 5B3, Canada; pierre.ayotte@inspq.qc.ca
Bai, G.: USDA-ARS Hard Winter Wheat Genetics Research Unit, Manhattan, KS 66506, USA
Bamberg, S.M.*: Federal University of Lavras, Lavras, Minas Gerais, Brazil; sorayabamberg@gmail.com
Bañuelos, Gary S.*: USDA-ARS Agricultural Research, San Joaquin Valley Agricultural Sciences Center, 9611 South Riverbend Avenue, Parlier CA 93648-9757 USA; Gary.Banuelos@ARS.USDA.GOV
Bao, Z.Y.*: School of Earth Sciences, China University of Geosciences, Wuhan 430074, China; zybao@cug.edu.cn
Barcelos, J.P.Q.: UNESP – Univ Estadual Paulista, Postal Code 15385-000, Ilha Solteira-SP, Brazil
Batista, K.D.*: Embrapa Roraima, Boa Vista, Roraima, Brazil; karine.batista@embrapa.br
Bauer, J.: Biology Department, Colorado State University, Fort Collins, CO, USA
Behra, Renata: Eawag, Swiss Federal Institute of Aquatic Science and Technology, CH-8600 Dübendorf, Switzerland; Institute of Biogeochemistry and Pollutant Dynamics, ETH Zürich, CH-8092 Zürich, Switzerland
Bertechini, A.G.*: Department of Animal Science, Federal University of Lavras, Lavras, Minas Gerais, Brazil; bertechini@ufla.br
Bissardon, Caroline*: ISTerre (Institut des Sciences de la Terre) – Université Joseph Fourier, Grenoble, France; caroline.bissardon@ujf-grenoble.fr
Blazina, Tim: Eawag, Swiss Federal Institute of Aquatic Science and Technology, 8600 Duebendorf, Switzerland; Institute of Biogeochemistry and Pollutant Dynamics, ETH Zürich, 8092 Zürich, Switzerland
Bobe, G.: Oregon State University, Corvallis, Oregon, USA
Bohic, Sylvain: Inserm GIN U836, & Nanoimaging ESRF Beamline ID16, Grenoble, France
Boldrin, P.F.: Federal University of Lavras, Lavras, Minas Gerais, Brazil
Boukhalfa, Inès*: Inserm U1096, Rouen, France; University of Rouen, Institute for Research and Innovation in Biomedicine, Rouen, France; ines.boukhalfa@gmail.com
Broadley, Martin R.*: School of Biosciences, University of Nottingham, Sutton Bonington, Loughborough, LE12 5RD, UK; martin.broadley@nottingham.ac.uk
Campos, L.L.: CENA/USP, Piracicaba/SP, Brazil
Cappa, J.J.: Biology Department, Colorado State University, Fort Collins, CO, USA

Cárdenas, M.E.: Global Conservation Agriculture Program, International Maize and Wheat Improvement Center, Mexico D.F. 06600, Mexico

Carneiro, M.A.C.: Federal University of Lavras, Lavras, Minas Gerais, Brazil

Carvalho, G.S.: Department of Soil Science, Federal University of Lavras (UFLA), Lavras (MG), Brazil

Chan, L.: Department of Biology, University of Ottawa, Ottawa, Ontario, K1N 6N5, Canada

Charlet, L.*: Earth Science Institute (ISTerre), University Grenoble Alps, CNRS, BP 53, 38041 Grenoble Cedex 9, France; charlet38@gmail.com

Chaudhary, Rohit J.: Christian Medical College & Hospital, Ludhiana (Pb), India

Chen, Huoyun: Institute of Application Technology for Crop Selenium Enrichment, Yangtze University, Jingzhou, Hubei, China; Agricultural College of Yangtze University, Jingzhou, Hubei, China

Chen, Shaozhan: Beijing Center for Disease Control and Prevention, Beijing 100013, China

Chen, X.L.: Zhejiang Institute of Geological Survey, Hangzhou, China

Chi, Feng-qin*: Soil Fertilizer and Environment Resources Institute, Heilongjiang Academy of Agriculture Sciences, Key Laboratory of Soil Environment and Plant Nutrition of Heilongjiang Province, Harbin 150086, China; fqchi2013@163.com

Chilimba, A.D.C.*: Ministry of Agriculture and Food Security, Department of Agricultural Research Services, Lunyangwa Research Station, P.O. Box 59, Mzuzu, Malawi; achilimba@gmail.com

Cochran, A.T.*: Biology Department, Colorado State University, Fort Collins, CO, USA; atcochra@rams.colostate.edu

Correa, L.B.: Department of Animal Science, Faculty of Animal Science & Food Engineering, University of São Paulo, Pirassununga, Brazil

Couture, R.M.: Norwegian Institute for Water Research, Gaustadallé en 21, 0349 Oslo, Norway

Cozzolino, S.M.F.: Department of Food Sciences, Faculty of Pharmaceutical Sciences, University of São Paulo, Sao Paulo, Brazil

Cui, Z.W.: College of Natural Resources and environment, Northwest A&F University, Yangling, Shaanxi 712100, China; Key Laboratory of Plant Nutrition and the Agri-environment in Northwest China, Ministry of Agriculture, Yangling, Shaanxi 712100, China

Cunha, J.A.: Department of Animal Science, Faculty of Animal Science & Food Engineering, University of São Paulo, Pirassununga, Brazil

da Silva Júnior, E.C.*: Department of Soil Science, Federal University of Lavras, Minas Gerais, Brazil; ediucarlos@gmail.com

da Silva, K.E.: Embrapa Amazônia Ocidental, Manaus, Amazonas, Brazil

da Silva, L.M.: Embrapa Acre, Rio Branco, Acre, Brazil

de Andrade, M.J.B.: Federal University of Lavras, Lavras, Minas Gerais, Brazil

de Figueiredo, M.A.: Federal University of Lavras, Lavras, Minas Gerais, Brazil; marislaine_alves@yahoo.com.br

de Lima, R.M.B.: Researcher at Embrapa Amazônia Ocidental, Manaus, Amazonas, Brazil

de Oliveira Júnior, R.C.: Embrapa Amazônia Oriental, Belém, Pará, Brazil

Deng, Yongjin: Sericulture Research Institute, Anhui Academy of Agricultural Sciences, Hefei, Anhui, China

Dhanjal, Noorpreet I.*: Department of Biotechnology, Thapar University, Patiala, India; pnoor85@gmail.com

Dhillon, Karaj S.*: Department of Soil Science, Punjab Agricultural University, Ludhiana 141004, India; dhillon_karaj@yahoo.com

Dhillon, Surjit K.: Department of Soil Science, Punjab Agricultural University, Ludhiana 141004, India

Domingues, C.R.S.: Federal University of Parana, Graduate Program of Soil Science, Curitiba, PR, Brazil

Domokos-Szabolcsy, Éva: Department of Agricultural Botany, Plant Physiology and Biotechnology, University of Debrecen, Debrecen, Hungary

Du Laing, Gijs*: Department of Applied Analytical and Physical Chemistry, Ghent University, Coupure Links 653, 9000 Gent, Belgium; Gijs.DuLaing@UGent.be

Du, Bin: Institute of Application Technology for Crop Selenium Enrichment, Yangtze University, Jingzhou, Hubei, China; Agricultural College of Yangtze University, Jingzhou, Hubei, China

Dumas, P.: Axe santé des populations et pratiques optimales en santé, Centre de recherche du CHU de Québec, Québec, Canada, G1V 2M2; Institut national de santé publique du Québec, Québec, G1V 5B3, Canada

Dumesnil, Anaïs: Inserm U1096, Rouen, France; University of Rouen, Institute for Research and Innovation in Biomedicine, Rouen, France

Ekholm, P.*: University of Helsinki, Department of Food and Environmental Sciences, Helsinki, Finland; paivi.ekholm@helsinki.fi

El Mehdawi, A.F.: Biology Department, Colorado State University, Fort Collins, CO, USA

El-Ramady, Hassan*: Soil and Water Dept., Faculty of Agriculture, Kafrelsheikh University, Egypt; hassanelramady@rocketmail.com

Eurola, M.: Natural Resources Institute, Jokioinen, Finland

Evangelista, A.: Center for Agricultural Research, COODETEC, Cascavel, PR, Brazil

Faisal, M.: Department of Microbiology and Molecular Genetics, University of the Punjab, Quaid-e-Azam Campus, Lahore-54590, Pakistan

Fan, S.X.: Beijing University of Agriculture, Beijing, China

Faquin, V.: Federal University of Lavras, Lavras, Minas Gerais, Brazil

Fári, Miklós: Department of Agricultural Botany, Plant Physiology and Biotechnology, University of Debrecen, Debrecen, Hungary

Faria, L.A.*: CENA/USP, Piracicaba/SP, Brazil; evazoot@yahoo.com.br

Fedirko, V.: Rollins School of Public Health, Emory University, Atlanta, GA, USA

Feng, P.Y.: College of Natural Resources and Environment, Northwest A&F University, Key Laboratory of Plant Nutrition and the Agri-environment in Northwest China, Ministry of Agriculture, Yangling, Shaanxi 712100, China

Fernandes, João Batista Kochenborger: Centro de Aquicultura da UNESP, Jaboticabal, SP, Brazil

Figueiredo, F.M.: Department of Animal Science, Federal University of Lavras, Lavras, Minas Gerais, Brazil

Fotovat, A.: Department of Soil Science, Ferdowsi University of Mahhad, Mashhad, Iran

Franco, F.A.: Center for Agricultural Research, COODETEC, Cascavel, PR, Brazil

Gautrin, L.: LCABIE, Université de Pau et des Pays de l'Ardour, Pau, 64053, France

George, G.N.*: Department of Geological Sciences, University of Saskatchewan, Saskatoon, Saskatchewan S7N 5E2, Canada; E-mail: g.george@usask.ca

Goda, J.: Advanced Centre for Treatment, Research and Education in Cancer, Kharghar, Navi Mumbai-410210, India

Gota, V.: Advanced Centre for Treatment, Research and Education in Cancer, Kharghar, Navi Mumbai-410210, India

Gu, A.Q.: School of Earth Sciences, China University of Geosciences, Wuhan, China; feiyu2008_@126.com

Guedes, M.C.: Embrapa Amapá, Macapá, Amapá, Brazil

Guilherme, L.R.G.*: Department of Soil Science, Federal University of Lavras (UFLA), Lavras (MG), Brazil; guilherm@dcs.ufla.br

Guo, Y.B.: College of Resources and Environmental Sciences, China Agricultural University, Beijing, China

Gupta, R.C.*: SASRD Nagaland University, Medziphema 797106, India; rameshguptal954@yahoo.com

Hall, J.A.*: Oregon State University, Corvallis, Oregon, USA; jean.hall@oregonstate.edu

Han, Dan: College of Resources and Environment, Huazhong Agricultural University, Wuhan, China

Han, Y.Y.: Beijing University of Agriculture, Beijing, China

Harouki, Najah: Inserm U1096, Rouen, France; University of Rouen, Institute for Research and Innovation in Biomedicine, Rouen, France

He, Wenjing: Institute of Application Technology for Crop Selenium Enrichment, Yangtze University, Jingzhou, Hubei, China; Hubei Collaborative Innovation Center for Grain Industry, Jingzhou, Hubei, China; Agricultural College of Yangtze University, Jingzhou, Hubei, China

Henry, Jean-Paul: Inserm U1096, Rouen, France; University of Rouen, Institute for Research and Innovation in Biomedicine, Rouen, France

Hesketh, J.E.: Institute of Cell and Molecular Biosciences, University of Newcastle, UK

Hoffmann, Peter R.*: University of Hawaii, John A. Burns School of Medicine, Honolulu, Hawaii, USA; peterrh@hawaii.edu

Hosseinkhani, B.: Ghent University, Ghent, Belgium

Hou, Q.Y.: China University of Geosciences, Beijing, China

Hou, Yuzhu: School of Nanoscience & Advanced Lab for Selenium and Human Health, USTC, China; Jiangsu Bio-Engineering Research Centre on Selenium, China

Huang, J.: College of Natural Resources and Environment, Northwest A&F University, Yangling, Shaanxi 712100, China; Key Laboratory of Plant Nutrition and the Agri-environment in Northwest China, Ministry of Agriculture, Yangling, Shaanxi 712100, China

Huang, Zhen*: School of Arts and Sciences, Georgia State University, Atlanta, USA; College of Life Sciences, Sichuan University, Chengdu, China; huang@gsu.edu

Hughes, D.J.*: Centre for Systems Medicine, Department of Physiology & Department of Epidemiology and Public Health Medicine, Royal College of Surgeons in Ireland, Dublin, Ireland; davidhughes@rcsi.ie

Hybsier, S.: Institute for Experimental Endocrinology, University Medical School Berlin, Germany

Imtiaz, Muhammad: College of Resources and Environment, Huazhong Agricultural University, Wuhan, China

Iwaoka, M.: Department of Chemistry, School of Science, Tokai University, Kitakanaem, Hiratsuka-shi, Kanagawa 259-1292, Japan

Jain, V.K.: Chemistry Division, Bhabha Atomic Research Centre, Trombay, Mumbai-400085, India

James Kirchner, Institute for Terrestrial Ecosystems, ETH Zürich, 8092 Zürich, Switzerland

Jamil, Zeenat: Fisheries and Aquaculture Laboratory, Department of Animal Sciences, Faculty of Biological Sciences, Quaid-i-Azam University, Islamabad, Pakistan

Jayne, T.S.: Michigan State University, East Lansing, Michigan, USA
Jenab, M.: Section of Nutrition and Metabolism, International Agency for Research on Cancer, Lyon, France
Jiang, Han: Key Laboratory of Agri-Food Safety of Anhui Province, School of Resources and Environment–School of Plant Protection, Anhui Agriculture University, Hefei 230036, Anhui, China
Jiang, Hao: College of Resource and Environment, China Agricultural University, Beijing, 100193, China
Johnson, Thomas M.: Department of Geology, University of Illinois at Urbana-Champaign, Urbana, IL 61801, USA
Jones, G.D.*: Eawag: Swiss Federal Institute of Aquatic Science and Technology, Ueberlandstrasse 133, P.O. Box 611, CH-8600 Duebendorf, Switzerland; Gerrad.Jones@eawag.ch
Jones, J.S.: Rollins School of Public Health, Emory University, Atlanta, GA, USA
Jones, L.: Environmental Sciences Program, Southern Illinois University, Edwardsville, Illinois, USA
Joy, Edward J.M.: School of Biosciences, University of Nottingham, Sutton Bonington, Loughborough, LE12 5RD, UK
Júnior, N.J.M.: Embrapa Amapá, Macapá, Amapá, Brazil
Kamogawa, M.Y.: ESALQ/USP, Piracicaba/SP, Brazil
Kaniz, Fatema: Soil and Environmental Sciences Department, University of Barisal, Barisal, 8200, Bangladesh
Karp, F.H.S.: ESALQ/USP, Piracicaba/SP, Brazil
Khan, Kifayat Ullah*: Fisheries and Aquaculture Laboratory, Department of Animal Sciences, Faculty of Biological Sciences, Quaid-i-Azam University, Islamabad, Pakistan; Centro de Aquicultura da UNESP, Jaboticabal, SP, Brazil; kifayat055@gmail.com
Khan, Ilyas: Regenerative Medicine Group, Swansea University Medical School, Swansea University, Wales
Korbas, M.: Science Division, Canadian Light Source Inc., Saskatoon, Saskatchewan S7N 2V3, Canada
Krone, P.H.: Department of Anatomy & Cell Biology, University of Saskatchewan, Saskatoon, Saskatchewan S7N 5E5, Canada
Kuang, En-jun: Soil Fertilizer and Environment Resources Institute, Heilongjiang Academy of Agriculture Sciences, Key Laboratory of Soil Environment and Plant Nutrition of Heilongjiang Province, Harbin 150086, China
Kumssa, Diriba B.: School of Biosciences, University of Nottingham, Sutton Bonington, Loughborough, LE12 5RD, UK
Kunwar, A.*: Radiation and Photochemistry Division, Bhabha Atomic Research Centre, Mumbai – 400085, India; kamit@barc.gov.in
Kwan, M.: Nunavik Research Center, Kuujjuaq, Québec, J0M 1C0, Canada
Lăcătuşu, Anca-Rovena: National Research & Development Institute for Soil Science, Agrochemistry and Environment Protection Bucharest, Romania
Lăcătuşu, Radu*: National Research & Development Institute for Soil Science, Agrochemistry and Environment Protection Bucharest, Romania; radu58rtl@yahoo.com
Lachat, Carl: Department of Food Technology & Nutrition, Coupure Links 653, 9000 Gent, Belgium
Läderach, Alexander: Institute for Atmospheric and Climate Science, ETH Zürich, 8092 Zürich, Switzerland
Laird, B.: School of Public Health and Health Systems, University of Waterloo, Waterloo, Ontario, N2L 3G1, Canada
Lavu, R.V.S.: Ghent University, Ghent, Belgium
Lemire, M.*: Axe santé des populations et pratiques optimales en santé, Centre de recherche du CHU de Québec, Québec, Canada, G1V 2M2; Institut national de santé publique du Québec, Québec, G1V 5B3, Canada; melanie.lemire@crchuq.ulaval.ca
Lessa, J.H.L.: Soil Science Department, Federal University of Lavras (UFLA), Lavras (MG), Brazil
Li, Fei: Jiangsu Bio-Engineering Research Centre of Selenium, Suzhou 215123, Jiangsu, China
Li, Hailan: College of Resources and Environment, Huazhong Agricultural University, Wuhan, China
Li, H.F.*: College of Resources and Environmental Sciences, China Agricultural University, Beijing 100193, China; lihuafen@cau.edu.cn
Li, J.: College of Natural Resources and Environment, Northwest A&F University, Key Laboratory of Plant Nutrition and the Agri-environment in Northwest China, Ministry of Agriculture, Yangling, Shaanxi 712100, China
Li, J.X.: College of Resources and Environmental Sciences, China Agricultural University, Beijing, China
Li, L.*: Robert W. Holley Center for Agriculture and Health, USDA-ARS, Cornell University, Ithaca, NY 14853, USA; ll37@cornell.edu
Li, M.: China Geological Survey, Beijing, China
Li, Miao*: Key Laboratory of Agri-Food Safety of Anhui Province, School of Resources and Environment–School of Plant Protection, Anhui Agriculture University, Hefei 230036, Anhui, China; miaoli@ustc.edu.cn
Li, Shengnan: College of Resource and Environment, China Agricultural University, Beijing, 100193, China
Li, Sumei: State Key Laboratory of Soil and Sustainable Agriculture, Institute of Soil Science, Chinese Academy of Sciences, Nanjing 210008, China

Li, T.*: College of Agronomy, Yangzhou University, Yangzhou 225009, China; taoli@yzu.edu.cn
Li, Z.: College of Resources and Environmental, Northwest A&F University, Key Laboratory of Plant Nutrition and the Agri-environment in Northwest China, Ministry of Agriculture, Yangling, Shaanxi 712100, China
Li, Zhibo: State Key Laboratory of Soil and Sustainable Agriculture, Institute of Soil Science, Chinese Academy of Sciences, Nanjing, China
Liang, D.L.*: College of Natural Resources and Environment, Northwest A&F University, Key Laboratory of Plant Nutrition and the Agri-environment in Northwest China, Ministry of Agriculture, Yangling, Shaanxi 712100, China; dlliang@nwsuaf.edu.cn
Lin, Z.-Q.*: Environmental Sciences Program, Southern Illinois University Edwardsville, Edwardsville, Illinois, 62026, USA; zhlin@siue.edu
Liu, M.: School of Earth Sciences, China University of Geosciences, Wuhan 430074, China
Liu, Dan: Chemical Synthesis and Pollution Control Key Laboratory of Sichuan Province, College of Chemistry and Chemical Engineering, China West Normal University, Nanchong 637000, China
Liu, Liping: Beijing Center for Disease Control and Prevention, Beijing 100013, China
Liu, Yan: Chemical Synthesis and Pollution Control Key Laboratory of Sichuan Province, College of Chemistry and Chemical Engineering, China West Normal University, Nanchong 637000, China
Loomba, Rinchu: Christian Medical College & Hospital, Ludhiana (Pb), India
Lopes, G.*: Soil Science Department, Federal University of Lavras (UFLA), Lavras (MG), Brazil; guilherme.lopes@dcs.ufla.br
Lui, Huan: Institute of Application Technology for Crop Selenium Enrichment, Yangtze University, Jingzhou, Hubei, China; Agricultural College of Yangtze University, Jingzhou, Hubei, China
Lungu, Mihaela: National Research & Development Institute for Soil Science, Agrochemistry and Environment Protection Bucharest, Romania
Luo, Y.Y.: School of Earth Sciences, China University of Geosciences, Wuhan, China
Luo, Yaomei: Institute of Application Technology for Crop Selenium Enrichment, Yangtze University, Jingzhou, Hubei, China; LiChuan Extension Center of Agricultural Technology, LiChuan, Hubei, China
Luo, Yongming: Yantai Institute of Coastal Zone Research, Chinese Academy of Sciences, Yantai, China
Luo, Z.: College of Resources and Environmental Sciences, China Agricultural University, Beijing 100193, China
Luxem, Katja E.: Eawag, Swiss Federal Institute of Aquatic Science and Technology, CH-8600 Dübendorf, Switzerland; Institute of Biogeochemistry and Pollutant Dynamics, ETH Zürich, CH-8092 Zürich, Switzerland
Lv, Y.Y.: China University of Geosciences, Beijing, China
Lyons, G.H.: School of Agriculture, Food & Wine, University of Adelaide, Waite Campus, Glen Osmond, SA 5064, Australia
Ma, B.: Earth Science Institute (ISTerre), University Grenoble Alps, CNRS, BP 53, 38041 Grenoble Cedex 9, France
MacDonald, T.C.: Toxicology Centre, University of Saskatchewan, Saskatoon, Saskatchewan S7N 5B3, Canada
Machado, M.C.: CENA/USP, Piracicaba/SP, Brazil
Magalhães, C.A.S.: Embrapa Agrossilvipastoril, Sinop, Mato Grosso, Brazil
Mahajan, R.: Environmental Sciences Program, Southern Illinois University, Edwardsville, Illinois, USA
Man, N.: College of Natural Resources and Environment, Northwest A&F University, Key Laboratory of Plant Nutrition and the Agri-environment in Northwest China, Ministry of Agriculture, Yangling, Shaanxi 712100, China
Mao, H.: College of Resources and Environment, Northwest Agriculture and Forestry University, Shaanxi 712100, China; Key Laboratory of Plant Nutrition and the Agri-environment in Northwest China, Ministry of Agriculture, China
Marques, J.J.: Department of Soil Science, Federal University of Lavras (UFLA), Lavras (MG), Brazil
Martinez, A. Fernandez: Earth Science Institute (ISTerre), University Grenoble Alps, CNRS, BP 53, 38041 Grenoble Cedex 9, France
Martinez, M.: LCABIE, Université de Pau et des Pays de l'Ardour, Pau, France, 64053
Martins, G.C.: Embrapa Amazônia Ocidental, Manaus, Amazonas, Brazil
Masinde, Peter Wafula: Department of Food Science & Nutrition, School of Agriculture & Food Science, P.O. Box 972-60200 Meru, Kenya
Méplan, C.: Institute of Cell and Molecular Biosciences, University of Newcastle, UK
Min, Ju: State Key Laboratory of Soil and Sustainable Agriculture, Institute of Soil Science, Chinese Academy of Sciences, Nanjing 210008, China
Minich, Waldemar B.*: Institute for Experimental Endocrinology, Charité – University Medicine Berlin, D-13353 Berlin, Germany; waldemar.minich@charite.de
Moraes, M.F.: Federal University of Mato Grosso, Postal Code 78600-000, Barra do Garças-MT, Brazil
Mulder, Paul: Inserm U1096, Rouen, France; University of Rouen, Institute for Research and Innovation in Biomedicine, Rouen, France

Musilova, L.: Department of Biochemistry and Microbiology, University of Chemistry and Technology in Prague, Prague, Czech Republic

Muyanga, M.: Department of Agricultural, Food, and Resource Economics, Michigan State University, East Lansing, Michigan, USA

Nazir, Samina: National Centre for Physics, Quaid-i-Azam University, Islamabad 45320, Pakistan

Netto, A. Saran: Department of Animal Science, Faculty of Animal Science & Food Engineering, University of São Paulo, Pirassununga, Brazil

Ngigi, Peter Biu*: Department of Applied Analytical and Physical Chemistry, Coupure Links 653, 9000 Gent, Belgium; peterbiu.ngigi@UGent.be

Nicol, Lionel: Inserm U1096, Rouen, France; University of Rouen, Institute for Research and Innovation in Biomedicine, Rouen, France

Oliveira, D.P.: Federal University of Lavras, Lavras, Minas Gerais, Brazil

Oliveira, T.F.B.: Department of Animal Science, Federal University of Lavras, Lavras, Minas Gerais, Brazil

Ortiz-Monasterio, I.*: Global Conservation Agriculture Program, International Maize and Wheat Improvement Center, Mexico D.F. 06600, Mexico; i.ortiz-monasterio@cgiar.org

Otieno, S.B.*: School of Public Health, Kenyattaa University, State Department for Livestock, Government of Kenya; samwelbotieno@yahoo.com

Ouellet, N.: Axe santé des populations et pratiques optimales en santé, Centre de recherche du CHU de Québec, Québec, Canada, G1V 2M7; Institut national de santé publique du Québec, Québec, G1V 5B3, Canada

Ouvrard-Pascaud, Antoine: Inserm U1096, Rouen, France; University of Rouen, Institute for Research and Innovation in Biomedicine, Rouen, France

Pascoalino, J.A.L.: Federal University of Parana, Graduate Program of Soil Science, Curitiba, PR, Brazil

Pavelka, S.*: Institute of Physiology, Academy of Sciences of the Czech Republic, Prague; Institute of Biochemistry, Faculty of Science, Masaryk University, Brno, Czech Republic; pavelka@biomed.cas.cz

Peng, Fang*: China Center for Type Culture Collection, College of Life Sciences, Wuhan University, Wuhan, Hubei, China; fangpeng2@yahoo.com.cn

Peng, Q.: College of Natural Resources and Environment, Northwest A&F University, Key Laboratory of Plant Nutrition and the Agri-environment in Northwest China, Ministry of Agriculture, Yangling, Shaanxi 712100, China

Pickering, I.J.*: Department of Geological Sciences, University of Saskatchewan, Saskatoon, Saskatchewan S7N 5E2, Canada; ingrid.pickering@usask.ca

Pilon, M.*: Biology Department, Colorado State University, Fort Collins, CO, USA; Marinus.Pilon@colostate.edu

Pilon-Smits, E.A.H.*: Biology Department, Colorado State University, Fort Collins, CO, USA; Elizabeth.Pilon-Smits@colostate.edu

Prakash, N. Tejo*: School of Energy and Environment, Thapar University, Patiala, India; tejoprakash@gmail.com

Pratti, V.L.: Ghent University, Ghent, Belgium

Priyadarsini, K. Indira*: Radiation & Photochemistry Division, Bhabha Atomic Research Centre, Trombay, Mumbai-400085, India; kindira@barc.gov.in

Prokisch, József: Bio- and Environmental Energetics Institution, Nano Food Lab, Debrecen University, Hungary

Qian, Wei: State Key Laboratory of Soil and Sustainable Agriculture, Institute of Soil Science, Chinese Academy of Sciences, Nanjing, China

Qiao, Yuhui*: College of Resource and Environment, China Agricultural University, Beijing, 100193, China; qiaoyh@cau.edu.cn

Qin, Hai-Bo: State Key Laboratory of Environmental Geochemistry, Institute of Geochemistry Chinese Academy of Sciences, Guiyang 550002, China

Qin, S.Y.: College of Natural Resources and environment, Northwest A&F University, Key Laboratory of Plant Nutrition and the Agri-environment in Northwest China, Ministry of Agriculture, Yangling, Shaanxi 712100, China

Qiu, Zhifang*: Barshop Institute for Longevity and Aging Studies, University of Texas Health Science Center at San Antonio, San Antonio, Texas 78245, USA; qiuzhifang1@gmail.com

Qu, Zhihao: Jiangsu Bio-Engineering Research Centre of Selenium, Suzhou 215123, Jiangsu, China; China Center for Type Culture Collection, College of Life Sciences, Wuhan University, Wuhan, Hubei, China

Rajinder Chawla*: Christian Medical College & Hospital, Ludhiana (Pb), India; rajinderchawla.cmc@gmail.com

Ralston, N.V.C.*: University of North Dakota, Grand Forks, ND 58202, USA; nick.ralston@und.edu

Ramos, S.J.: Vale Institute of Technology, Ouro Preto, Minas Gerais, Brazil

Rashid, Mohammad Mamunur: Environmental Sciences Program, Southern Illinois University Edwardsville, Edwardsville, Illinois, 62026, USA

Raymond, L.J.: University of North Dakota, Grand Forks, ND 58202, USA

Reis, A.R. dos*: UNESP-Univ Estadual Paulista, Postal Code 15385-000, Ilha Solteira-SP, Brazil; UNESP-Univ Estadual Paulista, Postal Code 17602-496, Tupã-SP, Brazil; andrereis@tupa.unesp.br

Reis, H.P.G.: UNESP – Univ Estadual Paulista, Postal Code 15385-000, Ilha Solteira-SP, Brazil
Rémy-Jouet, Isabelle: Inserm U1096, Rouen, France; University of Rouen, Institute for Research and Innovation in Biomedicine, Rouen, France
Ren, Wujie: Beijing Center for Disease Control and Prevention, Beijing 100013, China
Renko, Kostja: Institute for Experimental Endocrinology, Charité – University Medicine Berlin, D-13353 Berlin, Germany
Reynolds, R.J.B.: Biology Department, Colorado State University, Fort Collins, CO, USA
Riboli, E.: School of Public Health, Imperial College London, UK
Richard, Vincent: Inserm U1096, Rouen, France; University of Rouen, Institute for Research and Innovation in Biomedicine, Rouen, France.
Richterova, K.: Department of Biochemistry and Microbiology, University of Chemistry and Technology in Prague, Prague, Czech Republic
Righeto, P.P.: CENA/USP, Piracicaba/SP, Brazil
Santana, R.S.S.: Department of Animal Science, Faculty of Animal Science & Food Engineering, University of São Paulo, Pirassununga, Brazil
Santos, C.L.R.: Federal University of Mato Grosso, Graduate Program of Tropical Agriculture, Barra do Garças, MT, Brazil; Federal University of Parana, Graduate Program of Soil Science, Curitiba, PR, Brazil
Sarwar, Huda: Fisheries and Aquaculture Laboratory, Department of Animal Sciences, Faculty of Biological Sciences, Quaid-i-Azam University, Islamabad, Pakistan
Scheeren, P.L.: Embrapa Wheat, Passo Fundo, RS, Brazil
Schomburg, Lutz*: Institute for Experimental Endocrinology, Charité – University Medicine Berlin, D-13353 Berlin, Germany; lutz.schomburg@charite.de
Schuette, Andrea: Institute for Experimental Endocrinology, Charité – University Medicine Berlin, D-13353 Berlin, Germany
Schwiebert, Christian: Institute for Experimental Endocrinology, Charité – University Medicine Berlin, D-13353 Berlin, Germany
Sharma, Siddharth: Department of Biotechnology, Thapar University, Patiala, India
Shi, Weiming*: State Key Laboratory of Soil and Sustainable Agriculture, Institute of Soil Science, Chinese Academy of Sciences, Nanjing 210008, China; wmshi@issas.ac.cn
Silva Júnior, E.C.: Department of Soil Science, Federal University of Lavras (UFLA), Lavras (MG), Brazil
Silva, V.A.: Department of Animal Science, Federal University of Lavras, Lavras, Minas Gerais, Brazil
Singh, Shavinder: Christian Medical College & Hospital, Ludhiana (Pb), India
Siqueira, J.O.: Vale Institute of Technology, Belém, Pará, Brazil
Song, Jing: State Key Laboratory of Soil and Sustainable Agriculture, Institute of Soil Science, Chinese Academy of Sciences, Nanjing, China
Souza, G.A.: Department of Soil Science, Federal University of Lavras (UFLA), Lavras (MG), Brazil
Stanciu-Burileanu, Mihaela Monica: National Research & Development Institute for Soil Science, Agrochemistry and Environment Protection Bucharest, Romania
Su, Qing-rui: Soil Fertilizer and Environment Resources Institute, Heilongjiang Academy of Agriculture Sciences, Key Laboratory of Soil Environment and Plant Nutrition of Heilongjiang Province, Harbin 150086, China
Sun, H.: College of Life Sciences, Sichuan University, Chengdu, China
Sun, Xiaofei: College of Resource and Environment, China Agricultural University, Beijing, 100193, China
Sun, Zhenjun: College of Resource and Environment, China Agricultural University, Beijing, 100193, China
Sura-de Jong, M.: Department of Biochemistry and Microbiology, Institute of Chemical Technology in Prague, Prague, Czech Republic
Sztrik, Attila: Bio- and Environmental Energetics Inst., Nano Food Lab, Debrecen University, Hungary
Tack, F.M.G.: Ghent University, Ghent, Belgium
Tan, Genjia: Key Laboratory of Agri-Food Safety of Anhui Province, School of Resources and Environment–School of Plant Protection, Anhui Agriculture University, Hefei 230036, Anhui, China
Tang, Jiefu: School of Nanoscience & Advanced Lab for Selenium and Human Health, USTC, China; Jiangsu Bio-Engineering Research Centre on Selenium, Suzhou, China.
Tang, Z.Y.: Faculty of Materials Science and Chemistry, China University of Geosciences, Wuhan, China
Tardini, A.B.B.: Embrapa Agrossilvipastoril, Sinop, Mato Grosso, Brazil
Terry, Norman*: University of California, Berkeley, California 94720-3102, USA; nterry@berkeley.edu
Thuillez, Christian: Inserm U1096, Rouen, France; University of Rouen, Institute for Research and Innovation in Biomedicine, Rouen, France
Tian, H.: School of Earth Sciences, China University of Geosciences, Wuhan, China
Tu, Shuxin*: College of Resources and Environment, Huazhong Agricultural University, Wuhan, China; stu@mail.hzau.edu.cn

Van de Wiele, T.: Ghent University, Ghent, Belgium

Verma, P.: Radiation and Photochemistry Division, Bhabha Atomic Research Centre, Mumbai – 400085, India

Vriens, Bas: Eawag, Swiss Federal Institute of Aquatic Science and Technology, CH-8600 Dübendorf, Switzerland; Institute of Biogeochemistry and Pollutant Dynamics, ETH Zürich, CH-8092 Zürich, Switzerland

Wadt, L.H.O.: Embrapa Rondônia, Porto Velho, Rondônia, Brazil

Walker, Sue: Crops for the Future, The University of Nottingham Malaysia Campus, Jalan Broga, 43500 Semenyih, Selangor Darul Ehsan, Malaysia

Wan, Y.N.: College of Resources and Environmental Sciences, China Agricultural University, Beijing 100193, China; 2549354538@qq.com

Wang, A.: College of Agronomy, Yangzhou University, Yangzhou 225009, China

Wang, Dacheng*: Chemical Synthesis and Pollution Control Key Laboratory of Sichuan Province, College of Chemistry and Chemical Engineering, China West Normal University, Nanchong 637000, China; wangdacheng163@163.com

Wang, J.: Biology Department, Colorado State University, Fort Collins, CO, USA

Wang, J.: Environmental Sciences Program, Southern Illinois University, Edwardsville, Illinois, USA

Wang, Q.*: College of Resources and Environmental Sciences, China Agricultural University, Beijing, China; 805886968@qq.com

Wang, R.: College of Natural Resources and Environment, Northwest A & F University, Yangling, Shaanxi 712100, China; Key Laboratory of Plant Nutrition and the Agri-environment in Northwest China, Ministry of Agriculture, Yangling, Shaanxi 712100, China

Wang, X.F.: College of Resources and Environmental Sciences, China Agricultural University, Beijing, China

Wang, X.X.*: Beijing University of Agriculture, Beijing, China; 369198317@qq.com

Wang, Z.H.*: College of Resources and Environment, Northwest Agriculture & Forestry University, Shaanxi 712100, China; Key Laboratory of Plant Nutrition and the Agri-environment in Northwest China, Ministry of Agriculture, Yangling, Shaanxi 712100, China; zhwang@263.net

Wang, Zhaoshuang: Microelement Research Center, Huazhong Agricultural University, Wuhan 430070, China

Watts, Michael J.: Centre for Environmental Geochemistry, British Geological Survey, Keyworth, Nottingham, NG12 5GG, UK

Wei, C.H.: Faculty of Materials Science and Chemistry, China University of Geosciences, Wuhan, China

Wei, Dan: Soil Fertilizer and Environment Resources Institute, Heilongjiang Academy of Agriculture Sciences, Key Laboratory of Soil Environment and Plant Nutrition of Heilongjiang Province, Harbin 150086, China

Welsink, Tim: Institute for Experimental Endocrinology, Charité – University Medicine Berlin, D-13353 Berlin, Germany

Wernli, Heini: Institute for Atmospheric and Climate Science, ETH Zürich, 8092 Zürich, Switzerland

White, Philip J.*: The James Hutton Institute, Dundee DD2 5DA, United Kingdom; King Saud University, Riyadh 11451, Kingdom of Saudi Arabia; Philip.White@hutton.ac.uk

Winkel, L.H.E.*: Swiss Federal Institute of Technology (ETH), Institute of Biogeochemistry and Pollutant Dynamics, ETH Zürich, CH-8092 Zürich, Switzerland, Eawag: Swiss Federal Institute of Aquatic Science and Technology, Ueberlandstrasse 133, P.O. Box 611, CH-8600 Duebendorf, Switzerland; Lenny.Winkel@eawag.ch

Wu, Jie: Key Laboratory of Agri-Food Safety of Anhui Province, School of Plant Protection, Anhui Agriculture University, Hefei 230036, Anhui, China

Wu, Longhua: State Key Laboratory of Soil and Sustainable Agriculture, Institute of Soil Science, Chinese Academy of Sciences, Nanjing 210008, China

Wu, Zhilin: Key Laboratory of Agri-Food Safety of Anhui Province, School of Resources and Environment–School of Plant Protection, Anhui Agriculture University, Hefei 230036, Anhui, China

Xi, X.X.: China Geological Survey, Beijing, China

Xiao, Sen: Institute of Application Technology for Crop Selenium Enrichment, Yangtze University, Jingzhou, Hubei, China; Agricultural College of Yangtze University, Jingzhou, Hubei, China

Xiao, Y.: School of Earth Sciences, China University of Geosciences, Wuhan 430074, China

Xie, Y.H.: College of Natural Resources and Environment, Shanxi Agricultural University, Shanxi, China

Xie, Zhijian: College of Resources and Environment, Huazhong Agricultural University, Wuhan, China

Xing, Danying*: Institute of Application Technology for Crop Selenium Enrichment, Yangtze University, Jingzhou, Hubei, China; Hubei Collaborative Innovation Center for Grain Industry, Jingzhou, Hubei, China; Agricultural College of Yangtze University, Jingzhou, Hubei, China; pasufu@qq.com

Xiong, Shuanglian: College of Resources and Environment, Huazhong Agricultural University, Wuhan, China

Xu, Qiang: Northeast Agricultural University, Harbin 150030, China

Xu, Jianlong*: Hubei Collaborative Innovation Center for Grain Industry, Jingzhou, Hubei, China; Agricultural College of Yangtze University, Jingzhou, Hubei, China; Institute of Crop Sciences, Chinese Academy of Agricultural Sciences, Beijing, China; xujlcaas@126.com

Yang, S.I.: Department of Geological Sciences, University of Saskatchewan, Saskatoon, Saskatchewan S7N 5E2, Canada

Yang, Z.F.: China University of Geosciences, Beijing, China

Yao, Chunxia: Institute for Agro-Product Quality Standards and Testing Technologies, Shanghai Academy of Agricultural Sciences, Shanghai, China

Yasin, M.*: Department of Microbiology and Molecular Genetics, University of the Punjab, Quaid-e-Azam Campus, Lahore-54590, Pakistan; yasin_mmg@yahoo.com

Yin, Xuebin*: Advanced Lab for Selenium and Human Health, Suzhou Institute of USTC, Suzhou 215123, Jiangsu, China; Jiangsu Bio-Engineering Research Centre of Selenium, Suzhou 215123, Jiangsu, China; xbyin@ustc.edu.cn

You, Y.H.: Faculty of Materials Science and Chemistry, China University of Geosciences, Wuhan, China

Young, S.D.: University of Nottingham, School of Biosciences, Sutton Bonington Campus, Loughborough, LE12 5RD, UK

Young, Scott D.: School of Biosciences, University of Nottingham, Sutton Bonington, Loughborough, LE12 5RD, UK

Yu, T.*: China University of Geosciences, Beijing, China; yutao@cugb.edu.cn

Yuan, Linxi*: Jiangsu Bio-Engineering Research Centre of Selenium, Suzhou 215123, Jiangsu, China; yuanli@mail.ustc.edu.cn

Yuan, S.L.: College of Resources and Environmental Sciences, China Agricultural University, Beijing 100193, China

Zanetti, M.A.*: Department of Animal Science, Faculty of Animal Science & Food Engineering, University of São Paulo, Pirassununga, Brazil; mzanetti@usp.br

Zang, Huawei: Key Laboratory of Agri-Food Safety of Anhui Province, School of Resources and Environment–School of Plant Protection, Anhui Agriculture University, Hefei 230036, Anhui, China

Zhang, H.Y.*: School of Earth Sciences, China University of Geosciences, Wuhan 430074, China; hongyuzhang_cug@hotmail.com

Zhang, Jiu-ming: Soil Fertilizer and Environment Resources Institute, Heilongjiang Academy of Agriculture Sciences, Key Laboratory of Soil Environment and Plant Nutrition of Heilongjiang Province, Harbin 150086, China

Zhang, Nina*: Beijing Center for Disease Control and Prevention, Beijing 100013, China; znnms@sina.com

Zhang, W.: College of Life Sciences, Sichuan University, Chengdu, China

Zhang, Yu: Key Laboratory of Agri-Food Safety of Anhui Province, School of Resources and Environment–School of Plant Protection, Anhui Agriculture University, Hefei 230036, Anhui, China

Zhao, W.L.: College of Resources and Environmental, Northwest A&F University, Key Laboratory of Plant Nutrition and the Agri-environment in Northwest China, Ministry of Agriculture, Yangling, Shaanxi 712100, China

Zheng, Bao-Shan: State Key Laboratory of Environmental Geochemistry, Institute of Geochemistry Chinese Academy of Sciences, Guiyang 550002, China

Zhou, Jianli: Hubei Collaborative Innovation Center for Grain Industry, Yangtze University, Jingzhou, China

Zhu, Jian-Ming*: State Key Laboratory of Geological Processes and Mineral Resources, China University of Geosciences, Beijing 100083, China; State Key Laboratory of Environmental Geochemistry, Institute of Geochemistry Chinese Academy of Sciences, Guiyang 550002, China; zhujianming@vip.gyig.ac.cn

Zhu, Mei: School of Engineering, Anhui Agriculture University, Hefei 230036, Anhui, China

Zuberi, Amina: Fisheries and Aquaculture Laboratory, Department of Animal Sciences, Faculty of Biological Sciences, Quaid-i-Azam University, Islamabad, Pakistan

Biogeochemistry of selenium

The global biogeochemical cycle of selenium: Sources, fluxes and the influence of climate

Lenny H.E. Winkel
Eawag, Swiss Federal Institute of Aquatic Science and Technology, Dübendorf, Switzerland
Institute of Biogeochemistry and Pollutant Dynamics, ETH Zürich, Zürich, Switzerland

1 INTRODUCTION

Selenium (Se) is an essential element for humans and animals (Rayman, 2000; Fairweather-Tait et al., 2011). Selenium is present in the amino acids selenomethionine (SeMet) and selenocysteine (SeCys), which has been termed the 21st amino acid. SeCys is incorporated in selenoproteins that serve a wide range of biological functions, including oxidoreductions, redox signaling, antioxidant defense, thyroid hormone metabolism, and immune responses. (Lu & Holmgren, 2009). Due to the central role of Se in selenoproteins and the structural similarity of SeCys to the sulfur analogue cysteine, which may lead to the unspecific incorporation of selenocysteine instead of cysteine, Se only has a narrow range of safe concentrations for humans. Current estimates indicate that an intake of Se of 30 μg/day is inadequate, while intakes exceeding 900 μg/day are potentially harmful (Fairweather-Tait et al., 2011; Winkel et al., 2015).

2 DISTRIBUTION OF SE IN SOILS

The narrow range of safe concentrations in combination with an uneven distribution of Se in agricultural soils and crops has lead to environmental health problems, i.e. Se deficiency and toxicity (Winkel et al., 2012; Winkel et al., 2015). Food crops play a key role in dietary Se intake levels (Winkel et al., 2015; Steinnes, 2009) and the Se status of crops depends on factors such as Se concentrations and the presence of competing ions (Fernandez-Martinez & Charlet, 2009). Additionally, soil Se bioavailability and retention largely depend on speciation. The main inorganic species, selenite (SeO_3^{2-}) and selenate (SeO_4^{2-}) can be sorbed to mineral surfaces through pH-sensitive electrostatic interactions (Fernandez-Martinez & Charlet, 2009). Also, soil organic carbon is known to influence Se retention in soils, but the retention mechanisms are still poorly understood (Tolu et al., 2014). While the local distribution of Se in soils is governed by small-scale soil processes, it is unclear what processes determine the broad-scale spatial distribution of Se. These distributions are highly variable, which has resulted in geomedical problems in humans and livestock (Steinnes, 2009). For example, selenosis (Se poisoning) has been caused by excess Se levels in food crops in certain locations in China (Luo et al., 2004). For these cases, a close relationship was found between high Se concentrations in crops and underlying or adjacent bedrock geology (Fordyce, 2007). However, on a global scale, Se deficiency is much more prevalent than Se toxicity. While local Se enrichments in soils can be largely explained by geogenic sources (e.g., black shales) (Fernandez-Martinez & Charlet, 2009), bedrock geology has failed to explain the spatial Se distribution at a large scale (Blazina et al., 2014).

3 SOURCES, FLUXES AND SINKS OF SE

Deposition of Se transported in the atmosphere has been proposed as an alternative hypothesis to geogenic sources to explain large-scale patterns of environmental Se (Winkel et al., 2015; Wen & Carignan, 2007). Wet deposition is believed to be the dominant form of atmospheric deposition. Between 30°N and 90°N, Ross (1985) estimated Se fluxes of $3.6 - 10.0 \times 10^9$ g/yr and $0.55 - 2.6 \times 10^9$ g/yr, for wet and dry deposition respectively, and Cutter (1993) estimated a Se deposition of 2.2×10^9 g/yr over the N. Atlantic Ocean. Atmospheric sources of Se can be divided into anthropogenic (e.g., industrial emissions) and natural sources, which can further be divided into biotic (e.g., biomethylation) and abiotic (e.g., volcanic) sources (Wen & Carignan, 2007). Oceans are believed to be a major source of biotic Se to the atmosphere, accounting for 45 to 77% of global emissions and ranging between 0.4 and 9×10^9 g/yr (Nriagu & Pacyna, 1988; Nriagu, 1989). Marine organisms such as microalgae are thought to play an important role in these emissions via the production of volatile organic Se species, mainly dimethylselenide (DMSe) (Amouroux et al., 2001; Wen & Carignan, 2007). Significant positive correlations between concentrations of DMSe, dimethyl sulfide (DMS), marine plankton biomass or the amount of coccolithophores (a major group of marine phytoplankton) have been reported (Amouroux et al., 1996, 2001). Wen and Carignan (2007) proposed that coccolithophores could form DMSe similarly to DMS formation. Although the exact role of marine algae and other marine organisms is not clear in the production of volatile organic Se species, it is possible that volatile Se emissions from marine environments could contribute to broad-scale soil Se following atmospheric transport and transformation, and removal from the atmosphere via wet deposition. Recently, Blazina et al. (2014) used sediment sequences from the Loess Plateau in central China to show that over the last 6.8 million years rainfall likely controlled Se distributions in these sediments. This finding suggests that precipitation may be an important factor in determining the broad-scale Se distribution in soils in monsoonal China (Winkel et al., 2015). Further research is required to understand to what extent precipitation determines broad-scale soil Se patterns in other parts of the world. Currently, the relationship between Se and rainfall is being analyzed in high-resolution records of precipitation chemistry.

4 PREDICTIONS OF SE AND IMPACTS

Even though specific biogeochemical pathways of the Se cycle have been the focus of many studies, the relationships between these pathways are often not clear. To better understand global Se cycling, it is crucial to have mechanistic and quantitative understanding of not only different sources, sinks, and transport processes, but also on how these are linked. In this contribution, we will present new results on individual pathways of the global Se cycle and on the links between these pathways. Furthermore, we will show how information on these different pathways can be used to establish large-scale geospatial predictions of Se contents in soils using statistical models. The use of statistical models will be helpful in obtaining a better understanding of the processes driving large-scale distributions of Se and resulting predictions will help in the prevention of future health hazards related to unsafe levels of Se in soils.

REFERENCES

Amouroux, D., & Donard, O.F.X. 1996. Maritime emission of selenium to the atmosphere in eastern-Mediterranean seas. *Geophys. Res. Lett.* 23: 1777–1780.

Amouroux, D., Liss, P.S., Tessier, E., Hamren-Larsson, M., & Donard, O.F.X. 2001. Role of oceans as biogenic sources of selenium. *Earth Planet. Sci. Lett.* 189: 277–283.

Blazina, T., Sun, Y., Voegelin, A., Lenz, M., Berg, M., & Winkel, L.H.E. 2014. Terrestrial selenium distribution in China is potentially linked to monsoonal climate. *Nat. Commun.* 5: 4717.

Cutter, G.A. 1993. Metalloids in wet deposition on Bermuda: Concentrations, sources, and fluxes. *J. Geophys. Res.* 98(D9): 16777–16786.

Fairweather-Tait, S.J., Bao, Y., Broadley, M.R., Collings, R., Ford, D., Hesketh, J.R., & Hurst, R. 2011. Selenium in human health and disease. *Antioxid. Redox Signal.* 14: 1337–1383.

Fernandez-Martinez, A., & Charlet, L. 2009. Selenium environmental cycling and bioavailability: A structural chemist point of view. *Rev. Environ. Sci. Biotechnol.* 8: 81–110.

Fordyce, F.M. 2007. Selenium geochemistry and health. *Ambio* 36: 94–97.

Lu, J., & Holmgren, A. 2009. Selenoproteins. *J. Biol. Chem.* 284: 723–727.

Luo, K.L., Xu, L.R., Tan, J.A., Wang, D.H., & Xiang, L.H., 2004. Selenium source in the selenosis area of the Daba region, South Qinling Mountain, China. *Environ. Geol.* 45: 426–432.

Nriagu, J.O., & Pacyna, J.M. 1988. Quantitative assessment of worldwide contamination of air, water and soils by trace metals. *Nature* 333: 134–139.

Nriagu, J.O. 1989. A global assessment of natural sources of atmospheric trace metals. *Nature* 338: 47–49.

Rayman, M.P. 2000. The importance of selenium to human health. *Lancet* 356: 233–241.

Ross, H.B. 1985. An atmospheric selenium budget for the region 30°N to 90°N. *Tellus* 37B: 78–90.

Steinnes, E. 2009. Soils and geomedicine. *Environ. Geochem. Hlth.* 31: 523–535.

Tolu, J., Thiry, Y., Bueno, M., Jolivet, C., Potin-Gautier, M., & Le Hecho, I. 2014. Distribution and speciation of ambient selenium in contrasted soils, from mineral to organic rich. *Sci. Total Environ.* 479: 93–101.

Wen, H. & Carignan, J. 2007. Reviews on atmospheric selenium: Emissions, speciation and fate. *Atmos. Environ.* 41: 7151–7165.

Winkel, L.H.E., Johnson, C.A., Lenz, M., Grundl, T., Leupin, O.X., Amini, M., Charlet, L. 2012. Environmental selenium research: from microscopic processes to global understanding. *Environ. Sci. Technol.* 46 (2): 571–579.

Winkel, L.H.E., Vriens, B., Jones, G.D., Schneider, L.S., Pilon-Smits, E., & Banuelos, G.S. 2015. Selenium cycling across soil-plant-atmosphere interfaces: a critical review. *Nutrients* 7: 4199–4239.

Topsoil selenium distribution in relation to geochemical factors in main agricultural areas of China

T. Yu*, Z.F. Yang, Q.Y. Hou & Y.Y. Lv
China University of Geosciences, Beijing, China

X.X. Xi & M. Li
China Geological Survey, Beijing, China

1 INTRODUCTION

Selenium (Se) is an essential nutrient for human and animal health. The element often occurs in low concentrations in soils in many parts of the world (Sharma et al., 2015). In China, there is a low soil Se distribution zone, stretching from Heilongjiang Province of Northeast China to Yunnan Province of Southwest China, with soil Se concentrations of < 0.125 mg/kg (Tan et al., 1991). There are many epidemiological data on the Se-related human diseases, such as Kashin-Beck disease (KBD) and Keshan disease (KD) that were observed in the 1960s and 1970s from this low-Se belt reported. Studies showed a significant correlation between the concentration of Se in the environment and the incidences of the two diseases (Tan et al., 1991, 2002). Decades later, there is a strong need for collecting systematic information on the current soil Se status on the national scale and determining dominant factors influencing Se concentrations.

2 MATERIALS AND METHODS

2.1 Sample collection and preparation

A total of 376,679 topsoil samples were collected throughout the country in 1999 to 2012, including all main agricultural regions of China (Fig. 1). The sampling was conducted for the National Multi-Purpose Regional Geochemical Survey Project (NMPRGS) initiated by China Geological Survey (CGS). Systematic sampling using a regular grid was applied to ensure an even distribution of randomly selected sampling sites (Xi et al., 2005). The topsoils of 0–20 cm depth were collected at a density of 1 sample/km^2. Each sampling site was selected from the most representative land-use type of the most common soil type in each sampling grid. Specifically, topsoils were collected at 3–5 points within 50 m distance at each respective sampling site. In terms of sampling time, no sample was collected at the time of fertilization. The size of each soil sample was >1 kg. All the sampling sites were marked using GPS coordinates.

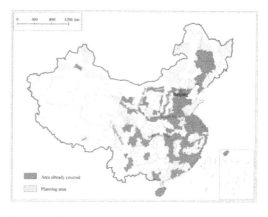

Figure 1. The sampling areas of the NMPRGS project in China. (Li et al., 2014).

2.2 Chemical analysis and accuracy

Air-dried soil samples were sieved with 20 mesh screen (<0.84 mm) and further with 200 mesh (<0.074 mm) for chemical analysis. Soil pH was measured after sieving with 10 mesh screen (<2 mm). Soil samples were acid-digested using a mixture of HF, HNO_3, $HClO_4$ and aqua regia until the digestion solution become clear. Concentrations of Se were measured using atomic fluorescence spectrometry (AFS, Model AFS-230E).

The chemical analysis precision and accuracy were validated using standard reference materials (SRM, including GBW7415, GBW7416, GBW7460, and GBW7461), spiked samples, internal and external duplicate samples and coded samples by cerified laboratories. In regards to soil samples, the accuracy of analysis was monitored by including 12 aforementioned SRMs for every 500 samples and analyzed simultaneously. Four SRMs were randomly inserted within every batch of 50 samples to assess the precision by the logarithmic standard deviations calculated between the analytical values and the recommended values. The accuracy and precision of the analyses of all samples satisfied the required specifications (Xi et al., 2005; Li et al., 2014).

Statistical analysis was conducted using Statics 19.0 software for calculating the mean, standard deviation, minimum value, maximum value, correlation coefficients and other statistical values.

3 RESULTS AND DISCUSSION

3.1 *Se concentration in topsoil*

The geometric and arithmetic means of Se concentration for all topsoil samples were 0.22 and 0.26 mg/kg, respectively, with the lowest value at 0.01 mg/kg and the highest value at 49.60 mg/kg. The arithmetic means of Se concentration in topsoils are higher than the corresponding geometric means due to the positively skewed distribution of the results.

The topsoil Se concentrations by geometric means in different administrative areas are in the following descending order: Central China> Southwest China> East China> North China> Northwest China> Northeast China. The highest means of topsoil Se concentration was 0.36 ± 0.26 mg/kg in Central China. North China has the lowest Se concentration at 0.19 ± 0.08 mg/kg. According to the threshold value of soil Se content reported by Tan et al. (2002), 11% of the farmland soil in China is Se-deficient, 21% is Se-insufficient, 55% is Se-sufficient, 13% is Se-high, and 0.04% of the soil is Se-excess.

3.2 *Factors influencing the distribution of soil Se*

Soil Se distribution can be affected different chemical processes (such as pH, redox potential, organic matter content, and the presence of other competitive ions), physical processes include sorption effects of soils and sediments, and biological processes, such as microbial Se transformation involving oxidation-reduction and alkylation or dealkylation (Lenz & Lens, 2009). The other geochemical factors that influence Se distribution in the topsoil include parent matereials, soil texture, soil organic matter and contents of Fe, Al, and other elements. These related factors were discussed by principal component analysis. Consequently, the parent rock is considered as one of the main factors lead to Se fractionation. The intrusive rocks and pyroclastic rocks are the main parent rocks that contribute high-Se soil, while the low-Se soil usually originated from the shallow marine deposits or the eolian deposits. The soil Se concentration in the agricultural area is also associated with soil Fe or Al content.

4 CONCLUSIONS

The results indicate that the highest soil Se concentrations tend to be observed in central China where Se is distributed with ultrabasic-basic intrusive rocks. According to the threshold value for the risk assessment of potential Se deficiency in Chinese agricultural soils, 11% of main agricultural areas of China are Se-deficient.

ACKNOWLEDGMENTS

This study was supported by the Major Programs of the Geological Survey of Land Resources, China Geological Survey (No. 12120114091901 and 12120113001800), National Natural Science Foundation of China (No. 41172326), and by the "Fundamental Research Funds for the Central Universities" (No. 2652015034).

REFERENCES

Fordyce, F.M. 2007. Selenium geochemistry and health. *Ambio* 36: 94–97.

Lenz, M. & Lens, P.N.L. 2009. The essential toxin: The changing perception of selenium in environmental sciences. *Science of The Total Environment* 407(12): 3620–3633.

Li, M., Xi, X.H., Xiao, G.Y., Cheng, H.X., Yang, Z.F., Zhou, G.H., et al. 2014. National multi-purpose regional geochemical survey in China. *Journal of Geochemical Exploration* 139: 21–30.

Sharma, V.K. McDonald, T.J., Sohn, M., Anquandah, G.A.K., Pettine, M., Zboril, R., et al. 2015. Biogeochemistry of selenium. A review. *Environmental Chemistry Letters* 13(1): 49–58.

Tan, J.A. & Huang, Y.J. 1991. Selenium in geo-ecosystem and its relation to endemic diseases in China. *Water, Air, and Soil Pollution* 57–58: 59–68.

Tan, J.A., Zhu, W.Y., Wang, W.Y., Li, R., Hou, S., Wang, D., et al. 2002. Selenium in soil and endemic diseases in China. *The Science of the Total Environment* 284: 227–235.

Xi, X.H. 2005. Multi-purpose regional geochemical survey and ecogeochemistry: new direction of quaternary research and application. *Quaternary Sciences* 25, 269–274 (English abstract).

Examining continental and marine sources of selenium in rainfall

Tim Blazina & Lenny H.E. Winkel*
Eawag, Swiss Federal Institute of Aquatic Science and Technology, Duebendorf, Switzerland
Institute of Biogeochemistry and Pollutant Dynamics, ETH Zürich, Zürich, Switzerland

Alexander Läderach & Heini Wernli
Institute for Atmospheric and Climate Science, ETH Zürich, Zürich, Switzerland

James Kirchner
Institute for Terrestrial Ecosystems, ETH Zürich, Zürich, Switzerland

1 INTRODUCTION

Atmospheric transport is important in the global biogeochemical cycle of Selenium (Se). The largest anthropogenic sources of atmospheric Se consist of fossil fuel combustion (1.8–3 × 10^9 g Se per year) and non-ferrous metal smelting (1.4–1.7 × 10^9 g Se per year) (Mosher & Duce, 1987; Nriagu, 1989). Similarly, there are various natural sources of atmospheric Se, including sea salt spray (0.04–1 × 10^9 g Se per year) and volcanic emissions (0.1–1.8 × 10^9 g Se per year) (Mosher & Duce, 1987; Nriagu, 1989). By and far the largest estimated natural atmospheric source of Se is the marine biosphere (0.6–17.7 × 10^9 g Se per year) (Wen & Carignan, 2007). This flux is thought to be driven by the production of volatile methylated Se compounds, such as dimethyl selenide (DMSe) and dimethyl diselenide (DMDSe) by marine organisms. However, the mechanisms behind the biogenic production of volatile organic Se species in the marine environment are still unclear.

Once emitted into the atmosphere, methylated Se species react rapidly with atmospheric oxidants such as OH, O_3 and NO_x and consequently have relatively short residence times, for example, <6 hour for DMSe (Atkinson et al., 1990). It has been proposed that biogenic methylated Se species may be adsorbed onto sub-micrometer sea salt particles, which can have atmospheric residence times of up to 60 hours and can be involved in long-range atmospheric transport (Wen & Carignan, 2009). However, the exact mechanisms of long-range atmospheric Se transport have yet to be explored. For example, the speciation of particle bound Se has yet to be directly measured, and it is not known if or how the speciation of Se would influence the likelihood of Se long-range transport in the atmosphere. Furthermore, despite the large amount of atmospheric Se that is estimated to be annually produced by the marine biosphere, it remains unclear how biogenic emissions of Se, in addition to other Se emissions (e.g., industrial emissions) affect atmospheric deposition and, ultimately, distributions of Se in terrestrial environments.

2 METHODS

In this study we investigate the marine and continental sources of Se by examining a high resolution (seven hourly) two-year data set of bulk deposition chemistry at the Plynlimon experimental research catchment in Wales, United Kingdom (Neal et al., 2013). This data set consists of time series spanning March 2007 through January 2009 for 50 separate analytes including major (e.g., Na, Mg, Cl) and trace elements (e.g., Se, As, Ni), and allows for detailed geochemical comparisons.

To investigate continental versus marine sources of atmospheric Se in rainfall we are employing the Lagrangian transport and dispersion model FLEX-PART (Stohl et al., 2005). The FLEXPART model uses the three-dimensional wind field, temperature, and specific humidity from ERA-Interim reanalysis data (Dee et al., 2011) to back-calculate where air travelled from and where moisture uptake occurred prior to arriving at the Plynlimon site. Given the maximum possible atmospheric lifetime of Se adsorbed onto sea salt aerosols (about 60 hours, see Wen & Carignan, 2009), we have analyzed two-day back trajectories.

3 RESULTS AND DISCUSSION

In this data set we find moderate positive correlations between Se and marine-derived ions such as chloride ($R^2 = 0.75$, $P < 0.0001$, n = 718) and bromide ($R^2 = 0.61$, $P < 0.0001$, n = 614). This positive correlation may indicate that there is a marine source for Se deposited at Plynlimon. The Plynlimon site is located <20 km from the coast and therefore sea salt aerosols could be a source of Se at this site. However, by comparing the average Se/Cl ratio in the rain water locations of moisture uptake for different rain events and elemental concentrations to explore the source locations of Se and other elements. To investigate whether marine biogenic emissions of Se influence Se concentrations in Plynlimon rainwater, we will examine the correspondence between air mass origin, moisture uptake location and Atlantic phytoplankton blooms (Fig. 1). With these analyses, we aim to better understand the sources of Se in Plynlimon rainfall and ultimately the influence of continental and marine sources of Se to soils.

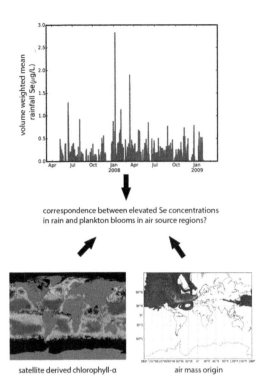

Figure 1. Schematic showing the 7 hourly volume weighted mean Se concentrations in Plynlimon rainwater from 2007 to 2009, and examples of satellite derived phytoplankton blooms (Feldman & McClain, 2015) and FLEXPART atmospheric back trajectory data.

(3.86×10^{-5}) to the average Se/Cl ratio in bulk seawater (4.13×10^{-9}) (Conde & Sanz Alaejos, 1997), we find that on average Se concentrations in the rainwater are enriched >9000 times relative to seawater, suggesting that there are other sources of Se in addition to sea salt inputs.

Here we will present results illustrating how Se concentrations at Plynlimon depend on the origin of air (e.g., distance/direction travelled) and the moisture uptake location. We will compare the traveled distances and directions of air masses as well as

REFERENCES

Atkinson, R., Aschmann, S.M., Hasegawa, D., Thompson-Eagle, E.T. & Frankenberger, W.T. 1990. Kinetics of the atmospherically important reactions of dimethyl selenide. *Environmental Science & Technology*, 24: 1326–1332.

Conde, J.E. & Sanz Alaejos, M. 1997. Selenium concentrations in natural and environmental waters. *Chemical Reviews*, 97: 1979–2004.

Dee, D & Coauthors. 2011. The era-interim reanalysis: configuration and performance of the data assimilation system. *Quart. J. Roy. Meteor. Soc.*, 137: 553–597,

Feldman, G.C. & Mcclan, C.R. 2015. *Ocean Color Web – Seawifs.* http://oceancolor.gsfc.nasa.gov/. NASA Goddard Space Flight Center. (Accessed on March 20 2015).

Mosher, B.W. & Duce, R.A. 1987. A global atmospheric selenium budget. *J. Geophys. Res.*, 92: 13289–13298.

Neal, C., Reynolds, B., Kirchner, J.W., Rowland, P., Norris, D., Sleep, D., et al. 2013. High-frequency precipitation and stream water quality time series from Plynlimon, Wales: An openly accessible data resource spanning the periodic table. *Hydrological Processes*, 27: 2531–2539.

Nriagu, J.O. 1989. A global assessment of natural sources of atmospheric trace metals. *Nature*, 338: 47–49.

Stohl, A., Forster, C., Frank, A., Seibert, P. & Wotawa, G. 2005. Technical note: the lagrangian particle dispersion model flexpart version 6.2. *Atmos. Chem. Phys.*, 5: 2461–2474.

Wen, H. & Carignan, J. 2007. Rreviews on atmospheric selenium: emissions, speciation and fate. *Atmospheric Environment*, 41: 7151–7165.

Wen, H. & Carignan, J. 2009. Ocean to continent transfer of atmospheric se as revealed by epiphytic lichens. *Environmental Pollution*, 157: 2790–2797.

The role of phytoplankton in marine selenium cycling

Katja E. Luxem, Bas Vriens, Renata Behra & Lenny H.E. Winkel*
Eawag, Swiss Federal Institute of Aquatic Science and Technology, Dübendorf, Switzerland
Institute of Biogeochemistry and Pollutant Dynamics, ETH Zürich, Zürich, Switzerland

1 THE MARINE BIOGEOCHEMISTRY OF SELENIUM

Selenium (Se) has a complex marine chemistry. It is one of the few chemical elements where anionic species with different oxidation states, such as selenite [Se(IV)] and selenate [Se(VI)], coexist at similar concentrations. The thermodynamically stable form of Se in oxic seawater is Se(VI), while Se(IV) and organic Se are kinetically stable in the surface ocean (Cutter & Bruland, 1984). In the surface ocean, organic Se species often dominate the Se speciation but these organic Se species are poorly characterized and only practically defined as non-anionic Se.

The presence of the thermodynamically disfavored species, Se(IV) and organic Se, could be explained by external inputs or production in the surface ocean. However, mass balance calculations have indicated that inputs into the surface ocean via upwelling, riverine input, or wet and dry atmospheric deposition, are too small to explain the observed concentrations of Se(IV) and organic Se (Cutter & Cutter, 1998). The marine depth profiles of Se (Fig. 1A) are frequently nutrient-like and correlated with profiles of phosphate, silicate or nitrate (Sherrard et al., 2004). The nutrient-like depletion of inorganic Se in the surface ocean has been attributed to biological uptake and export into the deep ocean (Cutter & Cutter, 2001, 1998). The production of Se(IV) and organic Se through biological activity is believed to contribute significantly to the observed Se concentrations in the surface ocean (Cutter & Cutter, 1998; Sherrard et al., 2004).

Within the surface ocean, marine microalgae (phytoplankton) can internalize, transform, and excrete Se. These transformations are an important component of the marine Se biogeochemical cycle (Fig. 1B) (Zhang & Gladyshev, 2008). Furthermore, the fraction of practically defined organic Se species in the surface ocean has been correlated to free amino acid concentrations (Aono et al., 1990), suggesting that these organic Se fractions are produced *in-situ* by marine microalgae. Many marine microalgae require nM levels of Se to sustain optimal growth (Araie & Shiraiwa, 2009).

Marine microalgae may also be able to produce methylated, volatile Se compounds that can be emitted to the atmosphere. Picomolar concentrations of methylated Se compounds (e.g., dimethyl selenide

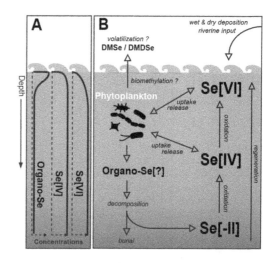

Figure 1. Depth profiles of the major Se species [i.e., practically defined organo-Se, Se(IV) and Se(VI)] in the oceanic water column with arbitrary concentration scales (Section A), and an overview of the major biogeochemical processes and Se species within the marine Se cycle, illustrating the central role of marine microalgae (Section B). (Cutter & Bruland, 1984).

and dimethyl diselenide) have been quantified in the Atlantic Ocean and have been related to biological activity within an algal bloom (Amouroux et al., 2001). Mass balance calculations have indicated that emissions of marine Se to the atmosphere may be large enough to significantly affect the global atmosphere Se budget (Wen & Carignan, 2007).

Direct evidence for the production of methylated Se compounds by marine microalgae remains absent. It is unclear whether these compounds are produced directly, and if so, by which microbial species, or indirectly, by degradation. If production is direct, it is expected to depend on intracellular Se concentrations. This process is likely to vary highly among different marine micro-algal taxa.

2 MICROCOSM EXPERIMENTS WITH MARINE MICROALGAE

To study the role of microalgae in the marine Se cycle and the production of methylated volatile Se compounds, we are performing microcosm experiments in which axenic cultures of marine microalgae are exposed to different concentrations of Se(VI) or Se(IV) (1 nM to 1 μM) in artificial seawater. Selenium uptake and metabolism is being investigated in two globally abundant, photosynthesizing unicellular marine microalgae, the silicifying diatom *Thalassiosira oceanica* and the calcifying coccolithophorid *Emiliania huxleyi*. Using a combination of analytical speciation methods (Vriens *et al.*, 2014, 2015), we are quantifying intracellular Se concentrations, concentrations of inorganic Se species in the exposure media, and volatile methylated Se compounds in the gaseous phase during algal growth. Parameters related to the physiology of cells (such as cell size and photosynthetic yield) are also being monitored.

3 ACTIVE SELENIUM UPTAKE BY MICROALGAE

After 24 h of exposure to concentrations between 1 nM and 1 μM of Se(VI) or Se(IV), we observed little change in culture growth rate, cell size or cellular ultrastructure in *T. oceanica* and *E. huxleyi*. Although the photosynthetic yield of *E. huxleyi* was slightly reduced after exposure to 1 μM Se(IV), these results indicate that both marine microalgae were relatively insensitive to acute Se exposure within the selcted concentration range. No indications for severe essentiality or toxicity of Se to these algae were observed within that Se concentration range.

Quantification of the intracellular Se concentrations after 24 h of exposure to variable concentrations of Se indicate that both algae accumulated Se(IV) up to a factor 10 more efficiently than Se(VI). Intracellular Se concentrations after 24 h reached up to 650 μmoles per liter cell volume and bioaccumulation factors exceeded 10,000 in experiments with low (<10 nM) Se concentrations. Results from the experiments with different concentrations of phosphate and silicate indicate that phosphate and silicate can compete with Se(IV) for the uptake via shared membrane-transporters. We are currently measuring the production of volatile methylated Se and sulfur in the microcosms with cultures of *E. huxleyi* and *T. oceanica*.

4 ENVIRONMENTAL IMPLICATIONS

Our research has shown that different marine microalgae vary in their ability to take up and metabolize Se(IV) and Se(VI), which is significant for Se biogeochemistry in the oceans. The ability of marine microalgae to produce methylated Se compounds is the focal point of our ongoing research. If marine microalgae are able to actively transform inorganic Se into volatile Se, they may be responsible for the large transfer of Se from the oceans to the atmosphere.

REFERENCES

Amouroux, D., Liss, P.S., Tessier, E., Hamren-Larsson, M. & Donard, O.F.X. 2001. Role of oceans as biogenic sources of selenium. *Earth & Planetary Science Letters* 189: 277–283.

Aono, T., Nakaguchi, Y., Hiraki, K. & Nagai, T. 1990. Determination of seleno-amino acid in natural water samples. *Geochemical Journal* 24: 255–261.

Araie, H. & Shiraiwa, Y. 2009. Selenium utilization strategy by microalgae. *Molecules* 14: 4880–4891.

Cutter, G.A. & Bruland, K.W. 1984. The marine biogeochemistry of selenium: a re-evaluation. *Limnology and Oceanography* 29: 1179–1192.

Cutter, G.A. & Cutter, L.S. 1998. Metalloids in the high latitude north Atlantic ocean: Sources and internal cycling. *Marine Chemistry* 61: 25–36.

Cutter, G.A. & Cutter, L.S. 2001. Sources and cycling of selenium in the western and equatorial Atlantic ocean. *Deep Sea Research Part II: Topical Studies in Oceanography* 48: 2917–2931.

Sherrard, J.C., Hunter, K.A. & Boyd, P.W. 2004. Selenium speciation in subantarctic and subtropical waters east of New Zealand: Trends and temporal variations. *Deep Sea Research Part I: Oceanographic Research Papers* 51: 491–506.

Vriens, B., Ammann, A.A., Hagendorfer, H., Lenz, M., Berg, M. & Winkel, L.H.E. 2014. Quantification of methylated selenium, sulfur and arsenic in the environment. PLoS ONE 9(7): e102906.

Vriens, B., Mathis, M., Winkel, L.H.E. & Berg, M. 2015. Quantification of volatile alkylated selenium and sulfur in complex aqueous media using solid-phase microextraction. *Submitted to Journal of Chromatography A.* 1407: 11–20.

Wen, H. & Carignan, J. 2007. Reviews on atmospheric selenium: Emissions, speciation and fate. *Atmospheric Environment* 41: 7151–7165.

Zhang, Y. & Gladyshev, V.N. 2008. Trends in selenium utilization in marine microbial world revealed through the analysis of the global ocean sampling (GOS) project. *PLoS Genetics* 4(6): e1000095.

Biogenic volatilization of nanoscale selenium particles in the soil-*Stanleya pinnata* system

J. Wang, R. Mahajan, L. Jones & Z.-Q. Lin*
Environmental Sciences Program, Southern Illinois University, Edwardsville, Illinois, USA

Y.H. Xie
College of Natural Resources and Environment, Shanxi Agricultural University, Shanxi, China

1 INTRODUCTION

Due to the rapid development of nanotechnology in recent years, more selenium (Se) nanoparticles (SeNPs) have been applied in various medical, nutritional, industrial, and remediation processes (Huang et al., 2015; Wang et al., 2015; Tran & Webster, 2013; Vekariya et al., 2012). Nanoscale Se particles have a high ratio of surface area to volume, which provides great driving forces for diffusion and becoming more reactive in the environment. As one of emerging contaminants, environmental impacts of SeNPs could be significantly different from bulk elemental Se and other Se compounds. Soil microorganisms play an important role in controlling the transport and fate of Se in the environment. Although previous studies have demonstrated the toxicity of SeNPs to different microbes (Li et al., 2008; Shi et al., 2011), little is known about the biotransformation or particularly biogenic volatilization of SeNPs in a soil-plant system.

The specific objective of this study was to determine the flux of Se volatilization in the soil- Princes'Plume (*Stanleya pinnata*) system treated with SeNPs, with the special focus on potential effects of microbial volatilization of SeNPs in the soil.

2 MATERIALS AND METHODS

Chemical synthesis of SeNPs: SeNPs were produced using selenous acid and ascorbic acid following the method by Li and Hua (2009). SeNPs were characterized using Malvern Zetasizer nano and NanoSight LM10. The average SeNP size was 75 nm in this study. The bacterial strain was previously isolated from the soil in the *Stanleya pinnata* field. Seeds of *Stanleya pinnata* were grown in six pots and each pot contained 1.5 kg soil (DW) that was treated with 1 mg/kg selenate. Prior to the maturity stage, whole plants were removed from three of the pots.

Volatilization measurement: Rates of Se volatilization were determined daily using the volatilization chamber and volatile Se trap solution (4:1 mixture of NaOH and H_2O_2, v/v) previously described by Shrestha et al. (2008). Rates of Se volatilization were measured daily from the soil pots without plants and the pots with *Stanleya pinnata*. Prior to the SeNP treatment, Se volatilization was measured for 4 days to obtain the background level of Se volatilization in the soil-plant system or in the soil without plants. Then, the soils were treated with 3 mg/kg of SeNPs. The volatilization measurement continued for another 7 days. Concentrations of Se in the trap solution were analyzed using ICP-MS (Agilent 7500cx).

Soil microbial volatilization of SeNPs was conducted using a soil bacterial strain *Psuedomonas sp.* that was previously isolated from the *S. pinnata* field. The bacterial cultural solution was treated with 5 mg/L of SeNPs in a 500 ml flask. With a starting bacterial OD_{600} of 0.1, the flasks were placed on the shaker with 125 rpm at room temperature. Each flask stopper had an inlet and outlet, and the inlet glass tube was attached with a 0.22 μm filter to remove airborne bacteria and possible airborne particulates, while the outlet was connected with two 250 ml gas washing bottles containing 100 ml of the trap solution.

3 RESULTS AND DISCUSSION

The rates of Se volatilized from the soil and the soil-plant system increased significantly after the soil was treated with SeNPs at the end of Day 4 (Fig. 1A), compared to the volatilization rates during the first four days. The soil-plant pots tended to volatilize more Se than the soil pots, but no statistically significant difference was observed.

The cumulative Se mass volatilized from the soil and the soil-plant systems increased with increasing time after the soils were treated with SeNPs (Fig. 1B). At the end of this experiment (Day 11), the cumulative Se mass volatilized from the soil and the soil-plant systems was 5.6 ± 1.9 μg/pot and 8.8 ± 3.5 μg/pot, respectively. No statistically significant ($p > 0.05$) difference was observed on Day 11 between the soil and the soil-plant systems, even though the difference between the means increased with increasing time after the SeNP treatment. The result also suggests that soil

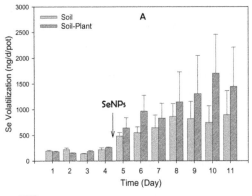

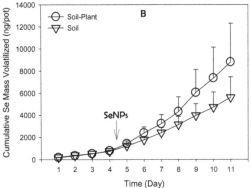

Figure 1. Rates of Se volatilization (A) and the cumulative Se mass volatilized (B) from the soil and the soil-*Stanleya pinnata* systems treated with SeNPs at the end of Day 4. Data shown are means and standard errors (n = 3).

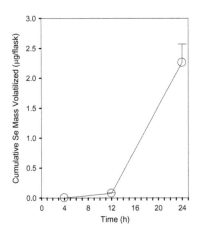

Figure 2. Cumulative Se mass volatilized from SeNP-treated cultural solution by *Psuedomonas sp.* during the first 24 hours from the inception of treatment. Data shown are means and standard errors (n = 3).

microbes play an important role in biotransformation or biogenic volatilization of SeNPs in the soil-plant system.

The bacterial Se volatilization process in the cultural solution treated with SeNPs showed a significant time delay (approximately 12 hours) from the inception of treatment (Fig. 2). The cumulative Se mass volatilized form SeNPs by *Psuedomonas sp.* that was associated with *Stanleya pinnata* increased significantly during the time period of 12 to 24 hours. The total amount of Se volatilized from each flask during that 12 hour time period was 2.26 µg, which accounts for 0.23% of the total SeNPs in each flask.

Using chemically synthesized SeNPs, this study demonstrated that, while bulk elemental Se is not bioavailable, elemental SeNPs can be bioavailable for biotransformation and volatilization in the soil-plant system and also in soil bacterial cultural solution. A 12-hour time-delay in volatilization of SeNPs was likely due to relatively slow bioprocesses of uptake and transformation of nanoscale elemental Se particles by the soil bacterium in the growth medium.

4 CONCLUSIONS

The results of this study demonstrate that SeNPs can be bioavailable in the soil and bacterial cultural solution for biotransformation and biogenic volatilization. Future studies need to explore potential effects of different chemical properties of SeNPs on Se volatilization, and to determine the importance of plant and soil microbial interaction in enhancing SeNP volatilization in the soil-plant system.

REFERENCES

Huang, L., Tong, X., Li, Y., Teng, J. & Bai, Y. 2015. Preparation of a novel supported selenium nanoparticles adsorbent and its application for copper removal from aqueous solution. J. Chem. Eng. Data, 60(1): 151–160.

Li, H., J. Zhang, T. Wang, W. Luo, Q. Zhou, & G. Jiang. 2008. Elemental selenium particles at nano-size (nano-Se) are more toxic to Medaka (*Oryzias latipes*) as a consequence of hyper-accumulation of selenium: A comparison with sodium selenite. Aquat. Toxicol., 89(4): 251–256.

Li, Z.L. & P.M. Hua. 2009. Mixed surfactant template method for preparation of nanometer selenium. E-Journal of Chemistry, 6, S1: S304–S310.

Shrestha, B., S. Lipe, K.A. Johnson, T.Q. Zhang, W. Retzlaff & Z.-Q. Lin. 2006. Soil hydraulic manipulation and organic amendment for the enhancement of selenium volatilization in a soil-pickleweed system. Plant Soil, 288: 189–196.

Tran, P.A. & Webster, T.J. 2013. Antimicrobial selenium nanoparticle coatings on polymeric medical devices. Nanotechnology, 24: 155101 doi:10.1088/0957-4484/24/15/155101.

Vekariya, K.K., J. Kaur, & K. Tikoo. 2012. ER alpha signaling imparts chemotherapeutic selectivity to selenium nanoparticles in breast cancer. Nanomedicine, 8: 1125–1132.

Wang, Q., Larese-Casanova, P., & Webster, T.J. 2015. Inhibition of various gram-positive and gram-negative bacteria growth on selenium nanoparticle coated paper towels. Int. J. Nanomedicine, 10: 2885–2894.

Global predictions of selenium distributions in soils

G.D. Jones*
Eawag: Swiss Federal Institute of Aquatic Science and Technology, Duebendorf, Switzerland

L.H.E. Winkel*
Swiss Federal Institute of Technology (ETH), Institute of Biogeochemistry and Pollutant Dynamics, Zürich, Switzerland
Eawag: Swiss Federal Institute of Aquatic Science and Technology, Duebendorf, Switzerland

1 GLOBAL SELENIUM DISTRIBUTION

Selenium (Se) is a vital trace element for human health. The risk of inadequate Se intake for good human health is governed in part by soil Se concentrations because Se intake occurs primarily through dietary consumption (e.g., grain). Soil Se is highly variable and typically ranges from ~0.01–2 mg/kg but can be >1,000 mg/kg in seleniferous soils. Standardized broad scale soil geochemical surveys have been conducted in few countries including China, Western Europe, and the United States (Zheng, 1994; Rawlins, 2012; USGS, 2014; Reimann, 2014), but for much of the globe, broad scale Se distributions are virtually unknown. Therefore, the health risks associated with Se deficiency/toxicity are also unknown.

The health risks associated with inadequate Se intake is at least in part related to the physicochemical properties that govern Se retention in soils. Several variables, including pH, soil organic matter, clay content, and others have been reported to influence the retention of Se in soils. While general trends have been reported to describe the relationship between these variables and Se in soils from small-scale experiments (Dhillon & Dhillon, 1999; Coppin et al., 2009; Winkel et al., 2015), such studies do not capture the environmental heterogeneity found in nature across broad scales. As a result, the dominant processes governing the biogeochemical cycling of Se in soils on larger scales are virtually unknown.

In addition to limited data availability across broad scales, many studies rely on traditional statistical analyses (e.g., linear regression, principal components analysis, discriminant function analysis) to describe biogeochemical cycles. These analyses, however, are often inappropriate for describing large scale cycles, due to inherent *a priori* assumptions (e.g., linearity). While on micro and small scales, relationships between predictor and dependent variables have been linear, as spatial scales increase these relationships becomes increasingly nonlinear and complex (Winkel et al., 2015), thus invalidating the use of traditional models. This complexity is likely a result of increased environmental heterogeneity on a large scale, due to different Se source contributions for example, resulting in highly nonlinear patterns. More sophisticated analyses are required to adequately handle the complexities of large-scale data to make accurate predictions of global soil Se concentrations.

For the first time, we attempt to predict soil Se concentrations on a global scale using national soil geochemical databases from North America, Europe, and Asia coupled with nonparametric statistical analyses to overcome the previously described limitations. The goals of this research are to (1) identify the dominant climatic, geologic, and soil physicochemical properties that govern spatial Se distributions in the soil and (2) develop a model to predict current soil Se concentrations on a global scale.

2 METHODS

Soil Se data were obtained from ~38,000 data points in China, Western Europe, and the USA from various geochemical surveys (USGS, 2014; Reimann, 2014; Zheng, 1994; Rawlins, 2012). To identify the dominant variables governing Se distributions, 18 global datasets describing climatic (e.g., aridity, temperature, evapotranspiration), soil (e.g., pH, clay content, bulk density), and other properties (e.g., lithology, land use) were collected and imported into ArcMAP. The predictor-variable data were extracted for each known Se data point.

To overcome the previously described analytical limitations, artificial neural network (ANN) models were used to identify the dominant variables governing Se distributions in the combined dataset. The main advantage of using ANN models over traditional techniques is that ANNs have no *a priori* data assumptions and are highly flexible non-linear models that can detect intricate patterns between dependent and independent variables. As a result, ANNs are ideally suited for predicting the global distribution of Se as a function of soil and climatic variables.

Data obtained from the geochemical surveys are inherently biased due to differences in sampling and analytical procedures. To minimize the bias of any particular dataset, the entire dataset was resampled using four different techniques: equal data representation, equal sampling density per data set, equal scale (all points within a 0.5° cell were averaged and represented by a single value), and randomly. The resampling procedure that resulted in the highest R^2 for the cross validation dataset was deemed the most appropriate procedure. All ANN models were developed with the free software R using the 'nnet' package.

3 RESULTS

The results of the ANN model indicate that climate variables were the most important for governing soil Se concentrations followed by soil chemical properties. Of the variables analyzed, categorical variables (e.g., geology, soil type, land use, etc.) were the least important in terms of governing the broad scale distribution of Se and were discarded from the dataset. Soil Se concentrations were predicted on a global scale with ∼70% accuracy. The resulting ANN model was used to quantitatively predict soil Se on a global scale. Such predictions could be used to evaluate the current status of Se in agricultural soils where no previous measurements have been made. Given that the most important variables are comparatively dynamic (e.g., climate vs. geology), these results suggest that the spatial distribution of Se in soils is likely to change through time.

REFERENCES

Barrow, N.J., Whelan, B.R. 1989. Testing a mechanistic model. viii. The Effects of time and temperature of incubation on the sorption and subsequent desorption of selenite and selenate by a soil. J. Soil Sci. 40: 29–37.

Coppin, F.; Chabroullet, C.; Martin-Garin, A. 2009. Selenite interactions with some particulate organic and mineral fractions isolated from a natural grassland soil. Eur. J. Soil Sci., 60: 369–376.

Dhillon, K.S. & Dhillon, S.K. 1999. Adsorption–desorption reactions of selenium in some soils of India. Geoderma, 93: 19–31.

Rawlins, B.G., McGrath, S.P., Scheib, A.J., Breward, N., Cave, M., Lister, T.R. 2012. The Advanced Soil Geochemical Atlas of England and Wales; British Geological Survey: Keyworth, UK. p. 227.

Reimann, C., Birke, M., Demetriades, A, Filzmoser, P., OConnor, P. 2014. Chemistry of Europe's agricultural soils. Part a: methodology and interpretation of the GEMAS data set; Bundesanstalt für Geowissenschaften und Rohstoffe: Hannover, Germany. p. 528.

United States Geologic Survey. 2014. Mineral resources online spatial data. Available online: Http://mrdata.usgs.gov/ds-801/.

Winkel, L.H.E., Vriens, B., Jones, G.D., Schneider, L.S., Pilon-Smits, E., Bañuelos, G.S. 2015. Selenium cycling across soil-plant-atmosphere interfaces: a critical review. Nutrients, 7: 4199–4239.

Zheng, C. (ed.). 1994. Atlas of Soil Environmental Background Value in the People's Republic of China; China Environmental Science Press: Beijing, China. p. 196.

Selenium in agroecosystems in tropical areas: A focus in Brazil

L.R.G. Guilherme*, G.S. Carvalho, E.C. Silva Júnior, L.B. Abreu, G.A. Souza & J.J. Marques
Department of Soil Science, Federal University of Lavras (UFLA), Lavras (MG), Brazil

1 INTRODUCTION

Selenium (Se) is an essential element to humans and other animals in trace amounts, but is harmful in excess. In fact, Se has one of the narrowest ranges between dietary deficiency (<40 μg/d) and toxic levels (>400 μg/d), which reinforces the importance of a better understanding of the relationships between environmental exposure and health (Fordyce, 2007). Research on Se content and behavior in soils has received special attention in recent times, particularly in countries where the role of Se on human and animal health has already been well recognized. However, the content of this element as well as its fate may vary considerably from region to region depending on the type of soil and environmental conditions. While some regions are well known to have low Se contents in the soil, others are classified as regions with high Se due to its availability. A better understanding of such aspects is key to define ways of increasing Se absorption by food crops, e.g., via agronomic biofortification, if applicable, or to avoid excessive transfer of this element into the food chain. This paper presents an overview on Se content and its behavior in tropical regions, providing insights on the particular situation related to Se research in Brazilian soils.

2 SELENIUM IN SOILS: FOCUS ON TROPICAL AGROECOSYSTEMS

Selenium is not homogeneously distributed in the earth's crust. Some particular countries, such as China and Brazil, possess Se deficient soils (<0.1 mg/kg) and soils with excess Se (>0.5 mg/kg), and often separated by a distance of only 20 km (Fordyce, 2007). The world's surface soils contain an average Se concentration of 0.33 mg/kg, but this concentration varies depending on the type of soil (Kabata-Pendias & Murkherjee, 2007).

Although the geological sources determine the Se content in soils, mobility and absorption of Se by plants and animals (i.e., Se bioavailability) is determined by a number of bio-physico-chemical parameters. These include soil pH and redox potential, forms or chemical species of Se, soil texture and mineralogy, organic matter (OM) content and the presence of competing ions (Singh et al., 1981; Vuori et al., 1994). An understanding of these parameters is essential for predicting Se deficiency or for remediating the health risks associated with excessive Se. Selenium occurs naturally in the soil environment in amounts that may cause nutritional toxicity or deficiency to livestock in areas throughout the world (Girling, 1984). Yet, even in soils containing sufficient Se, Se deficiency in animals may arise if the element is not in a readily available form (Fordyce, 2007).

In one of the earlier Brazilian studies 90 soil samples were collected from the Brazilian Cerrado, one of the most representative agricultural areas in Brazil. Carvalho (2011) observed a great variation in total Se concentration either in the surface (32–82 μg/kg) or in the subsurface horizons (22–63 μg/kg). The maximum Se concentration was less than the global average of 0.44 mg/kg soil (Kabata-Pendias, 2011). In more recent work, Gabos et al. (2014a) found soil Se concentrations in the State of São Paulo on both surface (0–20 cm) and subsurface (higher portion of B horizon) samples ranging from <0.089 to 1.612 mg/kg (average of 0.191 mg/kg, median of 0.111 mg/kg) for 58 soil samples. Half of the soils had less Se than the quantification limit (0.089 mg/kg), and 75% of the soil samples had less than 0.223 mg/kg, a value close to the background Se content set by the Environmental Agency of the State of Sao Paulo (0.25 mg/kg). These authors reported slightly higher values of total Se for subsurface than for surface soil samples. They also reported that Se in the superficial layer was positively correlated with cation exchange capacity (CEC), as well as with clay, organic matter (OM), and oxide contents of the soils. In the subsurface samples, only pH and aluminum oxide content were correlated with Se content.

Understanding how Se behaves in the soil and its interactions in the soil-water-plant system is critical for assessing the natural availability of Se or the soils ability to respond to its addition via agricultural inputs (Vuori et al., 1994). Abreu et al. (2011) evaluated selenate adsorption and desorption capacity of typical Cerrado soils in Brazil, and found that in sandy soils, which represented most of that area, Se tended to be less retained and could be easily available, whereas those soils with both higher clay content and Al and Fe oxides (gibbsitic and goethitic soils) showed higher affinity for Se, i.e., reduced bioavailability.

Many Brazilian soils are rich in Fe and Al oxides, which are known for being able to adsorb anions (e.g., phosphate, sulfate, selenite, and selenite), especially under acidic conditions. In fact, selenite is similar to phosphate, and tends to be much more strongly adsorbed than selenate that resembles chemical behaviors of sulfate in Fe and Al oxides (Gabos et al., 2014b; Goldberg, 2014). Phosphate is widely used in oxidic soils in Brazil, and gypsum containing sulfate is also recommended for improving subsurface soil conditions in tropical agriculture. In addition, lime is also an important input for assuring adequate agricultural yields in acidic soils. Therefore, by liming or adding phosphate and sulfate through fertilization practices, the availability of Se in the soil changes (Singh et al., 1981).

3 CONCLUSIONS

The background levels of Se in soils from Brazil reported so far are very low, i.e., most Brazilian soils can be considered deficient with respect to Se. For most cases, Se levels in soil were largely independent of parent material. In general the Se content may be related to OM and Al and Fe oxide content. Although total Se content is important to assess the current soil status of Se, measuring plant Se availability is still a challenge in the world. Since Se occurs in tropical upland soils mostly as selenate, and given the similarity of selenate with sulfate, we might have to extend our surveys on Se status in Brazilian agroecosystems not only to surface soils (0–20 cm), but also to subsurface soil layers. Actually, this is the recommended procedure for assessing plant available S in Brazil, i.e. analyses of $S-SO_4^{2-}$ have to be carried out in surface and subsurface soil horizons, as sulfate tends to be displaced by phosphate in the A horizon.

This new information on Se is especially relevant today because the Ministry of Agriculture, Livestock, and Supply (MAPA) is trying to deliver a normative (IN 5/2015) regulating the use of Se in Brazilian fertilizers. The recommended minimum level of Se in mixed fertilizers (fertilizers with macro and micronutrients) is 0.003% for soil application. Considering an average application rate of 400 kg of mixed fertilizer per ha in Brazil, it is expected that a fertilizer could deliver at least 12 g of Se per ha, which is a rate close to that recommended in countries with soil conditions similar to those in Brazil.

ACKNOWLEDGEMENTS

The authors thank CNPq, CAPES, and FAPEMIG for financial support.

REFERENCES

Abreu, L.B., Carvalho, G.S., Curi, N., Guilherme, L.R.G., & Marques, J.J.G.S.M. 2011. Sorção de selênio em solos do bioma cerrado. *Rev. Bras. Ci. Solo*, 35:1995–2003.

Carvalho, G.S. 2011. Selênio e mercúrio em solos sob cerrado. 93p. Tese (Doutorado em Ciência do Solo) – Universidade Federal de Lavras, Lavras-MG.

Gabos, M.B., Alleoni, L.R.F., & Abreu, C.A. 2014a. Background levels of selenium in some selected Brazilian tropical soils. *J Geochem. Explor.*, 145:35–39.

Gabos, M.B., Goldberg, S., & Alleoni, L.R.F. 2014b. Modeling selenium (IV and VI) adsorption envelopes in selected tropical soils using the constant capacitance model. *Environ. Toxicol. Chem.*, 33:2197–2207.

Girling, C.A. 1984. Selenium in agriculture and the environment. *Agr. Ecosyst. Environ.*, 11:37–65.

Goldberg, S. 2014. Macroscopic experimental and modeling evaluation of selenite and selenate adsorption mechanisms on gibbsite. *Soil Sci. Soc. Am. J.*, 78: 473–479.

Fordyce, F. 2007. Selenium geochemistry and health. *Ambio*, 36:94–97.

Kabata-Pendias, A. 2011. *Trace Elements in Soil and Plants*, 4 ed. Boca Raton: CRC press. 505 p.

Kabata-Pendias, A., & Mukherjee, A.B. 2007. *Trace Elements from Soil to Human*. New York: Springer. 550 p.

Singh, M., Singh, N., & Relan, P.S. 1981. Adsorption and desorption of selenite and selenate selenium on different soils. *Soil Sci.*, 132:134–141.

Vuori, E., Vääriskoski, J., Hartikainen, H., Kumpulainen, J., Aarnio, T., & Niinivaara, K. 1994. A long-term study of selenate sorption in Finnish cultivated soils. *Agr. Ecosyst. Environ.*, 48:91–98.

Selenium soil mapping under native Brazil nut forests in Brazilian Amazon

K.D. Batista*
Embrapa Roraima, Boa Vista, Roraima, Brazil

K.E. da Silva & G.C. Martins
Embrapa Amazônia Ocidental, Manaus, Amazonas, Brazil

L.H.O. Wadt
Embrapa Rondônia, Porto Velho, Rondônia, Brazil

L.M. da Silva
Embrapa Acre, Rio Branco, Acre, Brazil

N.J.M. Júnior & M.C. Guedes
Embrapa Amapá, Macapá, Amapá, Brazil

R.C. de Oliveira Júnior
Embrapa Amazônia Oriental, Belém, Pará, Brazil

C.A.S. Magalhães & A.B.B. Tardini
Embrapa Agrossilvipastoril, Sinop, Mato Grosso, Brazil

1 INTRODUCTION

Brazil nut (*Bertholletia excelsa* Bonpl., Lechytidaceae family) is one of the symbol trees of the Amazon. It is an important non-timber forest product for human livelihoods in the Amazon region due to its social, ecological, and economical importance (Wadt & Kainer, 2009). It is a dominant species that occupies the upper canopy of the forest, and influences the dynamics of gaps and forest succession. Nevertheless, the traditional harvesting system of nuts in most producing areas is still characterized at a low technological level. The reason is due to the changes in the local level of social organization around harvesting, to land tenure/use, and the diversity of production systems, which need to be better characterized. The variation in nut productivity in different Amazon regions is due to different biotic and abiotic relationships at their natural growing sites. These relationships demand further studies for better understanding the biotic and abiotic relationships and their influence on fruit production.

Another relevant aspect of Brazil nut is its high level of selenium (Se) in its nuts (Lemire et al, 2010; Parekh et al., 2008). The species has been considered an important source of Se (Gabos et al., 2014) and its concentration in nuts is highly variable, with values from 5 to 512 mg/kg of Se (Dumont et al., 2006; Chang et al., 1995). Although there is a variation in Se content in Brazil nuts, it is known that the ingestion of a single nut per day is enough to meet the daily human requirement for Se (Thomson et al., 2008).

The Se content in food usually depends on Se concentrations in soil (Gabos et al., 2014). In soil, the content of the element can be explained by the mineralogy (Winkel et al., 2012) or other soil characteristics, e.g. organic carbon (Perez-Roca et al., 2010), or by human activities (Gabos et al., 2014).

Because of the great importance of Brazil nut, Embrapa approved in 2013 the project "Brazil Nut Mapping and Environmental, Social and Economic Characterization of Production Systems in Amazon (MapCast)". The objectives were: (1) generate database and methodologies for mapping and modeling native Brazil nut; (2) characterize production factors and nut's production systems; (3) determine the social, environmental, and economic factors that can affect the Brazil nut production.

Among the specific objectives from MapCast, we reference the study that evaluated physical and chemical soil characteristics, including the evaluation of Se content in six different regions of Brazilian Amazon. Describing variations in Se soil levels at the landscape scale is very important to understand the behavior of this element in the Amazon soils. This information will make it possible to correlate the results with Se concentrations in nuts, and greatly contribute to the enhancement of nuts production chain for the human population.

Without knowing the different relationships involved in the Brazil nut harvest, and population variations as a result of environmental diversity, there is no way to move forward on the recommendations of practical

Brazil nut management in each respective Amazon region. The multi-disciplinarily aspect of the MapCast project will provide results that will support many other research activities. This integration is critical for generating positive impact in the development of Brazil nut production.

2 MATERIALS AND METHODS

MapCast is led by Embrapa Amazônia Ocidental (Manaus-AM), with 59 members composed by academics and researchers. Its research area comprises six Amazon States: Acre, Amapá, Amazonas, Mato Grosso, Pará and Roraima. The project was started in 2014 and will extend through 2018, encompassing about 776 soil samples collected in the Amazon. Among its research, the Se study is related to the "Characterization and environment modeling of natural occurrence of Brazil nut".

Soil samples were collected in a regular grid of 30×50 m, at 0–20 cm depth, using the UTM-WGS84 coordinate system for spatial location. The samples were brought to the laboratory for drying in the air forced circulation stove at 40°C and sieved in 2 mm mesh. Afterwards they were sent to the Soil Analysis Laboratory of the Federal University of Lavras, Minas Gerais for Se analysis.

Using geostatistics, regionalized maps will be created for the variable soil Se concentrations, resulting in a realistic mapping of Se content. Thus, comparisons of changes in the levels of soil Se will be made among different sampled locations and correlation analyses will be performed between fruit production and Se content. All statistical analyzes will be made in the R program (R Core Team, 2015).

3 EXPECTED RESULTS

Regionalized maps for Se will be made for each sampled location, showing the Se behavior in natural Brazil nut of Amazon soils. In the MapCast project, results will support additional studies of correlation between fruit yield and Se levels. We expect (based upon results), to perform additional studies on Se content in Brazil nuts at the same level of accuracy and comprehension as this study. Results will allow an accurate evaluation of this important non-timber forest product, as well as the aspects related to food safety for human consumers.

REFERENCES

Chang, J.C., Gutenmann, W.H., Reid, C.M., & Lisk, D.J. 1995. Selenium content of Brasil nuts from two geografic locations in Brasil. Chemosphere 30: 801–802.

Dumont, E.L. De Pauw, F. Vanhaecke, & R. Cornelis. 2006. Speciation of Se in Bertholletia excelsa (Brazil nut): A hard nut to crack? Food Chemistry 95: 684–692.

Gabos, M. B., Alleoni, L.R.F., & Abreu, C.A. 2014. Background levels of selenium in some selected Brazilian tropical soils. Journal of Geochemical Exploration 145: 35–39.

Lemire, M., Fillion, M., Barbosa, F., Guimarães, J.R.D. & Mergler, D., 2010. Elevated levels of selenium in the typical diet of Amazonian riverside populations. Sci. Total Environ. 408: 4076–4084.

Parekh, P.P., Khan, A.R., Torres, M.A. & Kitto, M.E., 2008. Concentrations of selenium, barium, and radium in Brazil nuts. J. Food Compos. Anal. 21: 332–335.

R Core Team. 2012. R: A language and environment for statistical computing. R Foundation for Statistical Computing, Vienna, Austria. ISBN 3-900051-07-0, http://www.R-project.org/.

Roca-Perez, L., Gil, C., Cervera, M.L., Gonzálvez, A., Ramos-Miras, J., Pons, V., et al. 2010. Selenium and heavy metals content in some Mediterranean soils. J. Geochem. Explor. 107: 110–116.

Thomson, C.D., Chisholm, A., McLachlan, S.K., & Campbell, J.M. 2008. Brazil nuts: an effective way to improve selenium status. Am J Clin Nutr 87: 379–384.

Wadt, L.H.O & Kainer, K.A. 2009. Domesticação e melhoramento da castanheira. In: Borém, M.T.G.L. & Charles, R.C. (eds.), Domesticação e melhoramento: espécies amazônicas: 297–318. Viçosa, MG.

Winkel, L.H.E., Johnson, C.A., Lenz, M., Grundl, T., Leupin, O.X., Amini, M., et al. 2012. Environmental selenium research: from microscopic processes to global understanding. Environ. Sci. Technol. 46: 571–579.

Fate of selenium in soil and engineered suboxic and anoxic environments

L. Charlet*, B. Ma & A. Fernandez Martinez
Earth Science Institute (ISTerre), University Grenoble Alps, CNRS, France

R.M. Couture
Norwegian Institute for Water Research, Gaustadallé, Norway

M.R. Broadley
School of Biosciences, University of Nottingham, Sutton Bonington Campus, Loughborough, UK

A.D.C. Chilimba
Lunyangwa Agricultural Research Station, Mzuzu, Malawi

1 INTRODUCTION

Selenium (Se) deficiency is associated with various ill health effects including; increased risk of mortality, articular cartilage disease, and cognitive decline (Rayman, 2012) and is estimated to affect the health of up to 1 billion people worldwide (Haug et al., 2007). Dietary Se intake is less available in developing countries, with a 22% risk of Se deficiency in Africa (Hurst et al., 2013), 52% in the East African region, and up to 91–100% in Kenya (Joy et al., 2014). In Africa, consumed food products are mainly produced locally and a deficiency may be directly correlated to low soil micronutrient contents or micronutrients being unavailable for plant uptake. There is generally a strong relationship between Se status in crops and the soils on which they grow (Steinnes, 2009). Selenium concentrations in the soil under 0.1 mg/kg are considered Se deficient, while concentrations greater than 0.5 mg/kg are considered seleniferous (Dhillon & Dhillon, 2003). However, within this rather narrow range, great differences in soil Se bioavailability is observed, depending on Se sorption mechanism as will be discussed, with a special focus on Malawi soil and crop Se content.

2 METHODS

The speciation of Se in soils, engineer barriers, and their main constituents has been determined by an array of techniques such as (1) element specific X-ray absorption spectroscopy (XAS), combining X-ray absorption near edge structure (XANES) and extended X-ray absorption fine structure (EXAFS) spectroscopies, (2) Se-O bond specific ATR-FTIR, together with (3) solid Mössbauer and PDF of HE-XRD techniques (to follow for instance structural change in FeS/FeSe solid solution). These bulk techniques are now used in situ to determine on-time speciation change during the course, e.g. of an anion exchange reaction. In addition, micro-focused XAS coupled to μXRF, μXRD and μ-FTIR can be used on soil sections to better understand (and thereafter model with a geochemical model) the localization of Se on different soil organic or mineral fractions, as these techniques can provide detailed information at microscopic spatial scale. The porosity distribution and migration of Se in of clay rich solid phases has been investigated by X-ray μ-tomography. Soil pore water sample Se speciation is analyzed by HPLC-HG-AFS.

3 SOIL SELENIUM SORPTION MECHANISMS

The fate of Se in soil depends on an array of redox reactions, diffusion, adsorption and precipitation processes or interactions with organic matter and biota, as well as Se present in unweathered primary mineral or fertilizers. We shall discuss the following species, their formation mechanisms (Fernandez Martinez & Charlet, 2009; Winkler et al., 2011) and importance in various soil types:

- Se complexation to organic matter through the formation of Se metal-humic ternary complexes
- Outer-sphere (weak; OSC) and inner-sphere (strong; ISC) complexeation of respectively selenate and selenite ions with various oxides and clays, including aluminosilicate nanotubes. The relative importance of OSC and ISC depends also on pH, soil humidiy.
- Biotic and abiotic formation of nano Se(0) and of FeS/FeSe solid solutions
- Volatilization of methyl Se species
- Precipitation of $CaSeO_4$ and $CaSeO_3$, and at lower concentrations, substitution Se(IV) within calcite and in calcareous soils.

4 RESULTS

The fate of Se is much contrasted depending on whether soil is acidic or not. In acidic conditions, both selenate and selenite ions form little bioavailable bidentate innersphere surface complex with Fe oxyhydroxides, Al oxyhydroxides (gibbsite) or tubular clay mineral (imogolite). This results in Se being nearly 100% adsorbed, thus poorly bioavailable although KH_2PO_4-extractable. In contrast, when soil pH value increases beyond pH 7, the sorption isotherms show a marked decrease in Se being immobilized on soil particles. This "adsorption edge" behavior is common to clays, imogolite and oxides. However, as opposed to a classical "adsorption edge" where the amount sorbed goes from 100% to 0% within a few pH units, as a response of OH^- and HCO_3^- increased concentration in the pore water, the retention of selenite ions remains significant even at high pH. At extreme pHs, such as the values found in cemented engineered barriers, anion exchange minerals, such as ettringite and Afm, can sorb selenite ions in their interlayer in a reversible manner; this complex structure is under investigation. At equilibrium with lime one even observes the precipitation of $CaSeO_3$, which then is the solid controlling the large (bio)availability of Se in pore water.

Finally anoxia may drastically affect the fate of Se in soils. Anoxia not only occurs upon soil flooding but also within microorganisms or, locally in vertisols upon swelling during the rainy season. In reductive conditions, Se° nanoparticles formation is little reversible under oxic conditions, i.e. slowly bioavailable upon reestablishment of oxic conditions. However our experiments also showed this reductive precipitation mechanism not to occur when other electron acceptors were present in the flooded soil, in which case Se is not reduced and remains soluble and thus bioavailable (Couture et al., 2015). The on–off switch mobility of Se in soils during redox oscillations are well described by a mixed kinetic-equilibrium biogeochemical model (Couture et al., 2015). In the case of vertisols redox switch during swelling/shrinking cycles may act as a slow Se delivery system and lead to high Se maize content.

5 CONCLUSION

In Malawe, the high bioavailability of Se denoted by high Se content in maize, when maize is grown on calcareous alkaline vertisols – compared to acidic soils – may result from a variety of mechanisms, namely: (1) the weakening of inner-sphere complexes (e.g. from bi- to mono-dentate) as pH increases, (2) the local precipitation of calcium selenite and partially reversible reductive precipitation of Se° and FeS/FeSe, which result in (3) the on-off switch mobilization of Se upon oxic/anoxic cycles, i.e. during swelling/shrinking cycles of these clayey (e.g. vertic) and Ca rich soils.

REFERENCES

Broadley, M.R., Chilimba, A.D.C., Joy, E.J., Young, S.D., Black, C.R., Ander, L.E. et al. 2012. Dietary requirements for magnesium, but not calcium, are likely to be met in Malawi based on national food supply data. *Int. J. Vitam. Nutr. Res.*, 82(3): 192–199.

Charlet, L., M. Kang, F. Bardelli, R. Kirsch, A. Géhin, J.M. Grenèche & F. Chen. 2012. Nanocomposite pyrite-greigite reactivity toward Se(IV)/Se(VI). *Environmental Science and Technology*, 46: 4869–4876.

Chilimba, A.D., Young, S.D., Black, C.R., Rogerson, K.B., Ander, E.L., Watts, M.J. et al. 2011. Maize grain and soil surveys reveal suboptimal dietary selenium intake is widespread in Malawi. *Scientific Reports* 1: 72. doi: 10.1038/srep00072.

Couture, R.M., Charlet, L., Markelova, E., Madé, B. & Parsons, C. 2015 On-off mobilization of contaminants in soils during redox oscillations. *Environ. Sci. Technol.* 49(5): 3015–3023.

Dhillon, K.S. & S.K. Dhillon 2003 Distribution and management of seleniferous soils. In: D. Sparks (ed.), Advances in Agronomy, 1st edn. Elsevier, Amsterdam, pp. 119–18.

Fernández-Martínez, A. & L. Charlet, 2009 Selenium environmental cycling and bioavailability: A structural chemist point of view. *Reviews in Environmental Science and Biotechnology*, 8: 81–110.

Haug, A., Graham, R.D., Christophersen, O.A. & Lyons, G.H. 2007. How to use the world's scarce selenium resources efficiently to increase the selenium concentration in food. *Microb Ecol Health Dis.* 19:209–228.

Hurst R, E.W.P. Siyame, S.D. Young, A.D.C. Chilimba, E.J.M. Joy, C.R. Black, et al. 2013. Soil-type influences human selenium status and underlies widespread selenium deficiency risks in Malawi Sci. Rep. 3, 1425.

Joy J.M., E.J.M. Joy, E.L. Ander, S.D. Young, C.R. Black, M.J. Watts, A.D.C. Chilimba, et al. 2014 Dietary mineral supplies in Africa *Physiol. Plant.* DOI: 10.1111/ppl.12144.

Kang, M., B. Ma, F. Bardelli, F. Chen, C. Liu, Z. Zheng, et al. 2013. Interaction of aqueous Se(IV)/Se(VI) with FeSe/ FeSe2: Implication to Se redox process. *Journal of Hazardous Materials*, 248–249: 20–28.

Steinnes, E. 2009 Soils and geomedicine *Environ. Geochem. Health* 31: 523–35.

Rayman, M.P. 2012 Selenium and human health *Lancet*, 379: 1256–68.

Winkel, L.H.E., C.A. Johnson, M. Lenz, T. Grundl, O.X. Leupin, M. Amini & L. Charlet. 2011. Environmental Selenium Research: From Microscopic Processes to Global Understanding. *Environmental Science & Technology*, 46:571–579.

Soil selenium contents, spatial distribution and their influencing factors in Heilongjiang, China

Feng-qin Chi*, En-jun Kuang, Jiu-ming Zhang, Qing-rui Su & Dan Wei
Soil Fertilizer and Environment Resources Institute, Heilongjiang Academy of Agriculture Sciences, Key Laboratory of Soil Environment and Plant Nutrition of Heilongjiang Province, Harbin, China

Qiang Xu
Northeast Agricultural University, Harbin, China

1 INTRODUCTION

Selenium is unevenly distributed on the surface of the earth. Ure and Berrow (1982) reported a Se concentration range of 0.03–2.0 mg/kg and a mean of 0.40 mg/kg for 1623 soils throughout the world. In China Tan and Zhu (2002) suggested that there existed a low Se disease belt from northeast to southwest with a Se content range of 0.022–3.806 mg/kg. Heilongjiang Province is the national base for the major commodity grain in China, which just lies in the starting end of low Se belt. Until now, knowledge on the soil Se content at different regions or in soils in Heilongjiang Province are not well documented. The specific objectives of this study were to (1) evaluate the total Se content in soils collected from all parts of Heilongjiang Province, and (2) investigate factors in related to the physio-chemical properties of the soils.

2 MATERIALS AND METHODS

2.1 Study area

Heilongjiang Province is located in the northeast of China (43°26′–53° 33′N, 121° 11′–135° 05′E) with a total land area of 473000 km². The area of arable land is 395020 km², accounting for 83.5% of the total land area of the province. Heilongjiang Province is the coldest province of China with temperate continental monsoon climate (Heilongjiang Provincial Bureau of Land Management, 1992).

2.2 Analytical methods

In total, 223 topsoil samples were collected from main soil types in the Province. Soil samples free of plant roots were air-dried, passed through a 100-mesh sieve for total Se determination using the HJ680-2013 method (China National Environmental Protection Standards, HJ680-2013). Concentrations of Se in the solution were determined using atomic fluorescence spectrophotometer with hydride generation.

Table 1. General physio-chemical properties of the soil samples (n = 223).

Soil properties	A.M.	Range	S.D.	C.V. (%)
pH (H$_2$O)	7.15	5.08~11.44	1.18	16.50
SOC (g kg^{-1})	23.40	2.13~100.80	12.08	51.62
Clay (%)	14.55	5.93~24.49	3.53	24.26
Sand (%)	28.22	0.53~56.33	10.66	37.77

The certified reference materials analyzed along with each batch of samples in this study was GBW-07424 (GSS-10, China). Soil particle size distribution was measured by laser particle size analyzer and Soil organic carbon (SOC) was measured by HT1300 dry method with multi N/C 2100 Analyzer. Soil pH was determined in water extracts in a soil to water ratio (1:2.5). Soil properties are shown in Table 1. SPSS software was used for the descriptive statistics and the Pearson correlation coefficient between total Se and physio-chemical properties of the soil.

3 RESULTS AND DISCUSSION

3.1 Total se content

According to the Se content division of ecological landscape safety threshold by Tan (1996), the following concentrations are designated as follows: deficient Se soil (<0.125 mg/kg), marginal Se (0.125–0.175 mg/kg), sufficient Se (0.175–0.450 mg/kg), Rich Se (0.450–2.000 mg/kg), high Se (2.000–3.000 mg/kg), excess Se (>3.000 mg/kg). For all 223 soil samples, total Se content ranged from 0.008 to 0.475 mg/kg, with an arithmetic mean of 0.137 mg/kg. The arithmetic mean was comparable to previous study, which reported 0.1605 mg/kg for soils in Heilongjiang Province (Xu, 1986). This value was far lower than the average value of the country (0.29 mg/kg) and close to the average crustal abundance of 0.130 mg/kg (Liu, 1996; Rudnick & Gao,

Table 2. Total selenium content of different soils types in Heilongjiang Province.

Soil type	Samples (n)	A.M. (mg/kg)	Range (mg/kg)	S.D.	C.V. (%)
Meadow soil	14	0.169	0.026–0.254	0.070	41.42
Lessive	18	0.153	0.029–0.275	0.065	42.48
Phaeozem	133	0.145	0.013–0.475	0.077	53.10
Chernozem	21	0.093	0.008–0.286	0.065	69.89
Dark Brown soil	27	0.118	0.039–0.218	0.055	46.61
Lithological soil	4	0.132	0.083–0.201	0.050	37.88
Saline-alkali soil	6	0.089	0.046–0.139	0.031	34.83
Total	223	0.137	0.008–0.475	0.073	53.28

Table 3. Total Se content of different regions in Heilongjiang Province.

Location	Samples (n)	Mean (mg/kg)	Range (mg/kg)	S.D.	C.V. (%)
Heihe	17	0.212	0.118~0.341	0.062	29.25
Qiqihaer	16	0.186	0.082~0.475	0.102	54.84
Suihua	42	0.151	0.036~0.291	0.064	42.38
Yichun	3	0.172	0.093~0.230	0.071	41.28
Jiamusi	27	0.130	0.028~0.286	0.077	59.23
Shuangyashan	16	0.124	0.008~0.256	0.085	68.55
Daqing	18	0.128	0.046~0.202	0.040	31.25
Harbin	56	0.098	0.013~0.261	0.055	56.12
Mudanjiang	13	0.129	0.039~0.218	0.053	41.09
Jixi	10	0.138	0.043~0.265	0.069	50
Qitaihe	5	0.163	0.251~0.286	0.060	36.81

Table 4. Pearson correlations between total Se and soil properties (n = 223).

	Se (mg/kg)	SOC (g/kg)	pH	Clay (%)	Sand (%)
Se	1	0.164*	0.003	0.225**	−0.149*
SOC		1	−0.265**	0.012	0.049
pH			1	0.229**	−0.302**
Clay				1	−0.701**
Sand					1

*: $p < 0.05$; **: $p < 0.01$.

2003). Total soil Se content in Heilongjiang Province belonged to marginal Se and approaching the deficient Se soil category.

3.2 Total Se content in different soil type and region

ANOVA showed a significant difference of Se content among various soil types. The Se content ranged in the following descending order: Meadow soil (0.169) (arithmetic mean, mg/kg) > Lessive (0.153) > Phaeozem (0.145) > Lithological soil (0.132) > Dark brown soil (0.118) > Chernozem (0.093) > Saline-alkali soils (0.089) (Table 2). The Se contents in chernozem and saline-alkali soils were significantly lower than in other soils.

Differences were observed in soil Se content in different regions (Table 3). The Se contents in the different regions ranged in the order of Heihe (0.212) (arithmetic mean, mg/kg) > Qiqihar (0.186) > Yichun (0.172) > Qitaihe (0.163) > Suihua (0.151) > Jixi (0.138) > Jiamusi (0.130) > Mudanjiang (0.129) > Daqing (0.128) > Shuangyashan (0.124) > Harbin (0.098). Higher variation coefficient of soil Se was found in Qiqihar, Jiamusi, Shuangyashan and Harbin districts.

3.3 Relationship between soil Se content and soil properties

A correlation analysis indicated that total Se content was positively correlated with SOC and soil clay content, but negatively correlated with sand content ($P < 0.01$) (Table 4). No significant relation was observed between the total Se content and soil pH.

4 CONCLUSIONS

Total soil Se content in Heilongjiang Province ranged from 0.008 to 0.475 mg/kg, with the mean value of 0.137 mg/kg, was lower than the average value of 0.29 mg/kg in China. In terms of soil type, meadow soil had a relatively high Se content (0.169 mg/kg), while Chernozem and saline-alkali soils had relatively low Se content. Among different regions, Heihe District had the highest total Se content, averaging 0.212 mg/kg, while Harbin had the lowest Se content, averaging 0.098 mg/kg. Correlation analysis showed that SOC and soil clay content were the main factors affecting soil selenium content.

REFERENCES

Heilongjiang Provincial Bureau of Land Management. 1992. *Heilongjiang soil*. Compiled by the Soil Survey Office of Heilongjiang Province. Agricultural Press. (in Chinese).

Liu Zheng. 1996. *Chinese Soil Trace Elements*. Nanjing: Jiangsu Science and Technology Press. (in Chinese).

Rudnick, R.L. & Gao, S. 2003. Composition of the continental crust//Holland, H.D. & Turekian, K.K. (eds), *Treatise on Geochemistry*. Amsterdam: Elsevier. 1–64.

Tan Jian'an. 1996. *Environmental life elementsand Keshan disease*. Beijing: Chinese Medical Science and Technology Press. (in Chinese).

Tan Jian'an. & Zhu Wenyu. 2002. Selenium in soil and endemic diseases in China. *Science of the Total Environment*. 284(1): 227–235.

Ure, A.M. & Berrow, M.L. 1982. The elemental composition of soils. *In Environmental Chemistry*, Vol. 2, H.J.M. Bowen (ed), 94–204, Royal Soc. Chem., London.

Xu Qingchun. 1986. Selenium content and distribution in soil and feed in Heilongjiang Province. *Journal of Northeast Agricultural College*. 17(4):399–406. (in Chinese).

Selenium sorption in tropical agroecosystems

G. Lopes*, L.R.G. Guilherme, A.M. Araujo & J.H.L. Lessa
Soil Science Department, Federal University of Lavras (UFLA), Lavras (MG), Brazil

1 INTRODUCTION

Selenium (Se) in soils depends of the geological parent material, among other factors. Generally, sandy soils have a lower Se content compared to organic and calcareous soils (El-Ramady et al., 2014). The presence of Se in the soil is important because as plants take up elements from the soil it can transfer these elements to animals that consume it. Selenium is well known for its essentiality in animals and humans (Tinggi, 2003). However, Se can also be toxic to organisms in higher concentrations, although, essential and toxic characteristics appear within a narrow concentration range (Lenz & Lens, 2009).

The deficiency of Se in soils has been reported worldwide, including in Brazilian soils (Carvalho, 2011; Gabos et al., 2014). Soils with a low Se content are not enough able to produce food with adequate Se levels. This result can contribute to Se deficiency in humans. Therefore, efforts have been addressed to understand the Se behavior in agroecosystems, especially to determine which factors may alter its availability and mobility in soils. This knowledge allowing us to add more Se into soils, and increase Se contents in the crops at safe levels.

In this presentation, we will discuss Se adsorption on some Brazilian soils, which have different soil characteristics, such as organic matter and clay percentages. Also we will address the effect of soil management comparing cultivated soils with uncultivated soils on soil Se availability.

2 SELENIUM SORPTION ON SOILS

Sorption behavior of Se depends, among other factors, of the Se speciation. In this context, selenate is less sorbed than selenite, thus, the last form is strongly adsorbed on solids, having low mobility in soils. Several studies have showed that selenite is adsorbed much more strongly than selenate on different solid surfaces, such as in soils (Eich-Greatorex et al., 2010). The difference in adsorption behavior of selenate and selenite leads to different Se contents that are plant available, which are higher for selenate than selenite due to its lower adsorption capacity.

Selenate and selenite may compete with organic acids as well as with other anions, such as phosphate and sulfate for adsorption sites. This fact is relevant for agroecosystems, particularly, for tropical soils taking into account that these soils usually receive high amounts of phosphate and sulfate from fertilizers and gypsum.

3 INFLUENCE OF SOIL MANAGEMENT

Selenium sorption behavior differs for soils due to differences in the soil management or production systems. In most cases, soil management practices include liming and fertilizations. Besides change the pH, an important soil characteristic that influence Se adsorption, the soil management may also modify chemical properties of the soil, causing variation in its adsorption capacity. For example, Øgaard et al. (2006) reported that cattle manure affects selenium behavior in soils, decreasing the sorption of both Se species, selenate and selenite.

There are studies in the literature showing the effect of phosphate under Se adsorption on soils. Most of these studies showed that the Se become more plant available following phosphate addition, i.e., the Se sorption decreases, as it is well reported for Japanese soils (Nakamura et al., 2006; Nakamura & Sekine, 2008). It has to be mentioned that this trend is more pronounced for selenite than selanate. This occurs due to chemical similarities between phosphate and the selenite anion (Eich-Greatorex et al., 2010), while the anion selenate tends to compete with sulfate. These information may assist explain the fact of Nakamura and Sekine (2008) found that selenite sorption did not change after an increase of the sulfate concentration.

4 FINAL REMARKS

Selenium is an essential micronutrient for animals and humans that has a narrow concentration range between the essentiality and toxicity. Thus, Se needs to be consumed in adequate amounts to express its positive functions in the organisms. Therefore, a relevant way to increase Se levels in the diet is through the plants, which can be reached adding Se in the soils to be transferred to crops. In this context, understand sorption behavior of Se on soils previous the Se addition is critical to better know the amounts of Se that are safe to be applied in the soils, which depend of soil characteristics as well as of the Se speciation.

In the focus of the above statement, we will present the results for Se adsorption on some selected Brazilian soils. Besides comparing cultivated soils with uncultivated soils, the studies that we will address also evaluated the effect of Se concentration added, ionic strength as well as other soil characteristics upon selenate sorption.

REFERENCES

Carvalho, G.S. 2011. Selênio e mercúrio em solos sob cerrado. p93. Tese (Doutorado em Ciência do Solo), Universidade Federal de Lavras, Lavras-MG.

Eich-Greatorex, S., Krogstad, T. & Sogn, T.A. 2010. Effect of phosphorus status of the soil on selenium availability. Journal of Plant Nutrition and Soil Science, 173: 337–344.

El-Ramady, H.R., Domokos-Szabolcsy, É., Abdalla, N.A., Alshaal, T.A., Shalaby, T.A., Sztrik, A., et al. 2014. Selenium and nano-selenium in agroecosystems. Environmental Chemistry Letters, 12: 459–510.

Gabos, M.B., Alleoni, L.R.F. & Abreu, C.A. 2014. Background levels of selenium in some selected Brazilian tropical soils. Journal of Geochemical Exploration, 145: 35–39.

Lenz, M. & Lens, P.N.L. 2009. The essential toxin: The changing perception of selenium in environmentam sciences. Science of the Total Environment, 407: 3620–3633.

Nakamura, Y.M. & Sekine, K. 2008. Sorption behavior of selenium and antimony in soils as a function of phosphate ion concentration. Soil Science and Plant Nutrition, 54: 332–341.

Nakamaru, Y., Tagami, K. & Uchida, S. 2006. Effect of phosphate addition on the sorption-desorption reaction of selenium in Japanese agricultural soils. Chemosphere, 63(1): 109–115.

Øgaard, A.F., Sogn, T.A & Eich-Greatorex, S. 2006. Effect of cattle manure on selenate and selenite retention in soils. Nutrient Cycling in Agroecosystems, 76: 39–48.

Tinggi, U. 2003. Essentiality and toxicity of selenium and its status in Australia: a review. Toxicology Letters, 137: 103–110.

Ionic strength effects upon selenate adsorption in cultivated and uncultivated Brazilian soils

A.M. Araujo, J.H.L. Lessa, G.A. Souza, L.R.G. Guilherme & G. Lopes*
Soil Science Department, Federal University of Lavras (UFLA), Lavras (MG), Brazil

1 INTRODUCTION

Selenium (Se) is considered an essential micronutrient for animals and humans due to its great importance in the antioxidant system, and in the human's defense against free radicals (VIARO, 2001). Moraes (2008) reported that the deficiency of essential elements, such as Se, is a well-known concern for human health worldwide, especially for developing countries. In addition, the low Se intake seems to be related to the low Se concentration in the diet (Pedrero, 2006).

Selenium concentration in plants vary depending on the amount and availability of this element in the soil, as well as genetic characteristics of the plants. Thus, the concentration of Se in plants may be related to its concentration found in soils. However, it also depends on the Se chemical speciation and on the soil characteristics, such as pH, clay percentage, and presence of competing anions (Faria & Karp, 2015).

Studies assessing sorption behavior of Se in Brazilian soils are relevant to understand how available or adsorbed Se occurs in such soils. In this context, we aimed to evaluate the effect of ionic strength (IS) on selenate adsorption in cultivated an uncultivated Brazilian soils.

2 MATERIALS AND METHODS

Sixteen soil samples from 8 different locations in Brazil were tested for Se adsorption. At each sampling location we collected two soils: the first one from a cultivated area and the second one from an uncultivated area (in most cases a native area never cultivated). All soil samples were dried at room temperature and sieved through a 2-mm sieve for the adsorption experiment as well as for soil characteristics described by Embrapa (1997) (Table 1).

For the adsorption experiment, 2 g of each soil were weighed in centrifuge tubes of 50 mL. Each tube received 20 mL of solution containing 100 µg/L of Se as Na_2SeO_4. The sodium selenate solution previously had pH adjusted to approximately 5,5 and it was prepared in a solution of sodium chlorite (NaCl) in both IS tested, 15 and 150 mmol/L. The adsorption time was 72 h, alternating 12 h of shaking and 12 h of rest. Then, the samples were centrifuged for 20 min at 3200 rpm and the supernatant was analyzed for Se using graphite furnace atomic absorption spectrometry (GFAAS).

The quantity of Se adsorbed was calculated by the difference between Se concentration initially added and the Se concentration in the analyzed supernatants.

Table 1. Maximum, mean, and minimum values of attributes of the soil samples studied.

	pH	S	P	OM	Clay	Silt	Sand
		mg/dm³		%			
Cultivated soils							
Maximum	6.5	48.62	91.8	9.59	58	33.5	92.5
Mean	5.6	15.22	30.6	3.46	37.3	13.8	48.8
Minimum	4.7	2.97	4.32	1.35	6	1.5	8.5
Uncultivated soils							
Maximum	5.7	7.03	3.53	4.6	79	21	95
Mean	5.2	5.42	1.94	2.95	40.2	10	49.7
Minimum	4.7	3.32	0.84	1.29	5	0	5

3 RESULTS AND DISCUSSION

Figure 1 shows the amounts of selenate adsorbed in the soils studied. The amount of selenate adsorbed was higher for uncultivated soils (Fig. 1B) compared to cultivated soils (Fig. 1A). In this context, range of Se adsorbed in the soils was approximately 206 µg/kg to 900 µg/kg for uncultivated soils and 216 µg/kg to 450 µg/kg for cultivated ones. The smaller adsorption capacity observed for cultivated soils may be explained due to the presence of competing anions, such as sulfate and phosphate, which are included during the soil management to achieve high crop yields, as shown in Table 1.

Similar studies conducted also reported that Se adsorption decreased in the presence of the following competitive anions: SiO_3^{2-}, SO_4^{2-}, PO_4^{3-}, and CO_3^{2-}, mainly in the case of phosphate (Zhang et al., 2010). According Neal (1995), the preference for the adsorption of anions occurs as follow: phosphate>arsenate>selenite>silicate>sulfate>selenate>nitrate>chlorite. This relationship explains the lower adsorption

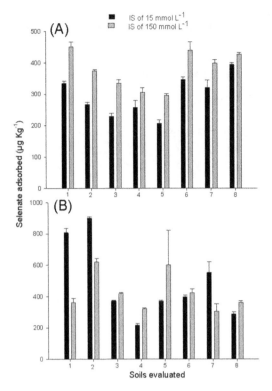

Figure 1. Amounts of selenate adsorbed (μg/kg) on cultivated (A) and uncultivated (B) soils for the two ionic strengths (IS) assessed.

of selenate found for cultivated soils in the present study.

As can be seen in Figure 1, the amounts of Se adsorbed also vary as a function of IS for both types of soils, cultivated (Fig. 1A) and uncultivated (Fig. 1B) soils. Because the adsorption capacity of Se was different in the two IS tested, we can infer that the main adsorption mechanism of selenate on these soils seems to be the outer-sphere complex or termed as non-specific adsorption. This fact agrees with other studies reported in the literature (Neal, 1995). The major adsorption mechanism may change depending on the pH, presence and concentration of competing ions, and mineral surface (PEAK, 2006).

4 CONCLUSIONS

Selenium adsorption, in the form of selenate, was lower in the cultivated soils compared to the uncultivated soils. Also, this study showed that IS affects selenate adsorption on the Brazilian soils.

ACKNOWLEDGEMENTS

The authors thank the CNPq, CAPES, and FAPEMIG for financial support. Also, we would like to acknowledge the Department of Soil Science for providing the structure used to perform this study.

REFERENCES

Embrapa. Centro Nacional de Pesquisa de Solos. 1997. Manual de métodos de análise de solos. 2 ed. rev. e atual.: EMBRAPA, 212p. Rio de Janeiro.

Faria, L.A. & Karp, F.H.S. 2015. Selênio: Um elemento essencial ao homem e aos animais e benéficos às plantas. Informações International Plant Nutrition Institute-Brasil (IPNI), informações agronômicas N° 149 ISSN 2311-5904.

Moraes, M.F. 2008. Relação entre nutrição de plantas, qualidade de produtos agrícolas e saúde humana. Informações Agronômicas, n. 123, p. 21–23. Piracicaba.

Neal, R.H. 1995. Selenium. In: ALLOWAY, B.J. Heavy metals in soils. 2 edition. New York, Wiley, p.260–283. New York.

Peak, D. 2006. Adsorption mechanisms of selenium oxyanions at the aluminum oxide/water interface, Journal of Colloid and Interface Science, New York, v. 303, p. 337–345.

Pedrero, Z., Madrid, Y. & Cámara, C. 2006. Selenium species bioaccesibility in enriched radish (*Raphanus sativus*): A potential dietary source of selenium. Journal of Agricultural and Food Chemistry 54: 2412–2417.

Viaro, R. S., Viaro, M.S. & Fleck, J. 2001. Importância bioquímica do selênio para o organismo humano. Ciências Biológicas e da Saúde, Santa Maria, v. 2, n. 1.

Zhang, N., Gang, D. & Lin, L. 2010 Adsorptive removal of ppm-level selenate using iron-coated GAC adsorbents. Journal Environ. Eng. 136: 1089–1095.

Soil cultivation affects selenate adsorption in Cerrado soils in Brazil

J.H.L. Lessa, A.M. Araujo, L.R.G. Guilherme & G. Lopes*
Department of Soil Science, Federal University of Lavras (UFLA), Lavras (MG), Brazil

1 INTRODUCTION

Selenium (Se) is essential for both animals and humans and enters the food chain via uptake from plants (Lisk, 1972). However, plants (e.g. foods) may contain different amounts of Se, depending on the Se content found in the soil where the plants were grown.

Regarding Se concentrations in Brazilian soils, recent studies evaluated some selected soils from São Paulo state and reported that the concentration of Se in these soils is low (Gabos et al., 2014). Also, Carvalho (2011) measured the concentrations of Se in samples of the Brazilian Cerrado soils and found low values, ranging from <22 to 81 μg/kg.

Selenium can occurs in soils mainly as selenate (SeO_4^{2-}) and selenite (SeO_3^{2-}), a form predominant in acid soils. Because of the negative charges, Se forms are attracted to positive charges, affecting Se mobility and availability to plants (Nakamaru et al., 2005). In the literature, there are few studies evaluating the adsorption capacity of Se in Brazilian soils (Abreu et al., 2011; Gabos et al., 2014). In this context, studies evaluating the Se behavior in terms of retention in Brazilian soils are still required. Additionally, more field information is needed to understand Se adsorption on tropical soils following additions of small amounts of Se to soil. Therefore, this study was aimed at assessing the adsorption of Se, in the form of selenate, in some selected, cultivated and uncultivated, Brazilian soils located in the state of Mato Grosso.

2 MATERIALS AND METHODS

Soil samples selected for this study were collected at the farm Agua Limpa, located in the state of Mato Grosso, Brazil. In the farm, we collected at the 0–20 cm depth two soil samples (cultivated and uncultivated) at three different locations. Thus, a total of 6 soil samples were tested in the present study and the uncultivated soils were used as a reference to assess the effect of soil management (e.g. fertilization) upon the Se adsorption capacity. Table 1 shows basic characterizations of both group of soils, cultivated and uncultivated.

For the adsorption experiment, 2 g of each soil were weighed in 50 mL centrifuge tubes. Then, 20 mL of a Se solution containing one of the following Se concentrations, as Na_2SeO_4 (μg/L): 0, 25, 50, 100, 200, 400, 800, 1200, 1600, and 2000 was added. The respective sodium selenate solution previously had the pH adjusted to approximately 5.5 and it was prepared in a solution with 15 mmol/L of NaCl. After adding the respective Se solution, an adsorption time of 72 h was allowed, alternating 12 h of shaking and 12 h of rest. Each treatment was assessed in triplicate. After sample preparation, the Se analyses were performed using graphite furnace atomic absorption spectrometry (GFAAS).

The data obtained from the adsorption tests were fitted to the Freundlich isotherm. For that, the Freundlich isotherm equation is as follows: $q_e = k_f C_e^{1/n}$, where q_e = Se adsorbed (mg/kg); C_e = equilibrium solution concentration (mg/L); k_f = measure of adsorptive capacity; and n = adsorption intensity.

For determining the Freundlich parameters to be applied in the equation mentioned before, we used previously the linear form of the Freundlich isotherm, as follows: $Log\ q_e = Log\ K_f + 1/n\ Log\ C_e$.

Table 1. Chemical and physical properties of the experimental soils.

Soils	pH	S	P	OM	Clay	Silt	Sand
		mg/dm³			%		
Location 1							
Uncultivated	5.4	4.0	2.0	2.9	51	27	22
Cultivated	5.4	18	32	2.9	33	11	56
Location 2							
Uncultivated	5.0	6.3	2.3	3.1	18	1	81
Cultivated	5.7	5.1	17.4	3.1	41	7	52
Location 3							
Uncultivated	4.7	5.5	2.3	2.1	16	6	78
Cultivated	5.6	3.0	18.3	2.6	11	5	84

3 RESULTS AND DISCUSSION

Adsorption data were well fitted to the Freundlich isotherm, as seen with the high correlation coefficients found for all six soils studied ($r^2 > 0.93$) (Table 2). Abreu et al. (2011) evaluated Se sorption on soils from the Cerrado biome also verified high correlation

Table 2. Freundlich isotherm parameters for the adsorption of Se on the soils studied.

Soils	n	K_f (L/kg)	r^2
Location 1			
Uncultivated	1.27	77.58	0.99
Cultivated	1.35	10.41	0.97
Location 2			
Uncultivated	1.26	9.18	0.99
Cultivated	1.20	7.94	0.99
Location 3			
Uncultivated	1.26	12.77	0.99
Cultivated	1.48	15.17	0.93

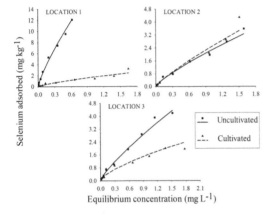

Figure 1. Selenium adsorbed (mg/kg) on the soils studied. The points indicate experimental data, while the lines were obtained with the Freundlich isotherm.

coefficients following fitting the data to the Freundlich model.

Figure 1 compares the adsorption of Se in cultivated and uncultivated soils in the three locations studied. The Se adsorption capacity was lower for cultivated soils compared to uncultivated soils, except in the location 2, where the soil used as reference (uncultivated) has much less clay than the cultivated soil, thus, reducing its potential to adsorb Se (Fig. 1). Therefore, soil cultivation can influence the Se adsorption ability. The reduction of Se adsorption following the soil cultivation may be explained due to the soil management applied to those soils. It is well known that cultivated soils, especially in tropical areas, receive high amounts of phosphate fertilizers due to the great importance of this nutrient for the crops and the deficient contents of P found in native areas (uncultivated soils). Thus, the presence of higher amounts of competing anions, such as phosphate, in the cultivated soils (Table 1) leads to the competition between those anions and the selenate for adsorption sites, decreasing Se retention. The presence of phosphate decreasing the Se adsorption capacity has been also reported for Japanese soils (Nakamura, 2006).

The greater amount of Se adsorbed among all soil samples was observed for the uncultivated soil of the location 1. This soil contains the greatest percentage of clay (Table 1), allowing it to adsorb more Se than other soil samples. Therefore, the soil clay percentage is an important soil characteristic to increase the adsorption capacity, since this amplifies the specific superficial area of the soil, which agrees with other studies (Abreu et al., 2011).

4 CONCLUSIONS

The soil characteristics, such as clay percentage and amount of competing anions, influence the adsorption of selenate on the soils in this study. Therefore, the amounts of Se adsorbed increased and decreased upon increasing the clay and competitive anion contents, respectively.

ACKNOWLEDGEMENTS

The authors thank the CNPq, CAPES, and FAPEMIG for financial support.

REFERENCES

Abreu, L.B., Carvalho, G.S., Curi N., Guilherme, L.R.G. & Marques, J.J.G.S.M. 2011. Sorção de selênio em solos do bioma cerrado. Revista Brasileira de Ciência do solo, 35: 1995–2003.

Carvalho, G.S. 2011. Selênio e mercúrio em solos sob cerrado.93p. Tese (Doutorado em Ciência do Solo) – Universidade Federal de Lavras, Lavras-MG.

Gabos, M.B., Alleoni, L.R.F.& Abreu, C.A. 2014. Background levels of selenium in some selected Brazilian tropical soils. Journal of Geochemical Exploration, 145: 35–39.

Lisk, D.J. 1972. Trace metals in soils, plants and animals. Advances in Agronomy, 24: 267–311

Nakamaru, Y., Tagami, K. & Uchida, S. 2005. Distribution coefficient of selenium in Japanese agricultural soils. Chemosphere, 58(10): 1347–1354.

Nakamaru, Y., Tagami, K. & Uchida, S. 2006. Effect of phosphate addition on the sorption-desorption reaction of selenium in Japanese agricultural soils. Chemosphere, 63(1): 109–115.

The effect of calcination on selenium speciation in selenium-rich rock

H.Y. Zhang*, M. Liu, Y. Xiao & Z.Y. Bao
School of Earth Sciences, China University of Geosciences, Wuhan, China

C.H. Wei
Faculy of Materials Science and Chemistry, China University of Geosciences, Wuhan, China

X.L. Chen
Zhejiang Institute of Geological Survey, Zhejiang, China

1 INTRODUCTION

Selenium (Se) is an indispensable nutrition element for humans and animals, but only within a very narrow concentration range between toxicity and deficiency (Schrauzer, 2000). Studying total Se and its distribution in rocks is insufficient for predicting how much Se will enter the food chains, because Se mobility and bioavailability are overwhelmingly dependent on Se chemical species. Recently, researchers have studied the transformation of Se from rocks into groundwater, soils, and, finally, the human body via the food chain; however, the Se speciation transformation within the rock itself before and after heating is still poorly understood.

Many solid sorbents can be used to trap Se during coal combustion, such as calcium materials (Diaz-Somoano et al., 2010; Xu et al., 2013). This study used a sequential extraction protocol to study the distribution and the transformation/speciation of Se in Se-rich rock samples before and after heating, using calcium material ($CaCO_3$) as a sorbent for trapping Se.

2 MATERIALS AND METHODS

Rock samples (slate) in this study were collected from Ziyang county in Shanxi Province, China. They were air-dried at 25°C, and then crushed to less than 200 μm diameter powder for chemical analysis. Then, 5 g of the sample and 0.5 g of calcium material ($CaCO_3$) were mixed, and placed in a pre-heated electric oven for 30 min at 400°C.

A five-step sequential extraction protocol was used according to Kulp & Pratt (2004) and Zhang et al. (1999) for Se speciation in Se-rich rock before and after heating, as described in Table 1. The Se concentrations in different species of Se were as follows: water soluble Se (F1), exchangeable Se (F2), organic bound Se (F3), sulfide/selenide Se (F4) and residue Se (F5). Selenium recovery from the five fractions ranged from 85.56 to 91.84% (average recovery of 87.49%) of the total Se value.

Table 1. Sequential extraction procedure.

Fraction	Extraction	Method
F1 Water soluble	DDI water	2 h, 25°C
F2 Ligand exchangeable	K_2HPO_4-KH_2PO_4	2 h, 25°C
F3 Organic bound	0.1 M NaOH	2 h, 90°C
F4 Sulfide/selenide	$KClO_3$ + HCl	1 h
F5 Residue	$HNO_3 + H_2O_2 + HF$	digestion

The total Se (Se_{total}) and the residual Se that remains after the sequential extraction of the complete rock sample was digested by a mixed acid digestion (i.e. HNO_3, HF, and $HClO_4$). The sample digestion was carried out in triplicate, as well as including two blanks and two standard reference materials to each digestion procedure. The Se concentration was measured by Hydride Generation-Atomic Fluorescence Spectrometry (HG-AFS) (Titan AFS-933, Beijing). The measured value of the reference materials were at least >90% of the certified value. The Se concentration in the whole rock sample was 39.53 ± 0.91 mg/kg.

3 RESULTS AND DISCUSSION

Figure 1 and Table 2 show Se concentrations before heating and after heating. In both situations (i.e. before and after heating), the total Se concentrations of the samples remained the same, and the highest Se speciation was still organic bound (F3). Furthermore, in both situations, the concentrations of Se in F1, F2 and F5 increased from 2.82 to 4.95%, 5.65 to 8.07%, and 8.64 to 19.32% of the total Se, respectively; whereas, the Se concentration of F3 and F4 decreased from 51.11 to 37.62% and 20.25 to 17.15% of the total Se, respectively. These results show that post the heating treatment stable Se species (F3 and F4) transform into less stable Se species (F1 and F2). However, the concentration of the most stable Se species, residue Se (F5), is higher than before.

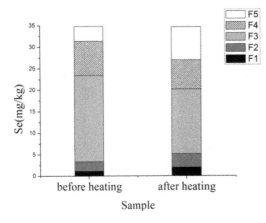

Figure 1. The change of speciation of Se before and after heating.

Table 2. Concentrations and speciation of Se before and after heating. STD: standard deviation.

	HP1			HP2		
	Mean	STD	Range	Mean	STD	Range
F1	1.11	0.08	1.06–2.21	1.97	0.25	1.80–2.25
F2	2.22	0.13	2.10–2.35	3.21	0.36	2.92–3.62
F3	20.09	0.94	19.20–21.07	14.96	0.40	14.55–15.34
F4	7.96	0.32	7.40–8.03	6.82	0.27	6.58–7.11
F5	3.40	0.46	2.90–3.80	7.68	1.29	6.19–8.38
Total Se Recovered:						
	34.51	0.61	33.90–35.11	34.63	0.41	34.17–34.92
Total Se Present:						
	39.31	1.07	38.23–40.36	39.75	0.89	39.20–40.77
Se Recovery Rate (%):						
	87.79		85.56–91.84	84.15		85.63–88.80

4 CONCLUSIONS

At a temperature of 400°C, organic bound Se (F3) and sulfide/selenide Se (F4) could partially transform to water soluble Se (F1), ligand exchangeable Se (F2) and residue Se (F5). In other words, the stable Se species can transform to less stable Se species. The increase of residue Se (F5) may be dependent on the composition of the rock and the temperature. In both situations (i.e. before and after heating), the total Se concentration of the sample remained the same and the highest Se speciation was still organic bound (F3), with calcium material ($CaCO_3$) as Se sorbent. This research is useful for result tracing geochemical processes of Se in rock under certain conditions.

REFERENCES

Diaz-Somoano, M., Lopez-Anton, M.A., Huggins, F.E. & Martinez-Tarazona, M.R. 2010. The stability of arsenic and selenium compounds that were retained in limestone in a coal gasification atmosphere. *J Hazard Mater,* 173: 450–4.

Kulp, T.R. & Pratt, L.M. 2004. Speciation and weathering of selenium in upper cretaceous chalk and shale from South Dakota and Wyoming, USA. *Geochimica et Cosmochimica Acta,* 68, 3687–3701.

Schrauzer, G.N. 2000. Selenomethionine: A review of its nutritional significance, metabolism and toxicity. *J. Nutr,* 130.

Xu, S.R., Shuai, Q., Huang, Y., Bao, Z. & Hu, S. 2013. Se capture by a CaO-ZnO composite sorbent during the combustion of Se-rich stone coal. *Energy & Fuels,* 27, 6880–6886.

Zhang, Z., Zhou, L.Y. & Zhang, Q. 1999. Speciation of selenium in geochemical samples by partial dissolution technique. *Rock and Mineral Analysis,* 16(4): 255–261 (in Chinese).

The fraction of selenium in cumulated irrigated soil in Gansu Province, China: Effect of aging

J. Li, S.Y. Qin, P.Y. Feng, N. Man & D.L. Liang*
College of Natural Resources and Environment, Northwest A&F University, Shaanxi, China
Key Laboratory of Plant Nutrition and the Agri-environment in Northwest China, Ministry of Agriculture, Yangling, Shaanxi, China

1 INTRODUCTION

Selenium (Se) is an essential trace element for humans and animals (Sager, 2006). In recently years, Se as fertilizer and industrial raw materials was applied in agriculture and industry, which make soil and water become the major sink of Se contamination, and eventually posing some potential problems such as toxicity to plants and animals. When Se is added into soils, its toxicity is not related to total Se content but largely depends on the Se fractions and their bioavailability in soil. However, the decrease in metal mobility and redistribution with time is thought to occur after metal addition to soil. And these processes was referred to as "aging" (Settimio et al., 2014). Aging plays a critical factor in determining metal bioavailability and toxicity in soil. The existence of aging process influences the metal bioavailability/toxicity between natural soil in field and in newly Se spiked soil (Tolu et al., 2014). However, these differences have been ignored by most of research, which may result in the overestimate on the exposure and ecological risk of Se. In addition, the aging of other metals in soil have been studied but little is known on the aging of Se in soil. Hence, it is therefore important to understand the fate and mobility of Se added into soil over time to (1) predict the long-term changes of Se fractionation and its transport in soil environment; (2) determine the equilibrium time of Se reaching steady at incubation periods; and (3) provide further insight into risk assessment of Se contaminated soil.

2 MATERIALS AND METHODS

Non-contaminated soil samples classified as cumulated irrigated soil was collected at the topsoil (0–20 cm) from Gansu province in China. The basic physiochemical properties of soil are shown in Table 1. The soil was air-dried and sieved < 0.15 mm. One kilogram of dried soil was weighed into plastic bag and spiked to the 2.5 mg/kg using Na_2SeO_3 in distilled water. The Se-spiked soil was mixed and then kept under incubation at 25°C. The soil moisture was maintained at 60% water holding capacity by weighing

Table 1. Physiochemical properties of tested soil.

pH	7.91
Organic matter (g/kg)	14.24
CEC (cmol/kg)	11.23
Carbonate (g/kg)	81.40
Amorphous iron (g/kg)	1.67
Amorphous aluminum (g/kg)	1.24
Total selenium (µg/g)	0.376
Clay (%)	20.6
Silt (%)	65.1
Sand (%)	14.3

and adding daily distilled water, during the incubation period. At the different incubation times (days) of 4, 10, 18, 33, 56, 85, 109, 141, 229, 285, 364 after Se was added, a subsample of soil was removed from plastic bag, air dried and then sieved, and then used for the Se fractionation analysis. A five-step sequential extraction method described by Hu et al. (2014) was applied for Se fractionation analysis. According to the method, Se is partitioned into soluble (SOL-Se), exchangeable (EX-Se), Fe/Mn oxide-bound (FMO-Se), and organic matter-bound (OM-Se). Data were analyzed by ANOVA and Duncan's multiple comparison tests in SPSS 20.0.

3 RESULTS AND DISCUSSION

After Se was added into soil, it would redistribute among different soil fractions. The distributions of Se fractionation with aging time are shown in Figure 1. Selenium was found predominantly in the exchangeable Se and Fe/Mn oxide-bound Se fractions, accounting for 56.4% and 26% of total Se content after 4 d incubation, followed by soluble Se, at 11.3%, and organic matter-bound Se was the lowest (at 7.0%). Moreover, after 364 d aging, the order of Se fractionation in the proportion of total Se was consistent with that first 4 d aging.

In addition, soluble Se and exchangeable Se in the proportion of total Se in soil decreased sharply within the first 56 d incubation, and then reduced

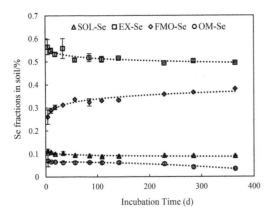

Figure 1. Soil selenium fractions changes with aging during 364 d incubation.

slightly with no significance decline after 56 d incubation ($P < 0.01$). The distribution of added Se into soil solid phase fractions appears to be consisted of two-phases over aging time was also observed in the metals of Cu, Zn, Cd and Pb aging (Jalali & Khanlari 2008; Liang et al. 2014). Otherwise, the ratio of organic matter-bound Se in total Se also diminished with increasing incubation time; but decreased lightly before a sharp decline in the first 229 d. In contrast to the first three fractionations, Fe/Mn oxide-bound Se in soil increased rapidly with the contact time between soil and pollutant, and then increased at slow rates. The amount of decrease in soluble, exchangeable and organic matter-bound Se fractions after 364 d incubation was 91.0%, 50.5%, 96.6%, respectively, compared with the 4 d aging, while Fe/Mn oxide-bound Se increased by 61.9%.

In general, those metals that can be readily absorbed and utilized by plants, can be defined as available fractionation. Soluble and exchangeable metals in soil are considered readily mobile and bioavailable for plant uptake (Tang et al. 2006). In irrigated soil, our results demonstrated that aging had a significant influence on Se fractionation distribution. With time, Se bioavailability decreases as soluble and exchangeable Se become more stable as Fe/Mn oxide-bound Se. Furthermore, these results also indicated that the transformation of both soluble Se and exchangeable Se in soil reached equilibrium within 56 d incubation.

4 CONCLUSIONS

The change of Se fractions was a result of aging in soil. A 56 d incubation could be the equilibrium time of Se aging in cumulated irrigated soil to provide the environmental quality criterion of Se. However, although the diminishing Se toxicity and bioavailability in soils during the aging process was verified by this study, the mechanism of Se aging still needs further study. Further studies also should testify how large the difference between the aging of short-term artificial Se-spiked soils in greenhouse compares to long-term naturally Se-contaminated fields.

REFERENCES

Jalali, M., & Khanlari, Z.V. 2008 Effect of aging process on the fractionation of heavy metals in some calcareous soils of Iran. *Geoderma* 143: 26–40.

Hu, B., Liang, D. L., Liu, J., Lei, L., & Yu, D. 2014. Transformation of heavy metal fractions on soil urease and nitrate reductase activities in copper and selenium co-contaminated soil. *Ecotoxicology and Environmental Safety* 110(0): 41–48.

Liang, S., Guan, D. X., Ren, J. H., Zhang, M., Luo, J., & Ma, L. Q. 2014. Effect of aging on arsenic and lead fractionation and availability in soils: Coupling sequential extractions with diffusive gradients in thin-films technique. *Journal of Hazard Material* 273C: 272–279.

Sager, M. 2006. Selenium in agriculture, food, and nutrition. *Pure and Applied chemistry* 78: 111–133

Settimio, L., McLaughlin M. J., Kirby, J. K., Langdon, K. A., Lombi, E., Donner, E., et al. 2014. Fate and lability of silver in soils: effect of ageing. *Environment Pollution* 191: 151–157.

Tang, X. Y., Zhu, Y. G., Cui, Y. S., Duan, J., & Tang, L. 2006. The effect of ageing on the bioaccessibility and fractionation of cadmium in some typical soils of China. *Environment International* 32: 682–689.

Tolu, J., Di Tullo, P., Le Hecho, I., Thiry, Y., Pannier, F., Potin-Gautier, M., et al. 2014. A new methodology involving stable isotope tracer to compare simultaneously short- and long-term selenium mobility in soils. *Analytical and Bioanalytical Chemistry* 406: 1221–1231.

Comparative study on the extraction methods and the bioavailability of soil available selenium

Zhaoshuang Wang, Shuxin Tu* & Dan Han
College of Resources and Environment, Huazhong Agricultural University, Wuhan, China
Hubei Collaborative Innovation Center for Grain Industry, Yangtze University, Jingzhou, China

1 INTRODUCTION

Practical food production shows that selenium (Se)-enriched soils may not produce Se-enriched agricultural products, which is due to no significant correlation between soil total Se and plant Se contents. For example, the relationship between the Se content in rice, wheat and potato and the total Se content of soil is poor (Zhang et al., 2012); while soil Se bioavailability was the main factor to determine the Se content of wheat (Zhao et al., 2009). Researchers have explored the analytical methods of soil available Se, but there is no uniform standard agreed upon for analyzing soil available Se (Tolu et al., 2011).

In this study, 14 different chemical extractants were employed to screen for the best chemical analytical method to determine available Se. The relationship between soil Se contents and the Se uptake of plants. We studied different plant species grown in Se-enriched soils in a seedling experiment.

2 MATERIALS AND METHODS

The soils tested were 10 different Se-enriched soils from Enshi, China, with the total Se concentrations ranging from 0.46 to 9.22 mg/kg. The 14 different chemical extractants for the measurement of the soil available Se were as follows: 0.5 mol/L $NaHCO_3$ (pH8.5), 0.03 mol/L NH_4F-0.025 mol/L HCl, 0.01 mol/L $CaCl_2$, 0.1 mol/L KH_2PO_4 (pH5.0), 0.25 mol/L KCl, 0.2 mol/L K_2SO_4, 0.05 mol/L EDTA, 0.01 mol/L $Ca(H_2PO_4)_2$ H_2O, deionized water, M3 extractants, 0.05 mol/L EDTA-0.05 mol/L NaOH, 0.05 mol/L NaOH, 0.05 mol/L KOH. The four plant species selected to grow in the soils for seedling Se uptake experiments were *Nicotiana tabacum*, *Brassica juncea*, *Oryza sativa*, and *Astragalus sinicus*. The total Se concentrations in plant tissues were determined based on dry biomass after sampling and sample preparation.

3 RESULTS AND DISCUSSION

3.1 *The content of soil available Se*

The soil available Se contents determined by sodium bicarbonate (including organic Se) were the highest among the 14 methods in the 10 soils. The available Se contents ranged from 50.76 to 292.94 μg/kg, and the average value was 132.58 μg/kg, accounting for 6.18% of the total soil Se. The average available Se contents by potassium dihydrogen phosphate, monocalcium phosphate, sodium bicarbonate, sodium hydroxide, potassium hydroxide were as follows (μg/kg): 28.10, 13.27, 72.19, 86.00, and 85.53, respectively, accounting for 1.13, 3.36, 4.01, 0.62 and 3.98% of the total soil Se, respectively. The available Se contents extracted via the other eight methods were very low, and could not be further studied (data not shown).

Soil Se fractionation showed that the average contents of soluble Se, exchangeable Se, and Fe-Mn-oxides-Se, organic sulfide binding state, and element Se and residual Seas follows (μg/kg): 34.94, 101.18, 33.3, 559.38, and 428.94 μg/kg, respectively, accounting for 1.63, 4.71, 15.51, 26.05, and 19.98% of the total soil Se, respectively.

Correlation analysis showed that there was no significant correlation between the available Se content and the total soil Se content ($p<0.05$), except the available Se content was determined by the sodium hydroxide method. This result indicated that total soil Se was not a good indicator for soil Se availability. However, with the exception of soluble Se content, the total soil Se content was positively correlated with other four chemical forms of Se. This finding was not similar to the research findings by Tan et al. (2002), who reported that there was significant correlation between total soil Se contents and soil soluble Se content, suggesting that total soil Se may be an index of soil slow-released Se.

3.2 *The correlation between plant Se and soil Se concentrations*

Correlation analysis between tobacco plant Se uptake, soil total Se content, and available Se content showed that there was no significant correlation with the different parts of the Se content in flue-cured tobacco. Available Se as determined by the method of extractants, i.e., potassium dihydrogen phosphate, sodium bicarbonate (including organic Se), sodium bicarbonate, had significant correlation with Se content in central, upper leaves and roots of flue-cured tobacco ($p < 0.05$). The correlation of sodium bicarbonate (including organic selenium) method was the best, and the correlation coefficients were 0.8189, 0.7371, and 0.8931, respectively. All the available Se analysis methods had extremely significant correlation with the Se content in flue-cured tobacco roots, but not with Se content in the lower leaves.

Soil available Se and Se content in *Indian mustard*, milk vetch, and rice were also significantly correlated, and the method of sodium bicarbonate (including organic Se) correlation was the best. Zhao et al. (2004) had also achieved similar results in their rye seedling experiments.

The available Se contents extracted by sodium bicarbonate (including organic Se) method had the best correlation with Se uptake of all four plants tested in this experiment. This may be due to the fact that the extractants include more soil available Se since this method adopts $K_2S_2O_8$ for decomposition of organic Se in the extractants to release more available Se from the soils.

4 CONCLUSIONS

The sodium bicarbonate (including organic Se) method was the best for extracting available soil Se, and it had the greatest correlative coefficient with Se contents in plants.

REFERENCES

Li, H.F., McGrath, S.P., & Zhao, F.J. 2008. Selenium uptake, translocation and speciation in wheat supplied with selenate or selenite. *New Phytologist*, 178 (1): 92–102.

Tan, J.A., Zhu, W.Y., Wang, W.Y., Li, R.B., Hou, S.F., Wang, D.C., et al. 2002. Selenium in soil and endemic diseases in China. *Science of the Total Environment*, 284 (1): 227–235.

Tolu, J., Le Hécho, I., Bueno, M., Thiry, Y., & Potin-Gautier, M. 2011. Selenium speciation analysis at trace level in soils. *Analytica ChimicaActa*, 684 (1): 126–133.

Zhang, H., Feng, X., Zhu, J., Sapkota, A., Meng, B., Yao, H., et al. 2012.Selenium in soil inhibits mercury uptake and translocation in rice (*Oryzasativa*L.). *Environmental Science & Technology*, 46, 10040–10046.

Zhao, F., Su, Y., Dunham, S., Rakszegi, M., Bedo, Z., & McGrath, S.P. 2009. Variation in mineral micronutrient concentrations in grain of wheat lines of diverse origin. *Journal of Cereal Science*, 49 (2): 290–295.

Zhu, Y.G., Pilon-Smits, E.A.H., Zhao, F.J., Williams1, P.N., & Meharg, A.A. 2009. Selenium in higher plants: understanding mechanisms for biofortification and phytoremediation. *Trends in Plant Sciences*.

Partitioning of SeNPs in the water soluble and the exchangeable fractions and effects of soil organic matter and incubation time

Mohammad Mamunur Rashid & Z.-Q. Lin*
Environmental Sciences Program, Southern Illinois University Edwardsville, Edwardsville, Illinois, USA

Fatema Kaniz
Soil and Environmental Sciences Department, University of Barisal, Barisal, Bangladesh

1 INTRODUCTION

Selenium (Se) nanoparticles have become one of the emerging contaminants in recent years due to their size-dependent unique properties and the limitless application in electronics and materials production, which may result in significant nanoselenium contamination in the environment with unknown consequences. In addition, elemental Se nanoparticles (SeNPs) can be produced through naturally occurring microbial metabolisms, but few studies have been conducted to determine the partitioning and transformation of biogenic and industrially produced SeNPs in the contaminated soil.

Nanoselenium of 20–60 nm has a similar bioavailability to sodium selenite (Zhang et al., 2001). Nanoselenium has comparable efficacy to selenite in up-regulating seleno-enzymes and Se levels in tissue, but is less toxic (Zhang et al., 2005, 2008). Previous studies have also indicated that soil organic matter plays an important role in Se mobility in soil by forming organo-mineral associations that may prevent adsorbed Se from leaching and/or create anoxic zones (aggregates) where Se is immobilized after its reduction (Tolu et al., 2014).

The present study investigated the potential effects of soil organic matter content and soil incubation time on the fractionation of nanoselenium in the soil and to determine the partitioning of bioavailable Se in the soil treated with nanoselenium particles as compared with selenate.

2 MATERIALS AND METHODS

Soil samples were collected from a local grassland area, air-dried and then sieved through a 2 mm mesh sieve. Part of the soil was amended with ground plant litter to increase soil organic matter content from 4.8% (low soil OM) to 20.3% (high soil OM). Nanoselenium particles (65 nm on average) were chemically synthesized from selenous acid (Li et al., 2009). The nanoparticle size was determined using Malvern Zetasizer Nano and the particle density was measured using NanoSight LM10. In each type of soil, the Se treatment was 5 mg/kg of SeNPs or selenate. Each treatment has five replicates.

The Se-treated soils were incubated at 25°C in a lab incubator with approximately 70% field capacity for two weeks. The soil samples were collected at 1st, 7th, and 14th day of incubation and then extracted with double deionized water (DDW) and phosphate buffer (1:25, w:w). One g (DW) fresh soil sample was placed into a 40 mL centrifuge tube and 25 mL aliquot DDW were mixed for the water soluble fraction, while 0.1 M (pH 7.0) K_2HPO_4-KH_2PO_4 was mixed for the exchangeable fraction. The centrifuge tubes were capped and shaken for 1 hour at room temperature on a horizontal shaker (200 rpm), centrifuged (10000 g, 20 min) and then decanted the supernatant. Five mL DDW were added to each sample and shaken for 2 minutes, centrifuged and supernatants were combined (Martens et al., 1997).

Total Se concentrations in both the fractions were determined by inductively-coupled plasma mass spectrometry (Agilent 7500cx).

3 RESULTS AND DISCUSSION

Previous researchers reported that bright and red nano-Se particles are water soluble and highly stable. Red elemental Se nanoparticles can aggregate into gray and black elemental Se particles that have a low bioavailability of <5 % (Garbisu et al., 1996). In our study, the size of nanoselenium particles increased significantly from initial 65 nm to 140 nm in the incubated soil after the nanoselenium treatment; the nanoparticle density of Se in the soil decreased with increasing soil incubation time. Large aggregates (approximately 140 nm) of nanoselenium particles were primarily found in the double deionized water extract from the soils that were incubated for 14 days (Fig. 1). The phosphate buffer extract contained similar large sized aggregates of elemental Se particles from the soil having a low organic matter content, but Se nanoparticles of <100 nm diameter were observed in the soil containing a high organic matter content.

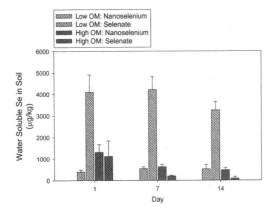

Figure 1. Water soluble Se in the soils treated with nano-selenium or selenate. Data shown are means and standard deviations (n = 5).

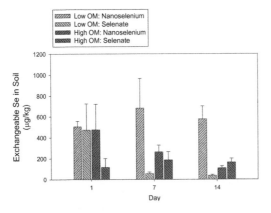

Figure 2. Exchangeable Se in the soils treated with nano-selenium or selenate

Water soluble nanoselenium concentrations were lower than the concentrations of water soluble selenate in the low OM soil, but higher than the selenate concentrations in the high OM soil after incubation for 7 or 14 days (Fig. 2). Exchangeable nanoselenium concentrations were higher than the exchangeable selenate concentrations in the low OM soil after incubation for 7 or 14 days, but no clear pattern was observed in the high OM soil.

Nanoselenium particles showed different chemical behaviors in the soil in comparison to selenate. Our study suggests that increasing soil incubation time of SeNPs decreases the potential environmental risk because of its tendency to increase in size and thus decrease surface reactive area. In high organic matter soil SeNPs gradually become stabilized. Less Se from SeNPs was found in the exchangeable when soil incubation time increased.

4 CONCLUSIONS

Our data show that the size of nanoselenium particles increased in the soil with increasing incubation time, while the nanoparticle density in the soil decreased with increasing soil incubation time. Soil organic matter contents significantly affect the bioavailable fractions of nanoselenium and selenate in the soils. Further study is needed to observe the fate of these nanoparticles in different levels of concentrations over longer time scale. Also, the chemical forms/species in the fractions should be determined to learn which species are going to pose significant risk of toxicity.

REFERENCES

Water soluble Se in the soils treated with nano-selenium or selenate. Data shown are means and standard deviations (n = 5).

Garbisu, C., Ishii, T., Leighton, T. & Buchanan, B.B. 1996. Bacterial reduction of selenite to elemental selenium. Chem. Geol., 132: 199–204.

Li, Z.L. & P.M. 2009. Mixed surfactant template method for preparation of nanometer selenium. E-journal of Chemistry. 6(s1), S304–S310.

Martens, D.A. & Suarez, D.L. 1997. Selenium speciation of soil/sediment determined with sequential extractions and hydride generation atomic absorption spectrophotometry. Environ. Sci. Technol. 31: 133–139.

Tolu, J., Y. Thiry, M. Bueno, C. Jolivet, M. Potin-gautier & I. Le Hécho. 2014. Distribution and speciation of ambient selenium in contrasted soils, from mineral to organic rich. Science of the Total Environment, 1: 479–480.

Zhang, J., Gao, X., Zhang, L. & Bao, Y., 2001. Biological effects of a nano red elemental selenium. Biofactors, 15: 27–38.

Zhang, J., Wang, H., Bao, Y. & Zhang, L. 2004. Nano red elemental selenium has no size effect in the induction of seleno-enzymes in both cultured cells and mice. Life Sci., 75(2): 237–44.

Zhang, J., Wang, X. & Xu, T. 2008. Elemental selenium at nano size (nano-Se) as a potential chemopreventive agent with reduced risk of selenium toxicity: comparison with Se-methylselenocysteine in mice. Toxicological Sciences, 101(1): 22–31.

A rapid analytical method for selenium species by high performance liquid chromatography (HPLC) coupled with inductively coupled plasma mass spectrometry (ICP-MS)

H. Tian & Z.Y. Bao*
School of Earth Sciences, China University of Geosciences, Wuhan, China

X.L. Chen
Zhejiang Institute of Geological Survey, Hangzhou, China

C.H. Wei & Z.Y. Tang
Faculty of Materials Science and Chemistry, China University of Geosciences, Wuhan, China

1 INTRODUCTION

The essential trace element selenium (Se) possesses many functions, such as anticarcinogenic activity, detoxification, and protection against oxidant damage or aging (Sunde, 2006). Human Se intake is mainly obtained from dietary sources, therefore consumption of Se-enriched products has generated a lot of new excitement in Se-deficient regions around the world (Wang et al., 2013). However, selenium's biochemical and molecular biology properties are dependent on its species. Inorganic Se compounds are not assimilated easily, and they can even be harmful to humans and animals. Therefore, investigating the Se species in Se-enriched products can provide useful information for human health.

In this study, the objectives are to investigate the effect of pH and methanol concentration of the mobile phase on the retention behaviors of five Se species, and to establish a rapid and efficient analytical method for Se species by HPLC-ICP-MS.

2 MATERIALS AND METHODS

An ELAN DRC-e HPLC-ICP-MS (PerkinElmer, USA) system was used for specific element detection. The parameters of ICP-MS were as follows: power 1100W, isotope monitored ^{82}Se, scanning mode: time-resolved mode. The HPLC was equipped with a Hamilton PRP-X100 column (250 mm × 4.1 mm × 10 μm) and a 50 μl loop. Numerous researchers have confirmed that citric acid is a well-suited mobile phase for Se compounds separation (Mechora et al., 2011; Fang et al., 2009). In our study, we selected 5 mM citrate buffer as mobile phase with the flow rate at 1.0 ml/min. Citric acid and methanol were of chromatographic grade, and other reagents were of at least analytical grade. All solutions were prepared in ultrapure water (18.2 MΩ/cm). Selenocystine (SeCys2, 98%), seleno-DL-methionine (SeMet, 99%), Methylselenocysteine (MeSeCys, 95%), sodium selenite [Se(IV), 98%], sodium selenate [Se(VI), 98%] and protease X_{IV} were all acquired from Sigma-Aldrich. Se(VI) and Se(IV) stock solutions were prepared in 2% HNO_3. Each of SeMet, SeCys2 and MeSeCys were dissolving in 0.1 M HCl, then stored at 4°C (Sanchez-Martinez et al., 2015). During the method optimization, the mixed standard solution of five Se species was diluted with mobile phase.

3 RESULTS AND DISCUSSION

3.1 *The effect of pH of mobile phase*

The pH of the mobile phase determines the retention behaviors of Se compounds (Stadlober et al., 2001). The retention times (Rt) of five Se species were shown with mobile phase pH (5 mM citric acid) (Fig. 1). The Rts of Se(VI) and Se(IV) showed a decrease as the pH increased from >25 min to 6.48 min and from 4.91 min to 2.86 min, respectively. However, the peaks of Se(IV) and MeSeCys overlapped despite the Rts of organic Se compounds were independent from mobile phase pH. Thus, pH 5.0–5.6 should be considered as the optimum acidity of mobile phase for rapid analysis of five Se species.

3.2 *The effect of methanol concentration*

The effect of methanol concentrations on the Rts of Se species was also investigated (Fig. 2). Some researchers had confirmed that directly introducing volatile carbon compounds (methanol, acetic acid) can enhance the signal of Se at m/z 82 (Kovačevič & Goessler, 2005). Our results showed that the signal of each species was enhanced considerably along with the addition of methanol. The elution of Se(IV) and Se(VI) were shifted to longer Rts, while shorter Rt of

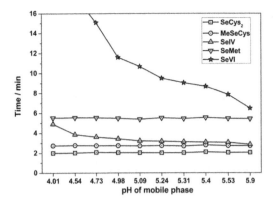

Figure 1. The effect of pH of mobile phase on the retention behaviors of five Se species.

Table 1. Linear ranges (μg/L), equations, correlation coefficients (r), and the detection limits (DL, μg/L) for the five selenium species.

Species	Ranges	Equations	r	DL
$SeCys_2$	20–200	$y = 380.6x - 1032.6$[†]	0.9991	1.45
MeSeCys	20–200	$y = 419.3x = 2316.8$	0.9974	1.33
Se(IV)	10–100	$y = 1242.2x = 348.7$	0.9999	1.73
SeMet	20–200	$y = 329.9x = 784.9$	0.9997	3.07
Se(VI)	10–100	$y = 144.5x = 2456.9$	0.9993	1.48

[†]y: peak area; x: concentration (μg/L)

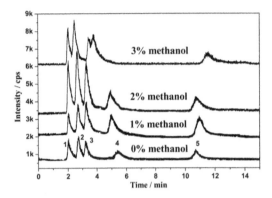

Figure 2. The effect of methanol concentrations on the retention behaviors of five Se species. Mobile phase: 5mM citric acid and methanol. $C_{Se(IV), Se(VI)} = 100\,\mu g/L$, $C_{SeCys_2, MeSeCys, SeMet} = 200\,\mu g/L$. 1, $SeCys_2$; 2, MeSeCys; 3, Se(IV); 4, SeMet; 5, Se(VI).

SeMet was observed when the methanol concentration increased. When the mobile phase contained 3% methanol, both Se(IV) and SeMet eluted with a Rt of 3.6 min. As a result, 2% was adopted as the optimal concentration of methanol in the mobile phase.

3.3 Calibration curves and the detection limits

To shorten the analysis time, different pHs of mobile phase of 5 mM citric acid with 2% methanol were also investigated. The results showed that the total separation time for five Se species were less than 10 min when the pH was greater than 5.31. However, the overlap of MeSeCys and Se(IV) was aggravated when the pH reached 5.65. Therefore, pH 5.5 was selected as the optimum acidity of mobile phase.

The calibration curves of five Se species were built under the optimum parameters (Table 1). The chromatographic detection limits were calculated by taking three times the standard deviation of nine replicates for the blank peak areas (3σ) divided by the slope of the calibration curve (IUPAC). Under the optimum conditions, the detection limits for Se ranged from 1.33 μg/L to 3.07 μg/L for the five Se species. Therefore, the method developed in our study can be used in the future for Se species in Se-enriched plants and food products.

4 CONCLUSIONS

This study demonstrates an efficient and rapid analysis method for five Se species in aqueous solution using HPLC-ICP-MS with [82]Se isotope as a signal monitor. The optimum mobile phase was 5 mM citric acid added 2% methanol at pH 5.5; the flow rate was 1.0 ml/min. The method developed in this study could be used to identify and quantify Se species in Se-enriched plants and food materials. This information is especially useful for implementation of the Se-biofortification strategy.

REFERENCES

Fang, Y., Zhang, Y., Catron, B., Chan, Q., Hu, Q. & Caruso, J. A. 2009. Identification of selenium compounds using HPLC-ICPMS and nano-ESI-MS in selenium-enriched rice via foliar application. Journal of Analytical Atomic Spectrometry, 24: 1657.

Kovačevič, M. & Goessler, W. 2005. Direct introduction of volatile carbon compounds into the spray chamber of an inductively coupled plasma mass spectrometer: Sensitivity enhancement for selenium. Spectrochimica Acta Part B: Atomic Spectroscopy, 60: 1357–1362.

Mechora, S., Cuderman, P., Stibilj, V. & Germ, M. 2011. Distribution of Se and its species in Myriophyllum spicatum and Ceratophyllum demersum growing in water containing Se (VI). Chemosphere, 84: 1636–41.

Sanchez-Martinez, M., Perez-Corona, T., Camara, C. & Madrid, Y. 2015. Preparation and characterization of a laboratory scale selenomethionine-enriched bread. Selenium Bioaccessibility. J Agric Food Chem, 63: 120–127.

Stadlober, M., Sager, M. & Irgolic, K. J. 2001. Effects of selenate supplemented fertilisation on the selenium level of cereals identification and quantification of selenium compounds by HPLC–ICP–MS. Food Chemistry, 73: 357–366.

Sunde, R.A. 2006. Selenium. In: Bowman, B.A. & Russell, R.M. (eds.), Present Knowledge in Nutrition, 9th ed. ILSI Press, Washington, D.C. 480–497.

Wang, J., Wang, Z., Mao, H., Zhao, H. & Huang, D. 2013. Increasing Se concentration in maize grain with soil- or foliar-applied selenite on the Loess Plateau in China. Field Crops Research, 150: 83–90.

Determination of selenium species in Se-enriched food supplement tablets by anion-exchange liquid chromatography-hydride generation-atomic fluorescence spectrometry

Nina Zhang*, Liping Liu, Wujie Ren & Shaozhan Chen
Beijing Center for Disease Control and Prevention, Beijing, China

1 INTRODUCTION

Selenium (Se) is an important element that exists in many different chemical forms in the environment and organisms. Selenium is a typical dual functional element because it is considered both an essential nutrient and a poisonous element. In particular, the beneficial health effects of Se are strongly dependent on the chemical form and the concentration range between essential and toxicity (Wrobel et al., 2005). The toxicity of Se is quite different among the different forms of Se. For example, inorganic Se species are more toxic than organic Se species, and Se(IV) is more toxic than Se(VI) (B'Hymer & Caruso, 2006).

The chemical analysis of different Se species requires the use of different instruments such as high performance liquid chromatography (HPLC) coupled with atomic absorption spectrometry (AAS) (Vale et al., 2010), ion chromatography (IC) coupled with inductively-coupled plasma-mass spectrometry (ICP-MS) (Guzmán Mar et al., 2009), or high performance liquid chromatography combined with inductively-coupled plasma mass spectrometry (HPLC-ICP/MS) (Iserte et al., 2004). HPLC is a high efficiency separation system, and ICP-MS has the high sensitivity of measuring trace amounts of Se. Thus, HPLC-ICP/MS has been demonstrated to be one of the most widely used methods for Se speciation analysis. However, the use of HPLC-ICP/MS also means high investment for instruments and high costs of operation. In this study, hydride generation (HG) atomic fluorescence spectrometry (HG-AFS) system has been used to determine the total Se concentration, while an anion-exchange liquid chromatography (LC) system can be coupled with AFS for Se speciation with relatively low costs.

2 MATERIALS AND METHODS

Anion-exchange LC-HG-AFS system was used for the analysis of Se species including; Se(IV), Se(VI), SeCys, and SeMet in Se-enriched food supplement tablets. Optimization of the chromatographic conditions led to a baseline separation of the four Se compounds in 12 minutes. To extract the inorganic Se, 0.6 mol/L hydrochloric acid was heated at 65°C and added to samples with ultrasonic for 5 minutes. Organic Se required an enzymatic reaction using Protease K. The extraction used 0.25 g of each sample, mixing 40 mg of Protease K and 4 mL ultra-pure water in a 37°C water bath for 12h. Afterwards, the mixture was centrifuged for 5 minutes at 3000 rpm and the supernatant was filtered through a 0.22 μm membrane filter. Inductively-coupled plasma/mass spectrometry (ICP-MS) was used to measure total Se concentrations, after microwave-assisted acid digestion, to quantify total Se in Se-enriched food supplement tablets and their extracts. The operating parameters are summarized in Table 1.

3 RESULTS AND DISCUSSION

Several mobile phases were tested including; citric acid (5 mM, pH 4.5 and 6.0, respectively), ammonium acetate (20 mM, pH 6.5), sodium bicarbonate (10 mM + 2% of acetonitrile, at pH 10.5), and $(NH_4)_2HPO_4$ (40 mM, pH 6.0). It was determined that $(NH_4)_2HPO_4$ (40 mM, pH 6.0) was the best mobile phase to achieve separation based on the retention times and peak resolution. The increase of $(NH_4)_2HPO_4$ from 10 mM to 50 mM had a significant effect on the retention time of inorganic Se species. Forty mM of $(NH_4)_2HPO_4$ was selected as the optimum $(NH_4)_2HPO_4$ concentration, based on a better sensitivity of the species and lower retention times. The effect of the pH on the separation of four Se species during the mobile phase was studied by varying the pH from 3.5 to 7.5 under test conditions, and the results showed that pH 6.0 provided better resolution than other pH values.

Several variables were studied to obtain optimal extraction procedures for the analyses to give satisfactory extraction efficiency. Due to the matrix of the tablets samples, 0.6 mol/L hydrochloric acid was found to be the most effective extractant for inorganic Se, and protease K was suitable for the organic Se. The extraction efficiencies of individual Se species were between 81% and 96%.

Table 1. Equipment and operating conditions

HG-AFS conditions:	
Selenium hollow cathode lamp	80 mA
Quartz furnace height	8 mm
Negative high voltage of photomultiplier	−300 V
Flow rate of 10% v/v HCl	2.5 mL/min
Flow rate of 2% m/v KBH$_4$– 0.5% m/v KOH-1% m/v KI	2.0 mL/min
Carrier gas flow rate	Ar, 400 mL/min
Shielding gas	Ar, 800 mL/min
LC conditions:	
Column	Hamilton PRP-X100 (250 × 4 mm, 10 μm)
Column temperature	Ambient
Injection volume	100 μL
Mobile phase	40 mmol/L (NH$_4$)$_2$HPO$_4$, pH 6.0
Mobile phase flow rate	1 mL/min
Operational Conditions of Agilent 7700 ICP-MS:	
RF power	1380
Carrier gas flow rate	Ar, 0.88 L/min
Makeup gas flow rate	Ar, 0.13 L/min
Sample depth	5.6 mm
Sample and skimmer cones	Ni, 1.0 and 0.4 mm, respectively
Isotopes monitored	^{82}Se

Table 2. Total Se and Se species with hydrochloric acid extraction from Se-enriched food supplement tablets by LC-HG-AFS (n = 3).

	Se Concentration (mg/kg)					
Sample	Se(IV)	Se(VI)	SeCys$_2$	SeMet	Sum[†]	Total[‡]
1	5.2	ND	ND	ND	5.2	5.5
2	13.6	ND	ND	ND	13.6	14.3
3	3.8	ND	ND	ND	3.8	4.2
4	22.5	ND	ND	ND	22.5	23.9
5	18.7	ND	ND	ND	18.7	19.5
6	56.1	ND	ND	ND	56.1	58.4
8	9.3	ND	ND	ND	9.3	9.8
9	38.9	ND	ND	ND	38.9	41.2
11	26.4	ND	ND	ND	26.4	27.8
12	31.3	ND	ND	ND	31.3	33.2

[†]Values are sum of Se species determined by LC-HG-AFS.
[‡]Values are total Se determined by ICP-MS.

The working curves of SeMet and other three Se species were in the concentration ranges of 50–2000 and 10–500 mg/L, respectively, with the regression coefficients greater than 0.999. The recovery rate was in the range of 75–104% for all the analyses.

The concentrations of Se species were determined by LC-HG-AFS and the total Se concentrations were also determined after microwave acid digestion by ICP-MS on Se-enriched food supplement tablets that were purchased from a local supermarket. The results

Table 3. Total Se and Se species with protease K extraction from Se-enriched food supplement tablets by LC-HG-AFS (n = 3).

	Se Concentration (mg/kg)					
Sample	Se(IV)	Se(VI)	SeCys$_2$	SeMet	Sum[†]	Total[‡]
7	0.5	ND	5.2	9.9	15.6	17.9
10	0.3	ND	3.4	7.2	10.9	13.4

[†]Values are sum of Se species determined by LC-HG-AFS.
[‡]Values are total Se determined by ICP-MS.

are shown in Tables 2 and 3. The sum of Se speciation obtained for 12 Se-enriched samples by anion-exchange LC-HG-AFS was in agreement with the results of total Se in the samples measured by ICP-MS.

4 CONCLUSIONS

A rapid, sensitive, and economic method has been developed for the determination of Se species, and the proposed method was applied for the measurement of Se species in 12 Se-enriched food supplement tablets obtained from local supermarket. Se(IV) was the predominant Se species identified in 10 samples and SeMet was the major species in the other two samples. We note that most of the chemical forms of extractable Se were inorganic Se in all 12 of the Se-enriched food supplement tablets.

REFERENCES

B'Hymer, C. & Caruso, J.A. 2006. Selenium speciation analysis using inductively coupled plasma-mass spectrometry. *J. Chromatogr. A.* 1114(1): 1–20.

Guzmán Mar, J.L., Reyes, L.H., Mizanur Rahman, G.M., & Skip Kingston, H.M. 2009. Simultaneous extraction of arsenic and selenium species from rice products by microwave-assisted enzymatic extraction and analysis by ion chromatography-inductively coupled plasma-mass spectrometry. *J. Agric. Food Chem.* 57: 3005–3013.

Orero Iserte, L., Roig-Navarro, A.F. & Hernandez, F. 2004. Simultaneous determination of arsenic and selenium species in phosphoric acid extracts of sediment samples by HPLC-ICP-MS. *Analytica Chimica Acta.* 527: 97–104.

Vale, G., Rodrigues, A., Rocha, A., Rial, R., Mota, A.M., Goncalves, M.L., et al. 2010. Ultrasonic assisted enzymatic digestion (USAED) coupled with high performance liquid chromatography and electrothermal atomic absorption spectrometry as a powerful tool for total selenium and selenium species control in Se-enriched food supplements. *Food Chemistry,* 21: 268–274.

Wrobel, K., Wrobel, K. & Caruso, J.A. 2005. Pretreatment procedures for characterization of arsenic and selenium species in complex samples utilizing coupled techniques with mass spectrometric detection. *Anal. Bioanal. Chem.,* 381(2): 317–331.

Selenium speciation in plants by HPLC-ultraviolet treatment-hydride generation atomic fluorescence spectrometry using various mobile phases

Dan Han, Shuanglian Xiong, Shuxin Tu*, Zhijian Xie, Hailan Li & Muhammad Imtiaz
College of Resources and Environment, Huazhong Agricultural University, Wuhan, China

Jianli Zhou & Danying Xing
Hubei Collaborative Innovation Center for Grain Industry, Yangtze University, Jingzhou, China

1 INTRODUCTION

Selenium (Se) is an essential element for human beings and animals. However, its toxic effects can be observed once it exceeds a critical concentration. The beneficial or toxic effect of Se on human beings and animals is not only dose-dependent, but also related to the chemical forms and the bioavailability of Se (Thiry et al., 2012). In living organisms, Se exists in non-volatile inorganic Se (selenite, selenate) and organic Se (SeMet) (Zhu et al., 2009). The objective of this study was to investigate systematically the relationship between Se retention and the eluent based on ammonium salt to address the limitations of the current determination of Se speciation in plant tissues. To achieve the objective, the following aspects were considered: (1) explore the relationship between Se retention and the mobile phases using a PRP-X100 column, (2) reduce the residue left on the sampler and skimmer cones of HPLC-AFS using an ammonium salt as the eluent, and (3) use the proposed method for the analysis of Se speciation in real samples.

2 MATERIALS AND METHODS

A soil pot experiment was conducted in the greenhouse of the Micro-element Research Center in Huazhong Agricultural University in Wuhan, China. There were two treatments: 2.2 and 22.2 mg/kg Se from sodium selenite. Each treatment was replicated three times. The leaves and roots of flue-cured tobacco were collected for the determination of Se speciation. The Se species were extracted from 0.2 g ground leaves or roots by using 10 mL 1:2 methanol water (Han et al., 2013) and being shaken for 15 min with ultrasonic wave at room temperature. The extracts were centrifuged at $5000 \times g$ for 30 min and the supernatant was collected in an eggplant shaped bottle. The methanol in solution was evaporated by rotary evaporation instrument (40°C). Finally, the supernatant was filtered through 0.22 μm Millipore filters before the analysis by HPLC-ultraviolet treatment-hydride generation atomic fluorescence spectrometry (HPLC-UV-HG-AFS).

3 RESULTS AND DISCUSSION

3.1 Selection of mobile phase

Four standard substances of SeCys, SeMet, Se (IV) and Se (VI) were separated well using a Hamilton PRP X-100 with a mobile phase containing 40mM $(NH_4)_2HPO_4$ at pH 6.00 as shown in Figure 1. However, when the mobile phase was 40mM $NH_4H_2PO_4$ at pH 4.48, only SeCys, SeMet and Se (IV) was detected (Fig. 1A); and only SeCys and Se (IV) were separated using 40 mM NH_4Ac at pH 6.94 and 40 mM NH_4NO_3 at pH 5.18 as the mobile phase as shown in Figure 1B and Figure 1C. The Se (VI) were not eluted because of their higher charge on the species under the eluent pH used, leading to their strong retained on the column (Chen et al., 2011; Chen et al., 2008). On the other hand, competing anion such as Ac^- and NO_3^- was too weak to elute SeMet and Se(VI) from the column (Chen et al., 2008).

3.2 Calibration curves and recovery by the proposed method

Linear calibrations were obtained over a concentration range of 0–236.4 μg/L with correlation coefficients ranging from 0.9991–0.9996 when 100 μL was injected. This was the basis for quantitative analysis of Se species, so the established method could be applied to Se speciation in plants. In 1:2 methanol water extraction conditions, the recoveries of SeCys, Se (IV), SeMet and Se(IV) in the Se-enriched samples were 95.72 ± 1.60%, 106.89 ± 2.20%, 93.28 ± 2.82%, and 91.38 ± 1.27%, respectively. The results showed that Se speciation remained stable in the extraction analysis process. Therefore, the proposed method could be used for the determination of Se species in contaminated plants.

3.3 Analysis of Se species in the leaves and roots of flue-cured tobacco

The proposed method was used for the determination of Se species in the leaves and roots of Se-enriched

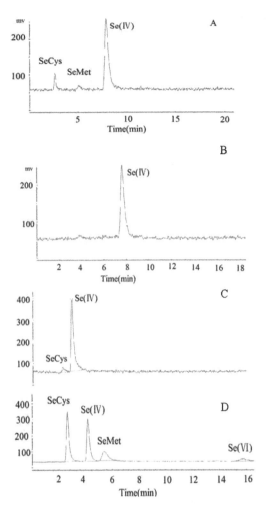

Figure 1. Typical chromatogram obtained when using four eluents containing 40 mM $NH_4H_2PO_4$ at pH 4.48 (A), 40 mM NH_4Acat pH 6.94 (B), 40 mM NH_4NO_3 at pH 5.18 (C) and 40 mM $(NH_4)_2HPO_4$ at pH 6.00(D). The concentration of SeCys 118.2 μg/L, Se(IV) 85.8 μg/L, SeMet 88.7 μg/L, Se(VI) 83.0 μg/L, CK was corresponding eluent.

flue-cured tobacco. It showed that the predominant Se species was Se(VI) in flue-cured tobacco leaves. The result is in accord with the findings of the previous study on leek (Lavu et al., 2012). When 2.2 mg/kg Se (sodium selenite) was added to soil, Se(VI) was the major Se species in leek (Lavu et al., 2012). The major Se forms were organic species (SeCys and SeMet) in roots of flue-cured tobacco, while Se(VI) was only detected at 22.2 mg/kg Se. The similar result was shown in the studies on carrot roots (Kápolna et al., 2009). When 100 mg/L Se (sodium selenite) was applied in foliar spray, the key Se species in carrot roots was SeMet, and only Se(IV) was detected among inorganic Se species (Kápolna et al., 2009).

4 CONCLUSIONS

A mobile phase containing 60 mM $(NH_4)_2HPO_4$ at pH 6.0 was found to be the most suitable for the separation of Se species by the PRP-X100 anion exchange column coupled with AFS detection. The retention of the target anion was affected by the competing ion and its concentration as well as the mobile phase pH. Furthermore, with the proposed method for the determination of Se species in the leaves and roots of flue-cured tobacco, four Se species, including SeCys, Se(IV), SeMet and Se(VI), were all detected at high Se levels.

REFERENCES

Chen, B., M. He, X. Mao, R. Cui, D. Pang, & B. Hu. 2011. Ionic liquids improved reversed-phase HPLC online coupled with ICP-MS for selenium speciation. Talanta 83: 724–731.

Chen, Z., K.F. Akter, M.M. Rahman, & R. Naidu. 2008. The separation of arsenic species in soils and plant tissues by anion-exchange chromatography with inductively coupled mass spectrometry using various mobile phases. Microchem. J., 89: 20–28.

Han, D., X. Li, S. Xiong, S. Tu, Z.G. Chen, J.P. Li, et al. 2013. Selenium uptake, speciation and stressed response of *Nicotiana tabacum* L. Environ. Exp. Bot., 95: 6–14.

Kápolna, E., P.R. Hillestrøm, K.H. Laursen, S. Husted & E.H. Larsen. 2009. Effect of foliar application of selenium on its uptake and speciation in carrot. Food Chem., 115: 1357–1363.

Lavu, R.V.S., G. Du Laing, T. Van De Wiele, V.L. Pratti, K. Willekens, B. Vandecasteele, et al. 2012. Fertilizing soil with selenium fertilizers: Impact on concentration, speciation, and bioaccessibility of selenium in Leek (*Allium ampeloprasum*). J. Agric. Food Chem., 60: 10930–10935.

Thiry, C., A. Ruttens, L. De Temmerman, Y.-J. Schneider & L. Pussemier. 2012. Current knowledge in species-related bioavailability of selenium in food. Food Chem., 130: 767–784.

Zhu, Y.G., E.A.H. Pilon-Smits, F.J. Zhao, P.N. Williams, & A.A. Meharg. 2009. Selenium in higher plants: understanding mechanisms for biofortification and phytoremediation. Trends Plant Sci., 14: 436–442.

Synchrotron studies of selenium interactions with heavy elements

I.J. Pickering* & G.N. George
Department of Geological Sciences, University of Saskatchewan, Saskatoon, Saskatchewan, Canada

T.C. MacDonald
Toxicology Centre, University of Saskatchewan, Saskatoon, Saskatchewan, Canada

P.H. Krone
Department of Anatomy & Cell Biology, University of Saskatchewan, Saskatoon, Saskatchewan, Canada

M. Korbas
Science Division, Canadian Light Source Inc., Saskatoon, Saskatchewan, Canada

1 SELENIUM AND ITS INTERACTION WITH HEAVIER ELEMENTS

The surprising interaction of selenium (Se) with a number of toxic heavy elements in biological systems has been known for decades. Selenium species, which are toxic at elevated levels, in some cases can interact with arsenic (As), mercury (Hg), or cadmium species to provide antagonism, or a significant reduction in toxicity compared with the individual species (e.g. Gailer *et al.*, 2007). These observations have led to the suggestion that Se may provide protection against environmental toxins such as these heavy metals. However, to date almost all studies reporting on such antagonistic effects have been carried out using exogenous Se at acute levels, whereas Se responding to environmental exposure is drawn from endogenous (*i.e.* essential) pools within the organism. Further research is needed on endogenous Se and its possible interactions with toxic elements.

2 SYNCHROTRON X-RAY METHODS TO STUDY SELENIUM-METAL BONDS IN BIOLOGY

Synchrotron X-rays can probe the binding of Se with heavier elements *in situ* within complex systems such as biological tissues. X-rays can probe the sample without substantial pretreatment of the tissue, which might modify chemical form or distribution. A suite of techniques enable the study of Se speciation and the microscopic co-localization of Se with the element.

2.1 *X-ray absorption spectroscopy (XAS)*

XAS is sensitive to the local structural and electronic environment of a given element such as Se. XAS can provide speciation of the type of chemical environment, though not the specific molecule, as

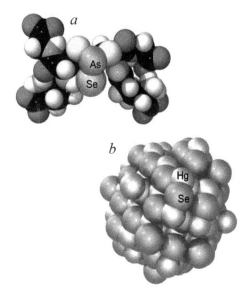

Figure 1. Seleno-bis(S-glutathionyl) arsinium ion (*a*) and nano-particulate mercuric selenide (*b*), deduced using XAS.

well as metrics, including interatomic distances within the local coordination sphere. Previously, XAS was used to investigate the antagonism between As and Se due to acute levels of arsenite and selenite, showing this to result from formation of an arsenic-selenium-glutathione molecule (Fig. 1a), which is excreted in bile (Gailer *et al.*, 2000b). The formation of nanoparticulate mercuric selenide was also revealed after acute co-exposure to selenite and inorganic mercury (Fig. 1b, Gailer *et al.*, 2000a).

2.2 *X-ray fluorescence imaging (XFI)*

XFI yields maps of multiple elements in a tissue sample (Pushie *et al.*, 2014), allowing visualization of Se localization in specific tissues, as well as co-localization of Se with another element. The sample is raster scanned through a microfocused synchrotron beam, with fluorescent X-rays collected using an energy-dispersive detector. In the conventional configuration, the thickness of each tissue section must be comparable with the beam size for optimum spatial resolution. A recent variant, confocal XFI, uses an additional focusing optic on the detector to enable high resolution maps of substantially thicker samples (Choudhury et al., 2015). XFI can provide graphic visualization of relationships between Se and heavier elements.

3 RECENT RESULTS IN MERCURY-SELENIUM INTERACTIONS

Zebrafish larvae, a model of vertebrate development, were used to study interactions of exogenous Hg with endogenous Se (MacDonald et al., 2015). In the absence of Hg, XFI showed that low levels of Se were present throughout the tissues. Selenium was more strongly concentrated in the eye lens, where selenomethionine substitutes for methionine in crystallin proteins (Choudhury et al., 2015), as well as in pigment cells, which are also rich in zinc.

Treatments both with methylmercury species and with inorganic mercury were investigated (MacDonald et al., 2015), since these Hg species show not only different mechanisms of toxicity but also different distributions (Korbas et al., 2012). In particular, methylmercury species are more toxic and are known to be neurotoxic, especially to the developing fetus. These species are present at low levels in edible marine fish, which has led to some concern over levels of such fish in the human diet.

In XFI of methylmercury treated larvae, Hg was widely distributed in the tissues, especially targeting the brain and the rapidly dividing cells of the eye lens. Nevertheless, Se distributions were largely unmodified from the Hg-free controls and did not correlate with distributions of Hg. In contrast, inorganic Hg treated larvae showed Se largely co-localized with Hg in tissues including the brain and the developing kidney. Moreover, Se was noticeably depleted from pigment spots (MacDonald et al., 2015). Mercury and Se were co-localized in very small regions consistent with nano-sized particles containing Hg, Se, and sulfur.

Previously, XAS and XFI of human brain following acute or chronic exposure to organomercury species showed that a portion of the total Hg and Se were bound as nanoparticulate mercuric selenide (Korbas et al. 2010). While such a co-localization of Hg and Se was not observed in methylmercury treated zebrafish larvae, it may suggest a slow demethylation of organic Hg prior to binding to Se. A longer time between exposure and death resulted in a larger proportion of Hg bound as nanoparticulate mercuric selenide. Moreover, Se levels in the brain after acute Hg exposures were substantially elevated compared with chronic or no exposure to Hg. Since mercuric selenide is extremely insoluble, it is possible that Se, under some circumstances, can play a protective role against Hg.

4 IMPLICATIONS

Since Se appears to be in a protective role, it is tempting to suggest that Se in the diet might play a palliative role against Hg toxicity. However, caution is appropriate since there are several observations (e.g. Cuvin-Aralar et al. 1991) of synergistic toxicological relationships between species of Hg and Se, in which together they are is substantially more toxic than the sum of the individual toxicities. Further work is needed to understand the complex interaction between Hg and Se in their different chemical forms.

REFERENCES

Choudhury, S., Thomas, J.K., Sylvain, N.J., Ponomarenko, O., Gordon, R.A., Heald S.M., et al. 2015. Selenium preferentially accumulates in the eye lens following embryonic exposure: a confocal X-ray fluorescence imaging study. *Environmental Science & Technology* 49(4): 2255–2261.

Cuvin-Aralar, M.L.A. & Furness, R.W. 1991. Mercury and selenium interaction: a review. *Ecotoxicology and Environmental Safety* 21: 348–364.

Gailer, J., George, G.N., Pickering, I.J., Madden, S., Prince, R.C., Yu, E.Y., et al. 2000a. Structural basis of the antagonism between inorganic mercury and selenium in mammals. *Chemical Research in Toxicology* 13(11): 1135–1142.

Gailer, J., George, G.N., Pickering, I.J., Prince, R.C., Ringwald, S.C., Pemberton, J.E., et al. 2000b. A metabolic link between arsenite and selenite: the seleno-bis(S-glutathionyl) arsinium ion. *Journal of the American Chemical Society*, 122: 4637–4639.

Gailer, J. 2007. Arsenic-selenium and mercury-selenium bonds in biology. *Coordination Chemistry Reviews* 251: 234–254.

Korbas, M., MacDonald, T.C., Pickering, I.J., George, G.N. & Krone, P.H. 2012. Chemical form matters: differential accumulation of mercury following inorganic and organic mercury exposures in zebrafish larvae. *ACS Chemical Biology* 7(2): 410–419.

Korbas, M., O'Donoghue, J.L., Watson, G.E., Pickering, I.J., Singh, S.P., Myers, G.J., et al. 2010. The chemical nature of mercury in human brain following poisoning or environmental exposure. *ACS Chemical Neuroscience* 1(12): 810–818.

MacDonald, T.C., Korbas, M., James, A.K., Sylvain, N.J., Hackett, M.J., Nehzati, S., et al. 2015. Interaction of mercury and selenium in the larval stage zebrafish vertebrate model. *Metallomics*, accepted.

Pushie, M.J., Pickering, I.J., Korbas, M., Hackett, M.J. & George, G.N. 2014. Elemental and chemically specific X-ray fluorescence imaging of biological systems. *Chemical Reviews* 114(17): 8499–8541.

Cellular and molecular functions of selenium

Selenium atom-specific modifications (SAM) of nucleic acids for human health

Rob Abdur*, W. Zhang, H. Sun & Zhen Huang*
School of Arts and Sciences, Georgia State University, Atlanta, USA
College of Life Sciences, Sichuan University, Chengdu, China

1 NATURALLY-OCCURRING SELENIUM NUCLEIC ACIDS

The reason that the selenium (Se) research has become an intense field of research is because this trace element of nutritional importance has unique sets of biological activities for human and animals. Twenty-five selenoproteins have been identified in human with different functions at various sub-cellular locations. Selenoproteins are mostly enzymes that constitute two major metabolic pathways, namely redox metabolism and thyroid hormone metabolism. The selenoproteins with redox activities are ubiquitous and help to maintain redox balance along with other pathways in the body. The selenoprotein GPx operates with a small ubiquitous peptide called glutathione (GSH) and forms one of the major pathways to neutralize or reduce reactive oxygen species (ROS) and reactive nitrogen species (RNS), which are generated continuously during metabolic and respiratory processes.

The natural occurrence of selenium in certain tRNAs was also reported in bacteria (Wittwer et al., 1984), archaea (Ching et al., 1984), mammals (Mizutani et al., 1999), and plants (Huang et al., 2001). The discovery of the presence of naturally-occurring selenium in tRNAs further provides the evidence for the nutritional role of selenium in life at the molecular level.

2 ARTIFICIALLY SELENO-MODIFIED NUCLEIC ACIDS (SENA)

The presence of the Se-modification in natural tRNA has encouraged Huang and co-workers to pioneer and incorporate the selenium functionalities at various selected positions of RNA and DNA, namely selenium-derivatized nucleic acids (SeNA) (Abdur et al., 2014; Carrasco et al., 2001; Lin et al., 2011; Salon et al., 2013; Sheng and Huang, 2010; Thompson et al., 2015; Wen et al., 2013). Their experimental data have demonstrated that nucleic acid oxygen in various positions can be stably replaced with selenium for structure and function studies, especially for the facilitation of phase solution, crystallization, and high-resolution structure determination in X-ray crystallography. Their studies imply that the selenium functionalities on nucleobases might improve the accuracy and efficiency of translation, RNA transcription and even DNA replication by enhancing base-pair interaction, selectivity, and fidelity. For example, the biochemical study of RNase H with the Se-modified DNA/RNA duplexes showed that the Se-modified nucleobase functionality can enhance the enzyme turnover rate by many folds (Abdur et al., 2014). The supporting evidences from the X-ray crystallographic structures have revealed that as a larger atom, the selenium functionality causes the local unwinding of DNA/RNA substrate duplex. The Se-functionalized DNA acts as a guiding sequence shifting the scissile phosphate of the RNA substrate closer to the active site, therefore further activating the attacking nucleophilic water molecule (Abdur et al., 2014). Previously others have reported that the selenium-modified nucleic acids can resist nuclease digestion (Brandt et al., 2006; Caton-Williams & Huang, 2008). Thus, this Se-atom-specific mutagenesis (SAM) has opened a new research area to further explore and design nucleic acid-based novel therapeutics.

Nucleotides and nucleic acids can be used to block or interfere specific pathological processes or pathways. The most common targets of these therapeutics are cancers and viruses. The potential oligonucleotide therapeutics, such as small interference RNAs (siRNA), antisense DNAs or CRISPR-RNA (crRNA) (Haurwitz et al., 2010; Li, 2015), are based on the sequence complementarity to the target RNAs (such as mRNAs) or DNAs (such as genomic DNA). This is an intensely advanced area, because the short oligonucleotides can target a specific RNA very precisely and hydrolyze the RNA by Dicer-Argonaut system (Castellano & Stebbing, 2013; Macrae et al., 2006; Schmitter et al., 2006) or target a specific DNA by crRNA and Case system (Haurwitz et al., 2010; Li, 2015; Liu et al., 2015). Potential application of the Se-modifications in short oligonucleotides could offer additional benefits to the oligonucleotides by improving the base-pair stability, selectivity and fidelity, by enhancing the catalytic rate, and by conferring resistance to nuclease digestion. Extensive *in vivo* research is, however, needed to take full advantage of all Se-modifications, especially the Se-nucleobase functionalities (Abdur et al., 2014; Salon et al., 2007; Sheng et al., 2012).

3 CONCLUSIONS

The selenium function in living systems and its effects on human health are essential. Moreover, selenium-atom specific modification (SAM) will allow the development and design of potential oligonucleotide therapeutics via antisense DNA, siRNA, miRNA and crRNA strategies, by improving the molecular specificity and substrate cleavage efficiency.

ACKNOWLEDGEMENT

The authors acknowledge the financial support provided by SCU and NIH (R01GM095881 and GM095086).

REFERENCES

Abdur, R., Gerlits, O.O., Gan, J., Jiang, J., Salon, J., Kovalevsky, A.Y., et al. 2014. Novel complex MAD phasing and RNase H structural insights using selenium oligonucleotides. Acta Crystallogr D Biol Crystallogr 70: 354–361.

Brandt, G., Carrasco, N., & Huang, Z. 2006. Efficient substrate cleavage catalyzed by hammerhead ribozymes derivatized with selenium for X-ray crystallography. Biochemistry 45: 8972–8977.

Carrasco, N., Ginsburg, D., Du, Q., & Huang, Z. 2001. Synthesis of selenium-derivatized nucleosides and oligonucleotides for X-ray crystallography. Nucleosides Nucleotides Nucleic Acids 20: 1723–1734.

Castellano, L., & Stebbing, J. 2013. Deep sequencing of small RNAs identifies canonical and non-canonical miRNA and endogenous siRNAs in mammalian somatic tissues. Nucleic Acids Res 41: 3339–3351.

Caton-Williams, J., & Huang, Z. 2008. Biochemistry of selenium-derivatized naturally occurring and unnatural nucleic acids. Chem Biodivers 5: 396–407.

Ching, W.M., Wittwer, A.J., Tsai, L., & Stadtman, T.C. 1984. Distribution of two selenonucleosides among the selenium-containing tRNAs from Methanococcus vannielii. Proc Natl Acad Sci U S A 81: 57–60.

Hassan, A.E., Sheng, J., Zhang, W., & Huang, Z. 2010. High fidelity of base pairing by 2-selenothymidine in DNA. J Am Chem Soc 132: 2120–2121.

Haurwitz, R.E., Jinek, M., Wiedenheft, B., Zhou, K., & Doudna, J.A. 2010. Sequence- and structure-specific RNA processing by a CRISPR endonuclease. Science 329: 1355–1358.

Huang, K.X., An, Y.X., Chen, Z.X., & Xu, H.B. 2001. Isolation and partial characterization of selenium-containing tRNA from germinating barley. Biol Trace Elem Res 82: 247–257.

Li, H. 2015. Structural Principles of CRISPR RNA Processing. Structure 23: 13–20.

Lin, L., Sheng, J., & Huang, Z. 2011. Nucleic acid X-ray crystallography via direct selenium derivatization. Chem Soc Rev 40: 4591–4602.

Liu, T., Li, Y., Wang, X., Ye, Q., Li, H., Liang, Y., et al. 2015. Transcriptional regulator-mediated activation of adaptation genes triggers CRISPR de novo spacer acquisition. Nucleic Acids Res 43: 1044–1055.

Macrae, I.J., Zhou, K., Li, F., Repic, A., Brooks, A.N., Cande, W.Z., et al. 2006. Structural basis for double-stranded RNA processing by Dicer. Science 311: 195–198.

Mizutani, T., Watanabe, T., Kanaya, K., Nakagawa, Y., & Fujiwara, T. 1999. Trace 5-methylaminomethyl-2-selenouridine in bovine tRNA and the selenouridine synthase activity in bovine liver. Mol Biol Rep 26: 167–172.

Salon, J., Gan, J., Abdur, R., Liu, H., & Huang, Z. 2013. Synthesis of 6-Se-guanosine RNAs for structural study. Org Lett 15: 3934–3937.

Salon, J., Sheng, J., Jiang, J., Chen, G., Caton-Williams, J., & Huang, Z. 2007. Oxygen replacement with selenium at the thymidine 4-position for the Se base pairing and crystal structure studies. J Am Chem Soc 129: 4862–4863.

Schmitter, D., Filkowski, J., Sewer, A., Pillai, R.S., Oakeley, E.J., Zavolan, M., et al. 2006. Effects of Dicer and Argonaute down-regulation on mRNA levels in human HEK293 cells. Nucleic Acids Res 34: 4801–4815.

Sheng, J., & Huang, Z. 2010. Selenium derivatization of nucleic acids for X-ray crystal-structure and function studies. Chem Biodivers 7: 753–785.

Sheng, J., Zhang, W., Hassan, A.E., Gan, J., Soares, A.S., Geng, S., et al. 2012. Hydrogen bond formation between the naturally modified nucleobase and phosphate backbone. Nucleic Acids Res 40: 8111–8118.

Thompson, A., Spring, A., Sheng, J., Huang, Z., & Germann, M.W. 2015. The importance of fitting In: Conformational Preference of Selenium 2' Modifications in Nucleosides and Helical Structures. Journal of Biomolecular Structure & Dynamics 33: 289–297.

Wen, Z., Abdalla, H.E., & Zhen, H. 2013. Synthesis of novel di-Se-containing thymidine and Se-DNAs for structure and function studies. Sci China Chem 56: 273–278.

Wittwer, A.J., Tsai, L., Ching, W.M., & Stadtman, T.C. 1984. Identification and synthesis of a naturally occurring selenonucleoside in bacterial tRNAs: 5-[(methylamino)methyl]-2-selenouridine. Biochemistry 23: 4650–4655.

Zhang, W., & Huang, Z. 2011. Synthesis of the 5'-Se-thymidine phosphoramidite and convenient labeling of DNA oligonucleotide. Org Lett 13: 2000–2003.

Comparative cytotoxicity and antioxidant evaluation of biologically active fatty acid conjugates of water soluble selenolanes in cells

A. Kunwar, P. Verma & K.I. Priyadarsini*
Radiation and Photochemistry Division, Bhabha Atomic Research Centre, Mumbai, India

K. Arai & M. Iwaoka
Department of Chemistry, School of Science, Tokai University, Kitakanaem, Hiratsuka-shi, Kanagawa, Japan

1 INTRODUCTION

Dihydroxy selenolane (DHS) and monoamine selenolane (MAS) are the synthetic water-soluble cyclic selenium (Se) analogues of dithiotheritol (Kumakura et al., 2010; Arai et al., 2014) (Fig. 1). These two compounds have been shown to exhibit wide range of biological activities such as free radical scavenging, mimicking the function of an antioxidant enzyme like glutathione peroxidise (GPx) and catalysing the oxidative folding of misfolded and/or denatured proteins (Singh et al., 2010; Kumakura et al., 2010; Chakraborty et al., 2012; Arai et al., 2011, 2014).

Recently others reported that the conjugation of a fatty acid of increasing alkyl chain length (C_6 to C_{14}) to parent molecules DHS and MAS not only increased their specificity towards substrates were projected to be better antioxidants compared to the parent compounds (Arai et al., 2014; Iwaoka et al., 2015). In the present study, fatty acid conjugates of DHS and MAS were evaluated for their cytotoxicity, associated mechanisms and antioxidant effects in cells.

2 MATERIALS & METHODS

The synthesis and characterisation of parent compounds DHS, MAS and their fatty acid conjugates of varying alkyl chain length (C_6–C_{14}) were reported previously (Arai et al., 2014; Iwaoka et al., 2015). Chinese Hamster Ovary (CHO) and Human breast carcinoma (MCF7) cells were obtained from National Centre for Cell Sciences Pune, India. The cells were cultured in Dulbecco Modified Eagle Medium (DMEM) supplemented with 10% fetal calf serum, 100 μg/ml streptomycin and 100 U/ml penicillin and maintained at 37°C under 5% CO_2 and humidified air. The cells were incubated with Se compounds for the desired time points prior to assay. All chemicals used in the study were of highest purity grade and procured from the local suppliers. The cytotoxicity of Se compounds was evaluated using (4,5-dimethylthiazol-2yl)-2,5-piphenyltetrazolium bromide (MTT) reduction assay. The mechanisms of toxicity, if any, was investigated by monitoring DNA fragmentation (propidium iodide (PI) staining), phosphatidylserine externalization (Annexin V staining), mitochondrial membrane potential (JC-1 staining) and membrane fluidity (1,6-diphenyl-1, 3,5-Hexatriene (DPH) anisotropy measurement) and integrity (lactate dehydrogenase (LDH) leakage). The aggregation behavior of the fatty acid conjugates of DHS and MAS were studies using fluorescence enhancement of a lipophilic fluorophore DPH (Plasek & Jarolim, 1987). DPH shows week fluorescence in aqueous solution, however, once it goes in to miceller structure or the aggregates, its fluorescence increases significantly. The incorporation/loading of DHS, MAS and their fatty acid conjugates in to cells was estimated in terms of the Se level using graphite furnace atomic absorption spectrometry. The non-toxic conjugates selected from above screening were analysed for antioxidant effects in terms of the expression of antioxidant selenoenzymes such as GPx1, GPx4 and thioredoxin reductase (TxR1) and the inhibition of 2,2′-Azobis (2-amidinopropane) dihydrochloride (AAPH) induced lipid peroxidation and protein carbonylation in CHO cells.

3 RESULTS & DISCUSSION

Parent molecules DHS, MAS and their C_6 fatty acid conjugates were nontoxic to CHO cells in the

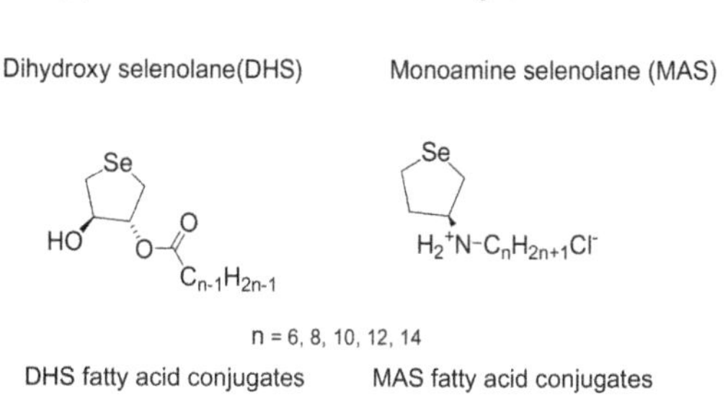

Figure 1. Chemical structures of DHS, MAS and their fatty acid conjugates.

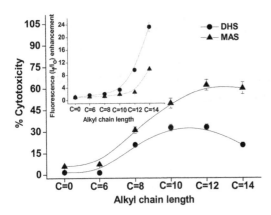

Figure 2. Cytotoxic effects of DHS, MAS and their fatty acid conjugates. The insert shows enhancement in the fluorescence intensity of a lipophilic fluorophore DPH in presence of the 30 μM aqueous solutions of DHS, MAS and their fatty acid conjugates. Data are mean ± SE (n = 3).

concentration range of 2–50 μM. Treatment with higher ($\geq C_8$) fatty acid conjugates in a similar concentration range showed significant chain length dependent increase in the cytotoxicity up to a treatment concentration of 10 μM. Above that concentration, cytotoxicity followed the order of $C_6 < C_8 < C_{10} \sim C_{12} > C_{14}$ (Fig. 2). Among the DHS and MAS conjugates, the former showed significantly lesser cytotoxicity than later at each chain length and treatment concentration.

The cytotoxicity evaluation of compounds in MCF7, a tumor cell line in the identical concentration range showed similar trend with respect to the alkyl chain length and did not exhibit differential toxicity as compared to CHO, the normal cell type. The mechanistic investigation revealed that the higher ($\geq C_8$) fatty acid conjugates of DHS and MAS led to the cell death by necrosis caused by cytolysis or membrane disintegration, as evidenced by the acute mitochondrial depolarization, increase in the plasma membrane fluidity and leakage of intracellular protein LDH followed by the $PI^{+ve} AnnxinV^{+ve}$ staining of cells (van Engeland et al. 1996). The aggregation behavior measured by the fluorescence enhancement of DPH indicated that the higher ($\geq C_{12}$) fatty acid conjugates of both DHS and MAS form aggregates as a function of concentration and this effect was found prominent in case of former than the later (Fig. 2). With all increase in alkyl chain length above C_{10}. There will be lesser availability of such compounds in solution to interact with cell membrane and thus less cytotoxicity. This result was supported by the fact that the loading/incorporation of Se by cells normalized with respect to amount of Se added in form of treatment with DHS and MAS conjugates (C_6–C_{14}), exhibited similar order as that of cytotoxicity. The differential toxicity induced by DHS and MAS conjugates and the decrease in toxicity after the alkyl chain length of $C_{12,}$ can be attributed to their aggregation behaviors. Further, treatment of CHO cells with the nontoxic C_6 conjugates of DHS and MAS, selected from above screening, showed significantly higher induction of antioxidant selenoenzymes such as GPx1, GPx4 and TxR1, both at mRNA and activity levels, as compared to the parent compounds. These conjugates also provided better protection against the AAPH induced lipid peroxidation and protein oxidation than the parent compounds, suggesting their potential antioxidant activities in cells.

4 CONCLUSIONS

The parent compounds DHS and MAS exhibit antioxidant effects in cells. Conjugating a fatty acid of variable chain length as a lipophilic unit to a hydrophilic antioxidant moiety may be an effective approach to enhance the antioxidant activity. However, the balance between the hydrophilicity and lipophilicity is important before the compound becomes toxic. In this study, such balance was observed at an alkyl chain length of C_6, showing increased antioxidant effects without compromising with the cytotoxicity.

REFERENCES

Arai, K., Dedachi, K., & Iwaoka, M. 2011. Rapid and quantitative disulfide bond formation for a polypeptide chain using a cyclic selenoxide reagent in an aqueous medium. Chem Eur J 17:481–5.

Arai, K., Moriai, K., Ogawa, A., & Iwaoka, M. 2014. An amphiphilic selenide catalyst behaves like a hybrid mimic of protein disulfide isomerase and glutathione peroxidase 7. Chem Asian J 9: 3464–71.

Chakraborty, S., Yadav, S.K., Subramanian, M., Priyadarsini, K.I., Iwaoka, M., & Chattopadhyay, S. 2012. DL-trans-3,4-dihydroxy-1-selenolane (DHS(red)) accelerates healing of indomethacin-induced stomach ulceration in mice. Free Radic Res 46: 1378–86.

Iwaoka, M., Sano, N., Lin, Y.Y., Katakura, A., Noguchi, M., Takahashi, K., et al. 2015. Fatty acid conjugates of water-soluble (±)-trans-selenolane-3,4-diol: effects of alkyl chain length on the antioxidant capacity. Chembiochem doi: 10.1002/cbic.201500047.

Kumakura, F., Mishra, B., Priyadarsini, K.I., & Iwaoka, M. 2010. A water-souble cyclic selenide with enhanced glutathione peroxidise-like catalytic activities. Eur J Org Chem 440–445.

Plasek, J., & Jarolim, P. 1987. Interaction of the fluorescent probe 1,6-diphenyl-1, 3, 5-hexatriene with biomembranes. Gen Physiol Biophys 6: 425–437.

Singh, B.G., Thomas, E., Kumakura, F., Dedachi, K., Iwaoka, M., & Priyadarsini, K.I. 2010. One-electron redox processes in a cyclic selenide and a selenoxide: a pulse radiolysis study. J Phys Chem A 114: 8271–8277.

van Engeland, M., Ramaekers, F.C., Schutte, B., & Reutelingsperger, C.P. 1996. A novel assay to measure loss of plasma membrane asymmetry during apoptosis of adherent cells in culture. Cytometry 24: 131–139.

Diselenodipropionic acid as novel selenium compound for lung radiotherapy

K. Indira Priyadarsini* & A. Kunwar
Radiation & Photochemistry Division, Bhabha Atomic Research Centre, Trombay, Mumbai, India

V.K. Jain
Chemistry Division, Bhabha Atomic Research Centre, Trombay, Mumbai, India

V. Gota & J. Goda
Advanced Centre for Treatment, Research and Education in Cancer, Kharghar, Navi Mumbai, India

1 INTRODUCION

Radioprotectors are used to minimize the damaging effects of ionizing radiation to normal cells during therapeutic procedures, wherein normal tissue protection is as important as the destruction of the cancer cells (Koukourakis, 2012). Initial studies on radioprotection was conducted on sulphur-containing amino acid, cysteine, and subsequently several synthetic sulfhydryl compounds were screened. The most effective radioprotective drug identified till date, and the only agent approved by the FDA for the use in clinic is amifostine. However, this drug is only specific to head and neck cancers and it shows considerable behavioral toxicity at radioprotective doses, which warranting the search for new effective and non-toxic radioprotectors.

Selenium (Se) belongs to the same group as sulfur in the periodic table and is an essential micronutrient. Se acts through selenoproteins like glutathione peroxidase (GPx) and thioredoxin reductase, which are necessary to maintain redox balance in cells (Papp et al., 2007). Sodium selenite, selenomethionine, selenium enriched yeast, etc. act as Se supplements. Sodium selenite was the first Se compound tested for radioprotection in mice. In a clinical study, selenite administered at a dosage of 200–400 μg per day increased the activity of serum GPx, and reduced the therapy-induced oxidative stress. However, selenite was also reported to exhibit behavioral toxicity in animals. With an aim to develop an alternate Se based radioprotector, our group initiated work on design, synthesis and development of GPx active Se compounds. A number of organoselenium compounds were synthesized in our laboratory. After screening all the water soluble selenium compounds, one lead compound, diselenodipropionic acid (DSePA) was found to exhibit promising radio-protective activity, which is now being developed as a radioprotector for lung radiotherapy.

2 MATERIALS AND METHODS

DSePA was synthesized in our laboratory according to the literature reported method (Arnold et al., 1986) and it was characterized by NMR, mass and IR spectroscopy. The GPx activity of DSePA was evaluated by estimating its mass equivalent for one GPx unit and kinetic parameter such as Michaelis-Menten constant (K_m) values for peroxide and GSH separately using the L-B plots. Irradiation of cells and mice was performed by using Cobalt-60 Bhabhatron facility with a dos rate of 1 Gy/minute. DSePA was supplemented at a dose of 2 mg/Kg body weight. Details of experiments in cells and in animals are given in references (Chaurasia et al., 2014; Kunwar et al., 2007, 2010, 2011, 2013)

3 RESULTS AND DISCUSSION

Study on GPx mimicking activity of DSePA indicated that it has more affinity towards the peroxide than the GSH and 12.84 mM of DSePA is equivalent to one unit of GPx. DSePA protected RBCs from peroxyl radical-induced lipid peroxidation, hemolysis and K^+ ions leakage (Kunwar et al., 2007). Before performing *in vivo* studies toxicological parameters of DSePA were performed using normal cells such as splenic lymphocytes and Chinese Hamster Ovary (CHO) cells and a tumor cell line of T lymphocytes (EL4). The results indicated that DSePA was non-toxic to both normal and tumor cells in the concentration range, where potent antioxidant activity was observed. The acute toxicity study of DSePA in mice revealed its maximum tolerable dose (MTD) as 8 mg/kg body weight for the intraperitoneal mode of administration. Further, DSePA administration at doses below MTD was shown not to cause any DNA damage or genotoxicity by measuring the induction of γ-H2AX foci and micronuclei frequency in the cell nuclei (Chaurasia et al., 2014). The radio-protective efficacy of DSePA was evaluated in mice after administering a non toxic and safe dose of 2 mg/kg body weight (i.p.) for 5 consecutive days prior to whole-body lethal doses of γ-irradiation by assessing the protection of hepatic tissue, hematopoietic organ (spleen), gastrointestinal (GI) tract and monitoring for any survival advantage. DSePA also protected against the depletion of endogenous antioxidants in the hepatic tissue of irradiated mice. Protection towards GI tract and hematopoietic system was confirmed by the restoration of radiation-induced reduction in villi height, number of crypt cells and spleen cellularity. DSePA ameliorated the radiation-induced intestinal inflammation and restored the immune balance (Th1/Th2 cytokines) in irradiated mice. In line with these observations, DSePA improved the 30-day survival of the irradiated mice by 35% with a dose reduction factor of about 1.10 (Kunwar et al., 2010, 2011).

Late lung tissue responses like pneumonitis and fibrosis are the most serious dose-limiting side effects of thoracic radiotherapy for several kinds of malignancies affecting organs in the thorax area. Whereas pneumonitis is an acute inflammatory response, fibrosis is characterized by progressive scarring of the lung, with vascular cell damage and collagen deposition in the interstitium. Since DSePA exhibited encouraging whole body radioprotection, the effect of DSePA in reducing the lung tissue response to thorax irradiation was also investigated. Whereas in previous work, DSePA was administered daily for 5 days before radiation exposure, in this study it was given three times a week, starting from the day of irradiation until the end of the experiment (i.e., until the presentation of lung disease). The results from this study indicated that despite the prolonged treatment, DSePA by itself did not cause any toxicity in the mice. Further, DSePA treatment increased the postirradiation asymptomatic survival time of mice by an average of 32 days, specifically reduced the levels of lipid peroxidation in the lungs, and increased GPx level, compared with mice receiving irradiation only. The intervention also decreased the extent of pneumonitis in mice, and decreased infiltration of mast cells and neutrophils in the lungs of irradiated mice (Kunwar et al., 2013) Recently, we also studied the bio-distribution of orally administered DSePA is in to different organs systems of tumor bearing mice. The results of this experiment showed the maximum bioavailability of DSePA in the lung followed by other organs such as kidney, liver and intestine. Interestingly, the bioavailability of DSePA in tumor tissue was found to be minimum. Encouraged by these results we propose to evaluate the efficacy of DSePA as lung radioprotector.

4 CONCLUSIONS

DSePA a non-toxic selenium compound has shown promising radioprotective effects in several radiosensistive organs and has ability to reduce late lung responses of radiation. With its preferential bioavailability in lungs it would be an effective radiopotector during lung radiotherapy.

REFERENCES

Arnold, A. P., Tan, K. S., & Rabenstein, D. L. 1986. Nuclear magnetic resonance studies of the solution chemistry of metal complexes, selenodryl containing amino acids and related molecules. *Inorg. Chem.* 25, 2433–2437.

Heilongjiang Provincial Bureau of Land Management. 1992. *Heilongjiang soil*. Compiled by the Soil Survey Office of Heilongjiang Province. Agricultural Press. (In Chinese)

Koukourakis, M.I. 2012. Radiation damage and radioprotectants: new concepts in the era of molecular medicine The British Journal of Radiology, 85: 313–330.

Kunwar, A., Mishra, B., Barik, A., Kumbhare, L.B, Jain, V.K., & Priyadarsini, K.I, 2007. 3,3′-Diselenodipropionic acid, an efficient peroxyl radical scavenger and a GPx mimic, protects erythrocytes (RBCs) from AAPH induced hemolysis *Chemical Research in Toxicology* 20: 1482–87.

Kunwar, A., Bansal, P., Kumar, S.J., Bag, P.P., Paul, P., Reddy, N.D., et al. 2010. In vivo radioprotection studies of 3,3′-diselenodipropionic acid, a selenocystine derivative. *Free Radic. Biol. Med.* 48: 399–410.

Kunwar, A., Bag, P.P., Chattopadhyay, S., Jain, V.K., & Priyadarsini, K.I. 2011. Anti-apoptotic, anti-inflammatory and immunomodulatory activities of 3,3′-diselenodipropionic acid in mice exposed to whole body γ-radiation. *Arch. Toxicol.* 85: 1395–1405.

Kunwar, A., Jain, V.K., Priyadarsini, K.I. & Christina, K.A. 2013. Selenocysteine derivative therapy affects radiation-induced pneumonitis in the mouse. Amer J. Resp. Cell Mol Biol., 49: 654–671.

Papp, L.A., Lu, J., Holmgren, A., & Khanna, K. 2007. From selenium to selenoproteins: synthesis, identity, and their role in human health. *Antioxidant & Redox Signaling*, 9: 775–806.

Association of selenoprotein and selenium pathway genetic variations with colorectal cancer risk and interaction with selenium status

D.J. Hughes*
Centre for Systems Medicine, Department of Physiology & Department of Epidemiology and Public Health Medicine, Royal College of Surgeons in Ireland, Dublin, Ireland

V. Fedirko & J.S. Jones
Rollins School of Public Health, Emory University, Atlanta, GA, USA

L. Schomburg & S. Hybsier
Institute for Experimental Endocrinology, University Medical School Berlin, Germany

C. Méplan & J.E. Hesketh
Institute of Cell and Molecular Biosciences, Newcastle University, UK

E. Riboli
School of Public Health, Imperial College London, UK

M. Jenab
Section of Nutrition and Metabolism, International Agency for Research on Cancer, Lyon, France

1 INTRODUCTION

Suboptimal intakes of the micronutrient selenium (Se) are found in many parts of Europe and experimental and observational evidence suggests that this contributes to the development of several cancers (Steinbrenner et al., 2013; Méplan & Hesketh, 2014; Hughes et al., 2015; Combs, 2015). Selenium is incorporated in 25 selenoproteins with roles which are thought to help prevent carcinogenesis largely due to the role of several of these proteins in cell protection from oxidative stress, redox control and the inflammatory response (Fairweather-Tait et al., 2011; Labunskyy et al., 2014). We recently reported in a nested cohort study of 966 CRC cases and 966 matched controls that a higher Se status (ascertained by serum levels of Se and Selenoprotein P, SePP) was associated with a lower colorectal cancer (CRC) risk within the European Prospective Investigation into Cancer and Nutrition (EPIC) (Hughes et al., 2015). Additionally, association studies have revealed that genetic variations in several of the selenoprotein genes may affect CRC risk (Méplan & Hesketh, 2014). We are currently examining the association of Se pathway gene variation (Méplan et al., 2012) with CRC risk, including the interaction of gene x Se status in disease risk modification.

2 MATERIALS AND METHODS

Candidate functional and common tagging SNPs (N = 999) in the Se pathway (selenoprotein genes, selenoprotein biosynthesis / transport genes, and genes in pathways sensitive to Se intake) were genotyped by *Illumina Goldengate* in DNA samples available for 1478 CRC case-control pairs matched within EPIC. 905 SNPs were successfully genotyped (94 SNPs with at least 20% missing data were excluded from analysis) in 149 genes comprising 154 variants in 24 of 25 selenoprotein genes and 751 variants in 125 Se pathway genes. Multivariable odds ratios and 95% confidence intervals were calculated using conditional logistic regression. In pathway analyses, genes were sorted into a primary best-known functional pathway and gene and pathway p-values were computed using the PIGE package Adaptive Rank Truncated test (Yu et al., 2009).

3 RESULTS AND DISCUSSION

The findings from this large prospective nested case-control study indicate that common genetic variation in 20 out of the 41 tested selenoprotein and Se biosynthesis genes, as well as variations in many genes related to the wider Se gene pathway, may affect CRC development risk.

There were 315 tagging SNPs analysed from 41 selenoprotein and Se transport/biosynthesis genes, and 590 variants from the other 108 wider Se interaction pathway genes. Among the 111 SNPs in 55 genes associated with an altered CRC risk ($p < 0.05$), 20 were located in 13 selenoprotein genes (Table 1), 17 in 7 other Se transport/biosynthesis genes and

Table 1. The 20 tagging SNPs in 13 of the selenoprotein genes significantly associated with CRC risk in at least one of the genetic models tested (p < 0.05). Genotypes were analysed by unconditional logistic regression with adjustment for age at recruitment, study centre and sex.

Selenoprotein Genes	No. SNP significantly associated with CRC risk	RS numbers
DIO1	1	rs11206244
GPX1	3	rs3448
		rs17080528
		rs9818758
GPX4	1	rs2074451
GPX6	1	rs974334
SEP15	1	rs472509
SELM	1	rs11705137
SELN	3	rs4659382
		rs11247710
		rs11247735
SELO	1	rs5771237
SELT	1	rs17214746
SELV	1	rs4802034
TXNRD1	3	rs11111979
		rs10778318
		rs7953266
TXNRD2	2	rs9606174
		rs17745314
TXNRD3	1	rs777238

another 74 variants in 35 wider Se interaction pathway genes (note that multiple testing corrections have not yet been applied). These 13 selenoprotein genes include those previously found to harbour variants associated with an altered CRC risk (*GPX1, GPX4, SEP15, TXNRD1, TXNRD2, TXNRD3*; for review, see Méplan & Hesketh, 2014) and also several with limited prior or no previous evidence of association with CRC risk (*DIO1, GPX6, SELM, SELN, SELO, SELT, SELV*). In addition, there were significant SNP x Se status interactions between Se and SePP levels with 54 and 41 tagging SNPs, respectively, that further modify CRC risk. Pathway analyses suggest that variations in the selenoprotein and biosynthesis pathway alone may alter CRC susceptibility risk, while risk associations attributed to the anti-oxidant and apoptosis pathways may depend more on gene x Se interactions.

4 CONCLUSIONS

Genetic variation in the selenoprotein gene pathway and in genes affected by Se intake may be associated with an increased or decreased risk of CRC development depending on genotype. Additionally, genetic variation is likely to alter risk associated with Se status measures (different genotypes were associated with increased and decreased risk modifications in interaction with Se status). Together with our previous findings for Se status markers (Hughes et al., 2015), we recommend that future Se supplementation trials for CRC prevention need to be conducted in populations with sub-optimal Se availability and should take account of selenoprotein pathway genotypes.

ACKNOWLEDGEMENTS

M. Jenab was involved in this study on behalf of the European Prospective Investigation into Cancer and Nutrition (EPIC). Funding for this study was provided by the Health Research Board of Ireland project grant HRA_PHS/2013/397 (to DJH). The EPIC study was supported by various funders (as detailed in Hughes et al., 2015).

REFERENCES

Fairweather-Tait, S.J., Bao, Y., Broadley, M.R., Collings, R., Ford, D., Hesketh, J.E., & Hurst, R. 2011. Selenium in human health and disease. *Antioxid. Redox Signal.* 14: 1337–1383.

Combs, G.F. 2015. Biomarkers of Selenium Status. *Nutrients* 7: 2209–2236.

Hughes, D.J., Fedirko, V., Jenab, M., Schomburg, L., Méplan, C., Freisling, H., et al. 2015. Selenium status is associated with colorectal cancer risk in the European Prospective Investigation of Cancer and Nutrition Cohort. *Int. J. Cancer* 136(5): 1149–1161.

Labunskyy, V.M., Hatfield, D.L., & Gladyshev, V.N. 2014. Selenoproteins: molecular pathways and physiological roles. *Physiol. Rev.* 94: 739–777.

Méplan, C., Rohrmann, S., Steinbrecher, A., Schomburg, L., Jansen, E., Linseisen, J., et al. 2012. Polymorphisms in thioredoxin reductase and selenoprotein K genes and selenium status modulate risk of prostate cancer. *PLoS One.* 7(11): e48709.

Méplan, C., & Hesketh, J. 2014. Selenium and cancer: a story that should not be forgotten-insights from genomics. *Cancer Treat. Res.*159: 145–66.

Steinbrenner, H., Speckmann, B., & Sies, H. 2013. Toward understanding success and failures in the use of selenium for cancer prevention. *Antioxid. Redox Signal.* 19: 181–191.

Yu, K., Li, Q., Bergen, A.W., Pfeiffer, R.M., Rosenberg, P.S., Caporaso, N., et al. 2009. Pathway analysis by adaptive combination of P-values. *Genet Epidemiol.* 33(8): 700–709.

Selenoenzymes iodothyronine deiodinases: 1. Effects of their activities in various rat tissues by administered antidepressant drug Fluoxetine

S. Pavelka
Institute of Physiology, Academy of Sciences of the Czech Republic, Prague
Institute of Biochemistry, Faculty of Science, Masaryk University, Brno, Czech Republic

1 INTRODUCTION

Selenoproteins iodothyronine deiodinases (IDs) are the key enzymes in the metabolism of thyroid hormones (THs). These enzymes catalyze the most important metabolic transformations of iodothyronine molecules through stepwise mono-deiodinations (Leonard & Köhrle, 2000). Three distinct types of IDs have been defined on the basis of substrate specificity, selectivity of the reactions they catalyze, sensitivity to inhibition by propylthiouracil, and response in activity that occurs *in vivo* with a change of thyroid status. All three IDs of types 1, 2 and 3 (D1, D2 and D3, respectively) are integral membrane proteins, which require thiols as a cofactor (Köhrle, 2002). Multiple biological effects of THs depend on intracellular levels of 3,5,3′-triiodo-L-thyronine (T_3) acting upon nuclear THs receptors. This active form of THs is mainly generated in peripheral tissues by outer-ring 5′-deiodination of prohormone thyroxine (T_4) by the action of IDs. Both D1 and D2 can catalyze this reaction. D1 is mainly expressed in the liver, kidneys, thyroid gland and pituitary, and is regarded as being the most important source of circulating T_3. However, D1 activity is broader and this ID can also perform inner-ring 5-deiodination of T_4 producing 3,3′,5′-triiodo-L-thyronine, reverse T_3 (rT_3), an inactive form of THs. D2 is mainly present in the brown adipose tissue, placenta, pituitary and muscle, and it is involved in local T_3 production in these tissues. Deactivation of THs is performed by D3 deiodinase, which catalyzes specifically inner-ring 5-deiodinations of T_4 and T_3 to produce rT_3 and diiodothyronine (T_2), respectively (Pavelka et al., 1997; Bianco et al., 2002; Macek Jilkova et al., 2010).

It is well known that THs can modulate concentration of neurotransmitters and metabolites in the CNS, among other effects. THs are supposed to control in this way the activity of some neurotransmitters (e.g., serotonin), which are hypothetically involved in the pathogenesis of depressive illness. On the other hand, many clinical studies indicated that various antidepressant treatments might influence the effects of hormones of hypothalamic-pituitary-thyroid axis in depressed patients (Kirkegaard & Faber, 1998; Sauvage et al., 1998). For example, Baumgartner et al. (1994) observed that subchronic administration of fluoxetine (Fluox) to rats produced significant effects on the enzyme activities of IDs in specific regions of the brain. Fluoxetine is the most frequently used drug today for the treatment of episodic depression and belongs to the group of non-tricyclic antidepressants known as selective serotonine re-uptake inhibitors. Importantly, inadequate activities of brain IDs can lead to local insufficient T_3 concentration and might be, therefore, one of the pathogenic factors of depression.

In the present studies, we followed in more details the effects of subchronic administration (for 25 days) of Fluox by itself to Wistar rats, or in combination with T_3, on the metabolism of THs in various rat tissues.

2 MATERIALS AND METHODS

Pharmacological experiments were performed on 90 young growing male Wistar rats, fed *ad libitum* a standard pelleted diet. During 25-day experimental period, the rats drank tap water with the addition of various amounts of fluoxetine (Ratiopharm), T_3 (Sigma), or their combinations. After the treatment, rats were decapitated under deep diethyl-ether anaesthesia. Serum total and free T_3 and T_4 levels were determined using commercial radio-immunoassay (RIA) kits. The thyroid glands, pituitary, different parts of the brain, liver, kidneys and several other tissues were also dissected and frozen. Samples of frozen tissues were used for the preparation of subcellular fractions – microsomes (liver, kidney), post-mitochondrial supernatants (brain, pituitary, thyroid gland) and cytosols (liver), used in the following radiometric enzyme assays.

Enzyme activities of D1, D2, and D3 were measured in principle as described previously (Pavelka, 2010), but with some modifications, e.g., of the thin layer chromatography (TLC) developing system and the way of radiochromatograms evaluation (Pavelka, 2015). Activities of conjugating enzymes iodothyronine sulfotransferases (in liver cytosols) and uridine 5′-diphospho-glucuronyltransferase (in liver microsomes) were also measured with the use of the adapted radiometric enzyme assays.

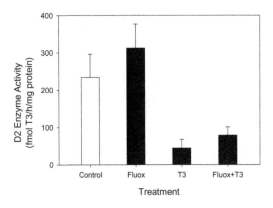

Figure 1. The influence of fluoxetine, T_3 and their combination on specific enzyme activities of D2 (fmol T_3/h/mg prot.) in rat pituitary.

3 RESULTS

Subchronic treatment of rats with Fluox by itself caused a moderate increase in D2 (55 ± 34 vs. 31 ± 9 fmol T_3/h/mg prot. in controls; mean $\pm$ SD for $n = 7$) and, in turn, a slight decrease in D3 activities in cerebellum and some other regions of the CNS. On the other hand, the administration of T_3 by itself caused, in accordance with our expectation, a substantial decrease in pituitary D2 activity (45 ± 23 vs. 234 ± 62 fmol T_3/h/mg prot. in controls; mean $\pm$ SD for $n = 7$) (Fig. 1) and a simultaneous increase in D1 and D3 activities (practically in all tissues studied). In contrast with the effects of T_3 by itself, the treatment of rats with the combination of Fluox plus T_3 caused only non-significant elevation of D1 and D3 activities.

4 CONCLUSIONS

With the use of our newly elaborated and adapted radiometric enzyme assays for IDs, we were able to quantify the effects of subchronic administration of xenobiotic Fluox, alone or in combination with T_3, on the metabolism of THs in the rats. Especially, we determined changes in deiodinating enzyme acti-vities in several specific tissues of the treated rats.

REFERENCES

Baumgartner, A., Dubeyko, M., Campos-Barros, A., Eravci, M. & Meinhold, H. 1994. Subchronic administration of fluoxetine to rats affects triiodothyronine production and deiodination in regions of the cortex and in the limbic forebrain. *Brain Research* 653: 68–74.

Bianco, A.C., Salvatore, D., Gereben, B., Berry, M.J. & Larsen, P.R. 2002. Biochemistry, cellular and molecular bio-logy, and physiological roles of the iodothyronine seleno-deiodinases. *Endocrine Reviews* 23: 38–89.

Kirkegaard, C. & Faber, J. 1998. The role of thyroid hormones in depression. *European Journal of Endocrinology* 138: 1–9.

Köhrle, J. 2002. Iodothyronine deiodinases. *Methods in Enzymology* 347: 125–167.

Leonard, J.L. & Köhrle, J. 2000. Intracellular pathways of thyroid hormone metabolism. In: L.E. Braverman & R.D. Utiger (eds) *Werner and Ingbar's the Thyroid – A Fundamental and Clinical Text* 8th edition. Philadelphia: Lippincott Williams & Wilkins. 136–173.

Macek Jilkova, Z., Pavelka, S., Flachs, P., Hensler, M., Kus, V. & Kopecky, J. 2010. Modulation of type I iodothyronine 5′-deiodinase activity in white adipose tissue by nutrition: possible involvement of leptin. *Physiological Research* 59: 561–569.

Pavelka, S., Kopecky, P., Bendlova, B., Stolba, P., Vitkova, I., Vobruba, V. et al. 1997. Tissue metabolism and plasma levels of thyroid hormones in critically ill very premature infants. *Pediatric Research* 42: 812–818.

Pavelka, S. 2010. Radiometric enzyme assays: development of methods for extremely sensitive determination of types 1, 2 and 3 iodothyronine deiodinase enzyme activities. *Journal of Radioanalytical and Nuclear Chemistry* 286: 861–865.

Pavelka, S. 2015. Selenoenzymes iodothyronine deiodinases: 2. Novel radiometric enzyme assays for extremely sensitive determination of their activities. Submitted *Abstract. 4th International Conference on Selenium in the Environment and Human Health*, Sao Paulo, Brazil.

Sauvage, M.F., Marquet, P., Rousseau, A., Raby, C., Buxeraud, J. & Lachatre, G. 1998. Relationship between psychotropic drugs and thyroid function: a review. *Toxicology and Applied Pharmacology* 149: 127–135.

Selenoenzymes iodothyronine deiodinases: 2. Novel radiometric enzyme assays for extremely sensitive determination of their activities

S. Pavelka
Institute of Physiology, Academy of Sciences of the Czech Republic, Prague
Institute of Biochemistry, Faculty of Science, Masaryk University, Brno, Czech Republic

1 INTRODUCTION

Selenium (Se), as an essential trace element is absolutely required for thyroid hormones (THs) synthesis and biotransformations. Three selenoenzymes, iodothyronine deiodinases (IDs) of types 1, 2 and 3 (D1, D2 and D3, respectively) have been identified with 5′ (outer-ring) and/or 5- (inner-ring) mono-deiodinating activities, which play an important role in the tissue-specific regulation of THs bioactivity (Bianco et al., 2002; Köhrle, 2002).

Recently, we have utilized radiometric methods for the extremely sensitive determination of IDs enzyme activities in homogenates of cultured mammalian cells (Pavelka, 2010). In general, the radiometric enzyme assays are based on the enzymaticaly catalyzed conversion of radioactively labeled substrates to labeled products, and on the measurement of radioactivity of either products or residual substrate after their quantitative separation. The availability of a simple and rapid method for quantitative separation of substrate and product is one of the two major requirements of a radiometric enzyme assay. The other requirement is the availability of a suitable labeled substrate of known specific radioactivity.

Most of the previous studies examining IDs in different tissues and species did not perform the measurement of enzyme activity under the optimum conditions (Baumgartner et al., 1994). In the present study, our objectives were, therefore, to determine proper assay conditions for measuring D1, D2 and D3 deiodinase activities in various sub-cellular fractions of rat and human tissues, including the concentrations of radioactively labeled substrates and thiol cofactor, the amount of total protein and enzyme concentration in the incubation mixtures, and appropriate incubation times.

2 MATERIALS AND METHODS

The newly elaborated radiometric methods for extremely sensitive determination of IDs enzyme activities in homogenates of cultured mammalian cells (Pavelka, 2010) were modified and adapted for other types of biological material (post-mitochondrial supernatants and microsomal frac-tions of various rat and human tissues). These radiometric enzyme assays were based on the use of high specific-radioactivity ^{125}I-labeled iodothyro-nines (NEN Radiochemicals) as substrates; sophisti-cated thin layer chromatography (TLC) separation of radioactive products from the unconsumed substrates; film-less auto-radiography of radiochro-matograms using storage phosphor screens (BAS-IP MS 2025, Fuji Photo Film Co., Japan); and quantification of the separated compounds with a BAS-5000 laser scanner (Fujifilm Life Science Co., Japan), using AIDA software (Raytest Isotopen-messgeräte, GmbH, Germany).

3 RESULTS

Modification of radiometric enzyme assays for IDs for determination of D1, D2 and D3 activities in post-mitochondrial supernatants and microsomal fractions of various rat and human tissues. Figure 1 shows an example of electronic image of radiochromatogram, which was prepared in the course of carrying out the radiometric enzyme assay for D1 deiodinase. The arrangement of the assays for D2 and D3 was similar. Duplicates of appropriate amounts of post-mitochondrial super-natants or microsomal fractions, isolated from samples of various rat tissues (usually 2–200 µg of protein in a volume of 20 µl, depending on types of tissue and ID measured) were incubated for appropriate time (usually 30 minutes) under optimum concentration conditions in the presence of the respective labeled substrate, partly at 37°C (samples) and partly at 0°C (corresponding blanks). The reaction was stopped on ice by adding 10 µl of a solution containing 10 µM T_3 and 10 µM T_4 in concentrated ammonium hydroxide (stop mix), followed by 30 µl of methanol, vortexing and centrifugation. Aliquots (4×2 µl) of supernatant extracts were analyzed by TLC on a silica gel plate using an optimized solvent system. Separated products of the reactions containing $\geq 0.5\%$ of the total radioactivity applied (corresponding to ≥ 25 cpm/spot) could be detected in this way. Our methodology enabled determination of IDs enzyme activities, in absolute terms, as low as 10^{-18} katals. On the other hand, the same system used under proper

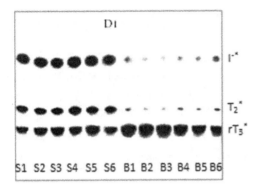

Figure 1. Electronic image (made with BAS-5000 laser scanner) of radiochromatogram of incubation mixtures in D1 enzyme assay in kidney microsomes (six samples: S1–S6 and six corresponding blanks: B1–B6). The radioactively labeled substrate was $[^{125}I]$-rT$_3$ and labeled products of its enzymatic conversion were $[^{125}I]$-3,3'-T$_2$ and $[^{125}I]$-I$^-$.

concentration conditions (e.g., a thousand-fold pre-dilution of samples of biological materials) enabled measurement of IDs activities as high as 10^{-12} katals.

4 CONCLUSIONS

We utilized novel, reliable radiometric methods for extremely sensitive determination of IDs enzyme activities in post-mitochondrial supernatants and microsomal fractions of various rat and human tissues, as well as in homogenates of cultured mammalian cells. The newly developed radiometric assays for IDs were based on the use of high specific-radioactivity ^{125}I-labeled iodothyronines as substrates; sophisticated TLC separation of radioactive products from the unconsumed substrates; film-less autoradiography of radiochromatograms using storage phosphor screens; and quantification of the separated compounds with a BAS-5000 laser scanner. This methodology enabled us to determine IDs enzyme activities as low as 10^{-18} katals. The assays proved to be very sensitive and rapid and, at the same time, reliable and robust.

REFERENCES

Baumgartner, A., Dubeyko, M., Campos-Barros, A., Eravci, M. & Meinhold, H. 1994. Subchronic administration of fluoxetine to rats affects triiodothyronine production and deiodination in regions of the cortex and in the limbic forebrain. *Brain Research* 653: 68–74.

Bianco, A.C., Salvatore, D., Gereben, B., Berry, M.J. & Larsen, P.R. 2002. Biochemistry, cellular and molecular bio-logy, and physiological roles of the iodothyronine seleno-deiodinases. *Endocrine Reviews* 23: 38–89.

Köhrle, J. 2002. Iodothyronine deiodinases. *Methods in Enzymology* 347: 125–167.

Pavelka, S. 2010. Radiometric enzyme assays: development of methods for extremely sensitive determination of types 1, 2 and 3 iodothyronine deiodinase enzyme activities. *Journal of Radioanalytical and Nuclear Chemistry* 286: 861–865.

The role of thioredoxin reductase 1 in cancer

E.S.J. Arnér
Division of Biochemistry, Department of Medical Biochemistry and Biophysics, Karolinska Institutet, Stockholm, Sweden

1 THIOREDOXIN REDUCTASE IS A SELENOPROTEIN WITH KEY FUNCTIONS FOR REDOX CONTROL IN CANCER CELLS

Thioredoxin reductase 1 (TrxR1, encoded by the *TXNRD1* gene in human) is a cytosolic selenoprotein with major importance for a wide range of redox processes carried out by thioredoxin 1 (Trx1), which is the main substrate for TrxR1. The active site disulfide of Trx1 is reduced to a dithiol motif by TrxR1 using NADPH. Reduced Trx, in turn, is well known for promoting growth and viability of cells through antioxidant and antiapoptotic functions, and by its support of synthesis of DNA precursors through ribonucleotide reductase. Several electrophilic anticancer drugs target the highly reactive Sec residue of TrxR1, which can explain parts of their cytotoxicity (Arnér, 2009; Eriksson et al., 2009).

We have found that the well-used anticancer agent cisplatin triggers the intracellular formation of covalently non-disulfide linked complexes of TrxR1 with both Trx1 and thioredoxin-related protein of 14 kDa (TRP14) (Prast-Nielsen et al., 2010). TRP14 is a Trx fold protein with an active site WCPDC motif previously shown to resemble Trx1 in being a ubiquitous cytosolic disulfide reductase. However, while being a substrate for TrxR1, others have shown that TRP14 does not reduce the typical protein substrates of Trx1 and it exhibits only 20% sequence identity with Trx1 (Woo et al., 2004). Interestingly, we recently discovered that TRP14 has a unique denitrosylation activity (Pader et al., 2014), implying that it is involved in cell signaling related to nitric oxide.

We have furthermore shown that selenium compromised forms of TrxR1 can induce cell death by a gain of function in the form of SecTRAPs (Se compromised thioredoxin reductase-derived apoptotic proteins). The formation of SecTRAPs possibly contributes to efficacy or side effects of certain electrophilic drugs used in cancer therapy, such as cisplatin (Anestal et al., 2008).

It is well established that cancer cells generate ROS at higher rates than normal cells as a result of oncogenic signaling, distorted metabolism, impaired mitochondrial function and other mechanisms (Luo et al., 2009). Their redox systems are thereby pushed to the limits of their capacity, which renders tumor cells more vulnerable than normal cells to agents that enhance oxidative stress, or reduce their capacity to cope with such stress. This fact may be utilized to selectively kill tumor cells (Wondrak, 2009; Gorrini et al., 2013). Indeed, many classical anticancer treatments - at least in part – are known to act by increasing ROS levels in cancer cells (Wondrak, 2009; Gorrini et al., 2013). It is thus not far-fetched to believe that TrxR1 targeting may be an important mechanism of action for anticancer drugs (Arnér & Holmgren, 2006; Urig and Becker, 2006; Lu et al., 2007; Yoo et al., 2007; Wang et al., 2008; Hedstrom et al., 2009; Shi et al., 2014).

2 ONGOING STUDIES AND UNRESOLVED QUESTIONS

To fully understand the properties of TrxR1 as a molecular target for anticancer therapy, we need to understand the full molecular details of the functions of this enzyme. To further probe this, we characterized novel low-activity tetrameric forms of the protein that may represent regulatory states (Rengby et al., 2009). We recently also crystallized and solved one such tetrameric structure, and identified a key Trp residue that plays an important role as a "redox sensor" affecting the catalysis of the enzyme in relation to redox perturbing conditions and cell death induction (Xu et al., 2015). In addition, we and others have shown that the same compounds that target TrxR1, or the genetic absence of TrxR1 in conditional knockout mouse models, will induce robust Nrf2 responses in normal cells (Locy et al., 2012; Prigge et al., 2012; Iverson et al., 2013). Indeed, it may be possible that TrxR1 is a potent regulator of Nrf2 and provide a link between cancer treatment using electrophilic anticancer drugs, and indicate both the importance of Nrf2 in cancer and the outcome of TrxR1 targeting for anticancer therapy (Cebula et al., 2015).

The importance of TrxR1 for cancer treatment and its links to Nrf2, however, needs to be confirmed experimentally. More studies are also needed to show whether novel inhibitors of TrxR1 can indeed be developed and evaluated for anticancer therapy.

3 CONCLUSIONS

Several studies suggest that TrxR1 may be a promising anticancer drug target. However, it remains to be shown which effects of its targeting are most important to yield efficacy, which include the formation of SecTRAPs, increased ROS levels in cancer cells and/or effects through its downstream substrates Trx1 and TRP14.

REFERENCES

Anestal, K., S. Prast-Nielsen, N. Cenas & E.S. Arner. 2008. Cell death by SecTRAPs: thioredoxin reductase as a prooxidant killer of cells. PLoS ONE 3(4): e1846.

Arnér, E.S.J. 2009. Focus on mammalian thioredoxin reductases – important selenoproteins with versatile functions. Biochim Biophys Acta, 1790(6): 495–526.

Arnér, E.S.J. & A. Holmgren. 2006. The thioredoxin system in cancer. Semin Cancer Biol 16(6): 420–426.

Eriksson, S.E., S. Prast-Nielsen, E. Flaberg, L. Szekely & E. S. Arner. 2009. High levels of thioredoxin reductase 1 modulate drug-specific cytotoxic efficacy. Free Radic Biol Med 47(11): 1661–1671.

Gorrini, C., I.S. Harris & T.W. Mak. 2013. Modulation of oxidative stress as an anticancer strategy. Nat Rev Drug Discov 12(12): 931–947.

Hedstrom, E., S. Eriksson, J. Zawacka-Pankau, E.S. Arner & G. Selivanova. 2009. p53-dependent inhibition of TrxR1 contributes to the tumor-specific induction of apoptosis by RITA. Cell Cycle 8(21): 3576–3583.

Iverson, S.V., S. Eriksson, J. Xu, J.R. Prigge, E.A. Talago, T.A. Meade, et al. 2013. A Txnrd1-dependent metabolic switch alters hepatic lipogenesis, glycogen storage, and detoxification. Free Radic Biol Med., 63: 369–380.

Locy, M.L., L.K. Rogers, J. R. Prigge, E.E. Schmidt, E.S. Arner & T.E. Tipple. 2012. Thioredoxin reductase inhibition elicits Nrf2-mediated responses in Clara cells: implications for oxidant-induced lung injury. Antioxid Redox Signal 17(10): 1407–1416.

Lu, J., E.H. Chew & A. Holmgren. 2007. Targeting thioredoxin reductase is a basis for cancer therapy by arsenic trioxide. Proc Natl Acad Sci USA. 104(30): 12288–12293.

Luo, J., N.L. Solimini & S.J. Elledge. 2009. Principles of cancer therapy: oncogene and non-oncogene addiction. Cell 136(5): 823–837.

Pader, I., R. Sengupta, M. Cebula, J. Xu, J.O. Lundberg, A. Holmgren, K. Johansson & E.S. Arner. 2014. Thioredoxin-related protein of 14 kDa is an efficient L-cystine reductase and S-denitrosylase. Proc Natl Acad Sci USA. 111(19): 6964–6969.

Prast-Nielsen, S., M. Cebula, I. Pader and E.S. Arner. 2010. Noble metal targeting of thioredoxin reductase–covalent complexes with thioredoxin and thioredoxin-related protein of 14 kDa triggered by cisplatin. Free Radic Biol Med 49(11): 1765–1778.

Prigge, J.R., S. Eriksson, S.V. Iverson, T.A. Meade, M.R. Capecchi, E.S.J. Arnér & E.E. Schmidt. 2012. Hepatocyte DNA replication in growing liver requires either glutathione or a single allele of txnrd1. Free Radic Biol Med 52: 803–810.

Rengby, O., Q. Cheng, M. Vahter, H. Jornvall & E.S. Arner. 2009. Highly active dimeric and low-activity tetrameric forms of selenium-containing rat thioredoxin reductase 1. Free Radic Biol Med 46(7): 893–904.

Shi, Y., F. Nikulenkov, J. Zawacka-Pankau, H. Li, R. Gabdoulline, J. Xu, 2014. ROS-dependent activation of JNK converts p53 into an efficient inhibitor of oncogenes leading to robust apoptosis. Cell Death Differ 21(4): 612–623.

Urig, S. & K. Becker. 2006. On the potential of thioredoxin reductase inhibitors for cancer therapy. Semin Cancer Biol 16(6): 452–465.

Wang, X., J. Zhang & T. Xu. 2008. Thioredoxin reductase inactivation as a pivotal mechanism of ifosfamide in cancer therapy. Eur J Pharmacol. 579(1–3): 66–73.

Wondrak, G.T. 2009. Redox-directed cancer therapeutics: molecular mechanisms and opportunities. Antioxid Redox Signal 11(12): 3013–3069.

Woo, J.R., S.J. Kim, W. Jeong, Y.H. Cho, S.C. Lee, Y.J. Chung, et al. 2004. Structural basis of cellular redox regulation by human TRP14. J Biol Chem. 279(46): 48120–48125.

Xu, J., S.E. Eriksson, M. Cebula, T. Sandalova, E. Hedstrom, I. Pader, et al. 2015. The conserved Trp114 residue of thioredoxin reductase 1 has a redox sensor-like function triggering oligomerization and crosslinking upon oxidative stress related to cell death. Cell Death Dis. 6: e1616.

Yoo, M.H., X.M. Xu, B.A. Carlson, A.D. Patterson, V.N. Gladyshev & D.L. Hatfield. 2007. Targeting thioredoxin reductase 1 reduction in cancer cells inhibits self-sufficient growth and DNA replication. PLoS ONE. 2(10): e1112.

Selenoprotein T: A new promising target for the treatment of myocardial infarction and heart failure

Inès Boukhalfa*, Najah Harouki, Lionel Nicol, Anaïs Dumesnil, Isabelle Rémy-Jouet, Jean-Paul Henry, Christian Thuillez, Antoine Ouvrard-Pascaud, Vincent Richard & Paul Mulder
Inserm U1096, Rouen, France;
University of Rouen, Institute for Research and Innovation in Biomedicine, Rouen, France

Youssef Anouar
Inserm U982, Mont Saint Aignan, France;
University of Rouen, Institute for Research and Innovation in Biomedicine, Rouen, France

1 INTRODUCTION

Selenium (Se) is an essential trace element and is considered as one of the most important in human health. In humans, Se deficiency is associated with numerous diseases, including Keshan disease. Selenium exerts its impact through Se-dependent proteins, i.e., selenoproteins, in which Se is incorporated into selenocystein (Sec) (Kryukov et al., 2003). In the heart, selenoproteins such as the mammalian glutathione peroxidases (GPxs) have been shown to play a major role in the antioxidant defense systems. In line with these observations, accumulating evidence from experimental studies indicates that SelT, an endoplasmic reticulum-resident thioredoxin-like protein, involves in many cellular processes such as cell growth and activity, in nervous, endocrine, and metabolic tissues (Prevost et al., 2013). However, SelT is re-expressed in cerebral ischemia and exerts neuro- protective properties. It is unknown whether SelT possesses similar protective properties in cardiovascular diseases, i.e. cardiac ischemia-reperfusion. Preliminary experiments performed in our laboratory clearly showed that myocardial infarction increased SelT protein expression, suggesting that SelT may play a yet unrevealed protective role in the cardiovascular system. Thus, the present study assessed the role of SelT in the development of left ventricular (LV) dysfunction using a myocardial infarction rat model.

2 MATERIALS AND METHODS

Animal model and treatment. Male Wistar rats (220 g, Janvier, France) were housed and experimented on in accordance with NIH guidelines. All animal experiments were approved by the local ethical commission. Myocardial infarction was induced by temporary occlusion of the left coronary artery (45 min ischemia, followed by reperfusion; I/R) in ketamine-xylazine anaesthetized animals. Five days before I/R, SelT (15 µg/kg/day, IP) or saline infusion was started by osmotic mini-pump (Alzet Model 2ML2, Cupertino, Calif., USA).

Cardiac evaluations. In sedated rats (sodium methohexital, 50 mg/kg IP), LV dimensions and function were assessed one week after I/R, using a Vivid 7 ultrasound echograph (GE Medical) equipped with a 14 MHz probe as described previously by Roche et al. (2015).

Myocardial perfusion was assessed by cardiac magnetic resonance imaging (MRI) in anesthetized rats using a 4.7 T horizontal bore scanner (4.7T, Bruker, France).

LV hemodynamics was assessed by determination of LV pressure-volume curves in anesthetized rats (Brietal™; 50 mg/kg, IP). The right carotid artery was cannulated for the recording of arterial blood pressure/heart rate, after which catheter was advanced into the LV. Finally, LV end-systolic as well as end-diastolic pressure–volume relations were measured as indexes of LV elastance and compliance, respectively.

Histological evaluations. Infarct size was determined using histological slices stained with Sirius Red and evaluated in 5 sections spanning the LV.

Statistical analysis. Data are presented as means ± sem. Sham, I/R and I/R+SelT groups were compared using one-way ANOVA followed, in case of significance, by Tukey's test for multiple comparisons. Infarct size in both I/R groups were compared using a Student's t test (2-tailed). A value of $p \leq 0.05$ was considered significant.

3 RESULTS

Mean infarct size was identical in I/R rats and I/R+SelT rats, as shown in Table 1.

LV remodeling. LV systolic and diastolic diameters were markedly increased and associated with a

Table 1. Effects of SelT administration on LV remodeling, hemodynamics and perfusion.

	Sham	I/R	I/R + SelT
LV remodeling			
LV syst. diam. (mm)	2.65 ± 0.15	6.32 ± 0.18†	4.80 ± 0.17‡
LV dias. diam. (mm)	5.92 ± 0.16	7.82 ± 0.13†	6.84 ± 0.08‡
LV weight (mg)	591 ± 3	743 ± 2†	658 ± 1‡
Infarct size (%)	–	11.6 ± 1.8	14.0 ± 2.3
LV hemodynamics			
Elastance (A.U.)	20.4 ± 1.4	13.3 ± 1.1†	17.9 ± 0.8‡
Compliance (A.U.)	0.91 ± 0.28	2.82 ± 0.43†	1.47 ± 0.21‡
Cardiac output (ml/min)	157 ± 7	116 ± 5†	145 ± 5‡
LV perfusion (ml/min/g)	8.1 ± 0.5	5.4 ± 0.1†	7.0 ± 0.2‡

† $p < 0.05$ Sham vs. I/R
‡ $p < 0.05$ I/R vs. I/R+SelT

reduced LV ejection fraction in I/R group as compared with sham-operated rats. The LV cavity remodeling was associated with an increase in LV weight. Administration of SelT resulted in significantly smaller LV systolic and diastolic diameters compared with saline-treated I/R animals.

LV hemodynamics. LV elastance was decreased and LV compliance was increased in I/R rats associated with a reduced cardiac output, illustrating LV systolic and diastolic dysfunctions. These changes were significantly ameliorated in the I/R+SelT group. Moreover, cardiac output was restored by SelT.

LV tissue perfusion. I/R resulted in a significant decrease of LV perfusion. SelT administration significantly ameliorated the perfusion of the 'viable' tissue of the LV.

4 CONCLUSIONS

Taken together, our experiments clearly show that SelT administration exerts beneficial effects after I/R. A pretreatment with SelT reduces I/R-related LV dilation, hypertrophy and impaired LV perfusion, resulting in an ameliorated LV function. However, additional experiments are warranted to (1) evaluate the mechanisms involved and (2) whether, when started after I/R, SelT-administration exerts similar beneficial effects. These data suggest that SelT might be a new therapeutic target for the treatment of cardiovascular diseases such as chronic heart failure.

REFERENCES

Kryukov, G. V., Castellano, S., Novoselov, S. V., Lobanov, A. V., Zehtab, O., Guigo, R., et al. 2003. Characterization of mammalian selenoproteomes. *Science,* 300: 1439–43.

Prevost, G., Arabo, A., Jian, L., Quelennec, E., Cartier, D., Hassan, S., et al. 2013. The PACAP-regulated gene selenoprotein T is abundantly expressed in mouse and human beta-cells and its targeted inactivation impairs glucose tolerance. *Endocrinology,* 154: 3796–806.

Roche, C., Besnier, M., Cassel, R., Harouki, N., Coquerel, D., Guerrot, D., et al. 2015. Soluble epoxide hydrolase inhibition improves coronary endothelial function and prevents the development of cardiac alterations in obese insulin-resistant mice. *Am J Physiol Heart Circ Physiol,* 308: H1020–1029.

Selenium status in humans: Measures, methods, and modifiers

Lutz Schomburg
Institute for Experimental Endocrinology, Charité – University Medicine Berlin, Berlin, Germany

1 SELENIUM INTAKE AND STATUS

Selenium (Se) is an essential trace element and it can become toxic at excessive concentrations. Cases of selenosis, both from veterinary and human medicine, are infrequently reported and highlight that there is an optimal range of Se supply. Daily Se intake differs widely among the world's populations, but symptoms of losing hair or nails are common (Lemire et al., 2012). In this regard, the body has developed tools and techniques for safeguarding its Se status.

Central to Se sufficiencies is the expression of selenoproteins. Two hierarchical principles exist in humans to reduce the risk of Se deficiency: (1) essential tissues are preferentially supplied and have developed pathways for limiting Se loss during deficiency, and (2) centrally house-keeping selenoproteins have priority over stress-related selenoproteins, with respect to translation. Hereby, the essential survival-supporting selenoproteins are always sufficiently expressed. With respect to organs, especially brain, bone, testes and other endocrine tissues reside high in the hierarchy of Se supply and retention, while liver, muscle or blood readily empty their Se reserves in times of poor supply (Schomburg & Schweizer, 2009).

2 HOW BEST TO DETERMINE SE STATUS IN HUMANS AND OTHER MAMMALS

The "Se status" is a term without clear definition. It may refer to (1) Se intake (assuming high intake will guarantee sufficiency), (2) Se excretion (assessment from 24 h urine samples), or (3) analyses by Se-dependent bio-markers. All these measures have advantages and limitations (Hoeflich et al., 2010; Hurst et al., 2010; Combs et al., 2011).

Several challenges are apparent: (1) It is virtually impossible to quantify Se intake in free-living subjects. A major problem lies in the varying Se content of identical-looking foodstuff. Most plants and animals accumulate Se according to its availability. (2) Determining Se status from excretion is cumbersome, displeasing and not reliable. Measurements under clinical conditions are not applicable to every-day life. (3) We need the analysis of specific seleno-compounds, metabolites or selenoproteins from bio-samples. In principle, every bio-sample may be analyzed. However, from an ethical and practical point of view, the choices are limited to hair, nails or body fluids (mainly blood, tear drops, and saliva). Hair and nails offer the inherent problem of being exposed to cosmetic manipulations (nail polish, shampoos, etc.), which can grossly affect quality and Se concentrations. Still, these matrices offer the theoretical advantage of providing insights into long-term Se availability instead of acute exposure. Similarly, measuring metabolites or seleno-compounds in body fluids might provide misleading data, as these short-lived biomarkers reflect acute exposure instead of average availability of Se.

From the considerations presented above, it is not surprising to find the most consistent data and interpretations from analyses of total Se or circulating selenoproteins from blood, plasma or serum. Yet, a consensus on the best biomarker has not been obtained. However, a most useful biomarker should be derived from tissues that are not preferentially supplied with Se but rather low in the hierarchy, e.g., the liver, and the selenoprotein should be Se-dependent and not preferentially supplied, as is the case for e.g. selenoproteins P, S, W or GPx1 or GPx3 (Fig. 1).

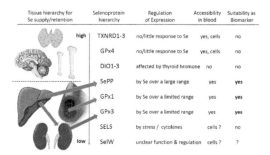

Figure 1. Overview on selenoproteins as potential biomarkers of Se status. Different tissues are characterized by a certain position within the hierarchy of Se supply and Se retention in times of deficiency with the thyroid, brain and bones being preferentially supplied. Similarly, selenoproteins are supplied with Se in a prioritized manner to preferentially express essential ones. Suitable biomarkers are thus rather originating from strictly Se-dependent organs like kidney or liver, with SePP displaying a larger range of responsiveness to Se supply than GPx1 or GPx3, qualifying it as probably the best and most suitable biomarker of Se status.

Selenium status determination from blood

The major fractions of Se in blood are Se-containing proteins. Small sized seleno-compounds such as Se-containing amino acids or their derivatives, selenite, selenate, Se-glutathione, the excretory seleno-sugars or other metabolites contribute only marginally to Se content of blood under normal conditions. Still, we hypothesize that these Se-containing molecules carry important health-relevant information under specific physiological conditions or diseases, but convincing data are presently lacking. We are thus left with total Se, or the circulating selenoproteins GPx1 (blood cells), GPx3 (mainly from kidney) and hepatic selenoprotein P (SePP) as possible choices.

Total Se is the most direct way for an assessment of Se status, and different techniques are available. However, there are good reasons to reconsider this choice and use selenoproteins, at least for the assessment of Se deficiency and sufficiency. Enzymatic tests are very sensitive to the preanalytical history of the sample, as the protein may be stable but its catalytic site often is sensitive. Hence, quantitative measures of protein abundance are preferred and offer high-throughput perspectives.

Selenoprotein P (SePP)

SePP is an actively secreted protein accounting for the majority of Se in human blood (Burk & Hill, 2015), and is derived mainly from liver, the organ controlling Se organification. Importantly, genetic inactivation of SePP leads to severe Se-dependent symptoms, including impaired growth and male infertility, neurological defects and endocrine derangements. The priority of supplying the brain with adequate Se is lost in the absence of SePP, highlighting its essential importance for tissue Se status. SePP thus represents a functional Se biomarker. The question remains whether measuring SePP is superior to total Se, or whether one is as good as the other.

Few studies in which both biomarkers have been determined indicate that SePP may be more relevant, e.g., Se and bone turnover (Hoeg et al., 2012), or colorectal cancer risk (Hughes et al., 2015). In both large studies, SePP correlated with higher significance to the clinical parameters than did total Se.

3 CONCLUSIONS

We find it important to monitor both the basal Se status and supplementation effects in clinical trials by robust biomarkers. At present, SePP appears as the best choice; especially in supplementation trials, Se determination alone is insufficient as total Se rather reflects the intake and not the effects of Se. Large scale clinical studies are needed to better define under what circumstances does total Se offer a sufficiently deep insight into Se status and when it will be mandatory to measure SePP as the functional biomarker of Se status and Se availability.

REFERENCES

Burk, R.F., & Hill, K.E., 2015. Regulation of selenium metabolism and transport. *Annual Review of Nutrition*, in press.

Combs, G.F., Jr., Watts, J.C., Jackson, M.I., Johnson, L.K., Zeng, H., Scheett, A.J. et al. 2011. Determinants of selenium status in healthy adults. *Nutrition Journal* 10:75.

Hoeflich, J., Hollenbach, B., Behrends, T., Hoeg, A., Stosnach, H., & Schomburg, L. 2010. The choice of biomarkers determines the selenium status in young German vegans and vegetarians. The *British Journal of Nutrition* 104: 1601–1604.

Hoeg, A., Gogakos, A., Murphy, E., Mueller, S., Kohrle, J., Reid, D.M., et al. 2012. Bone turnover and bone mineral density are independently related to selenium status in healthy euthyroid postmenopausal women. *The Journal of Clinical Endocrinology and Metabolism* 97: 4061–4070.

Hughes, D.J., Fedirko, V., Jenab, M., Schomburg, L., Meplan, C., Freisling, H., et al. 2015. Selenium status is associated with colorectal cancer risk in the European prospective investigation of cancer and nutrition cohort. *International Journal of Cancer.* 136: 1149–1161.

Hurst, R., Armah, C.N., Dainty, J.R., Hart, D.J., Teucher, B., Goldson, A.J., et al. 2010. Establishing optimal selenium status: results of a randomized, double-blind, placebo-controlled trial. The *American Journal of Clinical Nutrition* 91: 923–931.

Lemire, M., Philibert, A., Fillion, M., Passos, C.J., Guimaraes, J.R., Barbosa, F., Jr., et al. 2012. No evidence of selenosis from a selenium-rich diet in the Brazilian Amazon. *Environment International* 40: 128–136.

Schomburg, L., & Schweizer, U. 2009. Hierarchical regulation of selenoprotein expression and sex-specific effects of selenium. *Biochimica et Biophysica Acta* 1790: 1453–1462.

Selenoneine is the major Se compound in the blood of Inuit consuming of traditional marine foods in Nunavik, Northern Canada

M. Lemire*, A. Achouba, P.Y. Dumas, N. Ouellet & P. Ayotte
Axe santé des populations et pratiques optimales en santé, Centre de recherche du CHU de Québec, Université Laval, Québec, Canada;
Institut National de Santé Publique du Québec, Québec, Canada

M. Martinez
LCABIE, Université de Pau et des Pays de l'Ardour, Pau, France

L. Chan
Department of Biology, University of Ottawa, Ottawa, Canada

B. Laird
School of Public Health and Health Systems, University of Waterloo, Waterloo, Ontario, Canada

M. Kwan
Nunavik Research Center, Kuujjuaq, Québec, Canada

1 INTRODUCTION

Selenium (Se) is an essential element highly present in traditional marine foods consumed by Inuit (indigenous population inhabiting Artic regions of Greenland and North America), who exhibit one of the highest Se intake in the world (Lemire et al., 2015). In fish and marine mammal eating populations, there is increasing evidence suggesting that the high Se intake may play a role in offsetting some deleterious effects of methylmercury (MeHg) exposure (Fillion et al., 2013; Lemire et al., 2010, 2011; Valera & Dewailly, 2009; Ayotte et al., 2011). However, in other populations, elevated plasma Se concentrations have been recently associated to type 2 diabetes, hypercholesterolemia and/or hypertension (Rayman & Stranges, 2013). In addition to plasma Se levels, the most common biomarker of Se status, several other biomarkers (e.g. selenoproteins and small Se molecules such as selenoneine) have been identified (Xia et al., 2010; Yamashita et al., 2013; Klein et al., 2011) and these may help to better characterise Se status in the Inuit people.

2 METHODS

Archived plasma samples obtained from Inuit adults who participated to the Qanuippitaa health survey in 2004 (Rochette & Blanchet, 2004) were analysed for selenoproteins by affinity chromatography-inductively coupled plasma-mass spectrometry (ICP-MS) with quantification by post-column isotope dilution. Mercury (Hg) concentrations associated with selenoproteins were also quantified during the same analytical run.

We used our newly developed anion exchange liquid chromatography-ICP-MS method to determine the concentration of selenoneine in archived red blood cell samples from Qanuippitaa 2004 participants.

3 RESULTS

Concentrations (μg/L) of glutathione peroxidase, selenoprotein P and selenoalbumin were on average 25%, 52% and 23% of the total plasma Se concentration (n = 854), respectively. In addition, small concentrations of Hg co-eluted with each Se-containing protein. Plasma Se concentrations [median = 139 μg/L; interquartile range (IQR) = 23 μg/L] were markedly lower and less variable than corresponding blood Se levels (median = 261 μg/L, IQR = 166 μg/L]. We observed a non-linear relation between plasma and blood Se levels in Inuit, which is in contrast to the linear associations noted in other populations from Brazil and China displaying elevated Se status from plant-based food consumption (Lemire et al., 2012; Yang et al., 1983; Yang & Xia, 1995).

Our most recent speciation analyses confirm that selenoneine is the major Se compound found present in the blood cells of Inuit study participants with an elevated Se status.

4 CONCLUSIONS

Our results to date suggest that selenoneine accumulates in the blood cellular fraction of several Inuit adults relying on traditional marine foods. We are currently focusing our effort in studying the associations between selenoneine and selenoproteins status, as well as between biomarkers of Se status, MeHg exposure and cardiometabolic outcomes. These data will improve our capacity to assess the benefits and risks of Se intake and the traditional marine diet in this population.

REFERENCES

Ayotte, P., Carrier, A., Ouellet, N., Boiteau, V., Abdous, B., Sidi, E.A., et al. 2011. Relation between methylmercury exposure and plasma paraoxonase activity in inuit adults from Nunavik. Environ Health Perspect., 119(8): 1077–83.

Fillion, M., Lemire, M., Philibert, A., Frenette, B., Weiler, H.A., Deguire, J.R., et al. 2013. Toxic risks and nutritional benefits of traditional diet on near visual contrast sensitivity and color vision in the Brazilian Amazon. Neurotoxicology, 37: 173–81.

Klein, M., Ouerdane, L., Bueno, M., & Pannier, F. 2011. Identification in human urine and blood of a novel selenium metabolite, Se-methylselenoneine, a potential biomarker of metabolization in mammals of the naturally occurring selenoneine, by HPLC coupled to electrospray hybrid linear ion trap-orbital ion trap MS. Metallomics, 3(5): 513–20.

Lemire, M., Fillion, M., Frenette, B., Mayer, A., Philibert, A., Passos, C.J., et al. 2010. Selenium and mercury in the Brazilian Amazon: opposing influences on age-related cataracts. Environ Health Perspect., 118(11): 1584–9.

Lemire, M., Fillion, M., Frenette, B., Passos, C.J., Guimaraes, J.R., Barbosa, F., JR. et al. 2011. Selenium from dietary sources and motor functions in the Brazilian Amazon. Neurotoxicology, 32(6): 944–53.

Lemire, M., Kwan, M., Laouan-Sidi, A.E., Muckle, G., Pirkle, C., Ayotte, P., et al. 2015. Local country food sources of methylmercury, selenium and omega-3 fatty acids in Nunavik, Northern Quebec. Sci Total Environ., 509–510: 248–59.

Lemire, M., Philibert, A., Fillion, M., Passos, C.J., Guimaraes, J.R., Barbosa, F., JR. et al. 2012. No evidence of selenosis from a selenium-rich diet in the Brazilian Amazon. Environ Int., 40: 128–36.

Rayman, M.P. & Stranges, S. 2013. Epidemiology of selenium and type 2 diabetes: Can we make sense of it? Free Radic Biol Med., 65: 1557–64.

Rochette, L. & Blanchet, C. 2007. Methodological report. Inuit Health Survey 2004 Quanuippitaa? How are we? Quebec: Institut national de santé publique du Québec (INSPQ) & Nunavik Regional Board of Health and Social Services (NRBHSS). p. 338.

Valera, B., Dewailly, E., & Poirier, P. 2009. Environmental mercury exposure and blood pressure among Nunavik Inuit Adults. Hypertension, 54(5): 981–6.

Xia, Y.M., Hill, K.E., Li, P., Xu, J.Y., Zhou, D., Motley, A.K., et al. 2010. Optimization of selenoprotein P and other plasma selenium biomarkers for the assessment of the selenium nutritional requirement: A placebo-controlled, double-blind study of selenomethionine supplementation in selenium-deficient Chinese subjects. American Journal of Clinical Nutrition, 92(3): 525–31.

Yamashita, M., Yamashita, Y., Ando, T., Wakamiya, J., & Akiba, S. 2013. Identification and determination of selenoneine, 2-selenyl-N alpha, N alpha, N alpha-trimethyl-L-histidine, as the major organic selenium in blood cells in a fish-eating population on remote Japanese Islands. Biol Trace Elem Res., 156(1–3): 36–44.

Yang, G., Wang, S., Zhou, R., & Sun, S. 1983. Endemic selenium intoxication of humans in China. American Journal of Clinical Nutrition, 37(5): 872–81.

Yang, G.Q. & Xia, Y.M. 1995. Studies on human dietary requirements and safe range of dietary intakes of selenium in China and their application in the prevention of related endemic diseases. Biomed Environ Sci., 8(3): 187–201.

Selenium as a regulator of immune and inflammatory responses

Peter R. Hoffmann
University of Hawaii, John A. Burns School of Medicine, Honolulu, Hawaii, USA

1 INTRODUCTION

Experiments in animals together with clinical studies have supported an important role for sufficient dietary selenium (Se) in maintaining robust immunity against infections (Wang et al., 2009; Beck et al., 2003; Eze et al., 2013), effective responses to vaccines (Broome et al., 2004; Janbakhsh et al., 2013), and reducing pathology from chronic inflammatory disorders (Norton & Hoffmann, 2012; Speckmann & Steinbrenner, 2014; Nelson et al., 2011). However, the mechanisms by which Se intake regulates immunity are still not well understood, in part due to the wide variety of cellular processes affected by this micronutrient. The biological effects of dietary Se are exerted mainly through its incorporation into selenoproteins as the amino acid, selenocysteine (Sec). We have found that the endoplasmic reticulum (ER) transmembrane protein, selenoprotein K (Selk), is highly expressed in immune cells and its expression levels are sensitive to dietary Se in mice and Se levels in cell culture media (Verma et al., 2011). With this in mind, we undertook a series of experiments to more fully understand Selk's function in immune cells as a means to provide valuable insight into mechanisms by which Se influences immunity.

Some studies in cell lines have demonstrated a role for Selk and another ER transmembrane protein, selenoprotein S (Sels), in ER associated degradation of misfolded proteins and regulating ER stress (Shchedrina et al., 2011; Du et al., 2010). Investigations into the in vivo role of Selk have been facilitated by our novel Selk-/- mice. Immune cells from Selk-/- mice did not show indications of ER stress in the resting or activated state (Verma et al., 2011), perhaps due to overlapping roles of Selk and Sels related to ER stress. We instead found that receptor mediated Ca2+ flux was impaired in Selk-/- T cells, neutrophils, and macrophages. This impaired Ca2+ flux led to defects in a variety of immune cell functions including migration, proliferation, and oxidative burst that were associated with higher susceptibility to West Nile virus infection due to impaired viral clearance (Verma et al., 2011). Our recent work has uncovered a key molecular mechanism by which Selk regulates calcium flux and other cellular processes through its role in the palmitoylation of certain proteins. Palmitoylation is an important post-translational modification that involves the addition of a 16-carbon fatty acid moiety onto cysteine residues on target proteins that stablize these proteins, facilitates association with membranes, or regulates subcellular localization.

2 MATERIALS AND METHODS

Established techniques with acyl-biotin exchange experiments were used to visualize protein palmitoylation following previously published studies (40). This approach involves immunoprecipitation of the IP3R and exchange of palmitoyl groups with biotin, and Western blotting performed with detection of the biotin using streptavidinylated fluoresent reagents. The bands are then visualized using a Li-Cor Odyssey fluorescence scanner. IP3 concentration was measured in mouse bone marrow derived macrophages (BMDMs) and T cells using a HitHunter IP3 FP Assay (DiscoveRx). A key technique employed for these studies was the measurement of Ca2+ Flux in immune cells after stimulation through various immune receptors. Ca2+ flux was assayed for BMDMs, primary T cells, and Jurkat T cells via time-lapse video microscopy or via flow cytometry A detailed description of all experimental procedures can be found in (Fredericks et al., 2014).

3 RESULTS AND DISCUSSION

We have found that several proteins crucial for immune cell activation and function require Selk for palmitoylation, including the scavenger receptor CD36 and the calcium channel protein IP3R (Meiler et al., 2013; Fredericks et al., 2014). For palmitoylation to occur, Selk must complex with the palmitoyl acyl transferase (DHHC6) at the endoplasmic reticulum membrane and together this complex completes the transfer of palmitic acid to the cysteine residues on target proteins. These two proteins bind each other through SH3/SH3 binding domain interactions on the cytosolic side of the endoplasmic reticulum membrane. DHHC6 serves as the enzyme that catalyzes palmitoylation, and we believe that Selk functions as a coenzyme. Selk deficiency induced through low Se intake or genetic mechanisms leads to low levels of palmitoylation of target proteins, which in turn leads to low expression

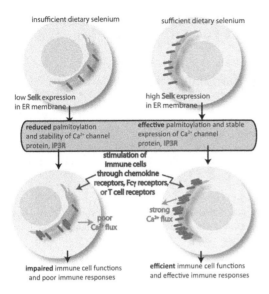

Figure 1. Our data have uncovered a role for Selk in receptor mediated Ca^{2+} flux in immune cells. Our new studies will definitively determine the role of Selk in palmitoylation/stabilization of the ER Ca^{2+} channel protein, IP3R, and other proteins in regulating immune cell function.

of the IP3R calcium channel (Fig. 1). Overall, we propose that Selk functions as a coenzyme for ZDHHC6 dependent palmitoylation and this represents a key mechanism by which Selk, and perhaps dietary Se levels, impact immune cell function. Moreover, this molecular role for Selk has effects on inflammation and immunity in vivo.

4 CONCLUSIONS

Since Se is observed to be in an apparent protective role, it is tempting to suggest that Se in the diet might play a palliative role against mercury toxicity. However, given that there are several observations of synergistic toxicological relationships between species of mercury and Se (*e.g.* Cuvin-Aralar et al., 1991), in that the sum of the two is much more toxic than the individual toxicity, caution is appropriate. The interaction of mercury and Se will depend in a complex manner on the chemical forms of both elements, with more work needed to better understand their relationships.

REFERENCES

Beck, M.A., O.A. Levander, & J. Handy. 2003. Selenium deficiency and viral infection. *J Nutr* 133 (5 Suppl 1):1463S–7S.

Broome, C.S., F. McArdle, J.A. Kyle, et al. 2004. An increase in selenium intake improves immune function and poliovirus handling in adults with marginal selenium status. *Am J Clin Nutr* 80 (1):154–62.

Du, S., J. Zhou, Y. Jia, & K. Huang. 2010. SelK is a novel ER stress-regulated protein and protects HepG2 cells from ER stress agent-induced apoptosis. *Arch Biochem Biophys* 502 (2):137–43.

Eze, J.I., M.C. Okeke, A.A. Ngene, J.N. Omeje, & F.O. Abonyi. 2013. Effects of dietary selenium supplementation on parasitemia, anemia and serum proteins of Trypanosoma brucei brucei infected rats. *Exp Parasitol* 135 (2):331–6.

Fredericks, G.J., F.W. Hoffmann, A.H. Rose, et al. 2014. Stable expression and function of the inositol 1,4,5-triphosphate receptor requires palmitoylation by a DHHC6/selenoprotein K complex. *Proc Natl Acad Sci U S A* 111 (46):16478–83.

Janbakhsh, A., F. Mansouri, S. Vaziri, et al. 2013. Effect of selenium on immune response against hepatitis B vaccine with accelerated method in insulin-dependent diabetes mellitus patients. *Caspian J Intern Med* 4 (1):603–6.

Meiler, S., Y. Baumer, Z. Huang, et al. 2013. Selenoprotein K is required for palmitoylation of CD36 in macrophages: implications for foam cell formation and atherogenesis. *J Leukoc Biol* 93 (5):771–80.

Nelson, S.M., X. Lei, and K.S. Prabhu. 2011. Selenium levels affect the IL-4-induced expression of alternative activation markers in murine macrophages. *J Nutr* 141 (9):1754–61.

Norton, R. L., and P. R. Hoffmann. 2012. Selenium and asthma. *Mol Aspects Med* 33 (1):98–106.

Shchedrina, V.A., R.A. Everley, Y. Zhang, S.P. Gygi, D.L. Hatfield, & V.N. Gladyshev. 2011. Selenoprotein K binds multiprotein complexes and is involved in the regulation of endoplasmic reticulum homeostasis. *J Biol Chem* 286 (50):42937–48.

Speckmann, B., & H. Steinbrenner. 2014. Selenium and selenoproteins in inflammatory bowel diseases and experimental colitis. *Inflamm Bowel Dis* 20 (6):1110–9.

Verma, S., F.W. Hoffmann, M. Kumar, et al. 2011. Selenoprotein K knockout mice exhibit deficient calcium flux in immune cells and impaired immune responses. *J Immunol* 186 (4):2127–37.

Wang, C., H. Wang, J. Luo, et al. 2009. Selenium deficiency impairs host innate immune response and induces susceptibility to Listeria monocytogenes infection. *BMC Immunol* 10:55.

Effects of selenium exposure on neuronal differentiation of embryonic and induced pluripotent stem cells

Miao Li & Zhifang Qiu*
Barshop Institute for Longevity and Aging Studies, University of Texas Health Science Center at San Antonio, San Antonio, Texas, USA

1 INTRODUCTION

Selenium (Se) is an essential trace element for human and animals. Through selenoproteins, this mineral participates in various biological processes such as antioxidant defence, thyroid hormone production, and immune responses (Zwolak & Zaporowska, 2012). Some reports indicate that a human organism deficient in Se may be prone to certain diseases. Adverse health effects following Se overexposure, although very rare, have been found in animals and people (Zwolak & Zaporowska, 2012).

In the last two decades, cell culture techniques for both mammalian embryonic stem cells and adult stem cells have been developed and improved, and are now widely available (Mori & Hara, 2013). These stem cells are either pluri- or multi-potent, which makes them favorable for use in vitro developmental toxicity assays. Recent studies have reported several applications for embryonic and adult stem cells in cytotoxicity and developmental toxicity testing (Mori & Hara, 2013). The applications have the potential to provide alternative assessment techniques for evaluating pharmacological agents, as well as environmental and chemical toxins, and possibly reveal novel toxic and developmental effects that are difficult to investigate in humans because of ethical considerations. With pluripotent stem cells, we are able to investigate in vitro the potential effects of pollutants on human health. The objective of this study was to evaluate toxicity effects of Se on embryonic and induced pluripotent stem cells-derived neuron.

2 MATERIALS AND METHODS

Three clonal lines of non-human primate marmoset iPS cells (B8, 88, 15) were grown in E8 medium supplemented with 10% fetal bovine serum. At the beginning of the differentiation protocols, cells were removed from the dish with Accutase. Cells were then transferred into differentiation medium as described below, or were incubated with sodium selenite prior to the differentiation treatment.

Following detachment of the cells with Accutase, cells were plated in differentiation medium containing various concentrations (0.05–1.0%) of sodium selenite. Differentiation medium comprised DMEM/F12 with 20% KSR, 0.32 μM dorsomorphin hydrochloride, 0.32 μM SB431542, 20 ng/ml FGF2, 25 μg/ml insulin, 1 nM retinoic acid (all-trans retinoic acid, Sigma), and 10 μM ROCK inhibitor Y-27632. Cells were plated in this medium on plates coated with Matrigel.

Following 24 hours of incubation in differentiation medium/Se, cells were removed from the dish with Accutase and resuspended in differentiation medium without Se. Cells were then permitted to form aggregates using 384-well hanging-drop plates. Each well of the hanging drop plate received 3000 cells in 30 μL differentiation medium. Plates were placed in a humidified incubator at 37.5°C for 72 h. Following this incubation period, the resultant aggregates were collected from the plates. They were then dissociated into single cells by incubation in 1 mL 0.25% trypsin/EDTA for 10 min at room temperature. Cells were transferred into DMEM/F12 medium containing 20% KSR to stop the digestion and mechanically dissociated. 100,000 cells were plated per 35 mm Matrigel-coated dish. Cells were plated in DMEM/F12 containing 20% KSR, 2% B27 supplement, 15 μg/mL transferrin (human, Sigma), 10 μM retinoic acid, 1 μM SB431542, 1 μM dorsomorphin, 3.2 μM SAG (Sonic hedgehog agonist; EMD Chemicals, Billerica, MA) and 10 μM Y-27632. Cells were maintained in the same medium for 72 h with medium changes at 24 h intervals. Following this period, cells were harvested for preparation of RNA.

Total RNA was isolated from cells using RNA Bee according to the manufacturer's instructions. A total of 2 μg of RNA was reverse-transcribed by using superscript II. Quantitative PCR was conducted using SYBR green detection and an ABI 7900HT system. Levels of mRNAs are reported as cycles versus β-actin, using marmoset gene-specific primers.

Statistical analyses were performed using Prism software. Responses to differentiation and Se pretreatment were analyzed by one-way ANOVA followed by Dunnett's test of multiple comparisons with a control.

3 RESULTS AND DISCUSSION

Marmoset ESCs and iPSCs derived neural cells with expressing of MAP2 and β-tubulin III markers by

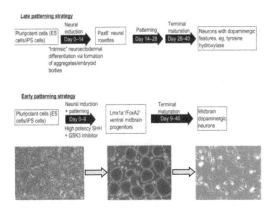

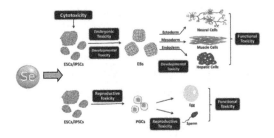

Figure 2. Stem cell-based toxicity of Se assays.

Figure 1. Differentiation of pluripotent cells to neural cells and then to dopaminergic neurons.

immunocytochemistry were treated for different concentrations of Se. ESCs and iPSCs were treated with Se from the time of EB formation (Day 0) to the end of post-plating (Day 6), or only during the 6 days of post-plating. The declining in outgrowth was observed in neural morphology in a dose dependent manner. The number of neurons emerging out of EBs gradually decreased with increasing concentrations of Se, the EBs did not plate and had a dark appearance. Inhibition of neuronal differentiation was determined to be the 45% reduction of EBs with neural extensions as compared to the control. Moreover, results of q-PCR at six days post-plating showed that in spite of there being no significant change in Nestin expression, MAP2 was significantly reduced, respectively (Fig. 1).

In the experiment, marmoset ESCs and iPSCs-derived neuron were used as an in vitro model for evaluation of Se-induced neurotoxicity. According to the present study, the results suggest that the early phase of neural development, which coincides with neural tube formation, is more sensitive than the later phase when neural precursor cells differentiate into mature neurons. Also, it appears that the toxicity increment in vitro had direct correlation with time in mESCs and iPSCs derived neural cells. Regarding ESCs and iPSCs abilities such as tissue-specific properties, especially on the formation of neuroectoderm and mesoderm, they could act as a system to evaluate inhibiting or inducing effects on the differentiation processes of early embryonic stages. Importantly, they provide a unique and widely applicable system for studies that are directly relevant to human health without the use of animal models.

In addition, stem cell toxicology of Se allows for the concurrent assessment of many forms of toxicity including acute, embryonic, developmental, organ, reproductive, and functional (Fig. 2). Importantly, iPSC technology will enable us to assess toxicity without the ethical issues associated with the derivation and use of iPSCs, but with the added benefit of the potential for personalized toxicology (Faiola et al., 2015).

4 CONCLUSIONS

This study suggests that stem cells are a suitable model for the assessment of Se toxicity, especially when assessment of Se in vivo is difficult and a large number of animals are required to be sacrificed to achieve a similar conclusion.

REFERENCES

Zwolak, I. & Zaporowska, H. 2012. Selenium interactions and toxicity: a review. Selenium interactions and toxicity. *Cell Biology and Toxicology*. 28(1): 31–46.

Mori, H. & Hara, M. 2013. Cultured stem cells as tools for toxicological assays. *Journal of Bioscience and Bioengineering*. 116(6): 647–652.

Faiola, F., Yin, N.Y., Yao, X.L. & Jiang, G.B. 2015. The Rise of stem cell toxicology. Environment Science & Technology. 49 (10): 5847–5848.

Functional deletion of brain selenoenzymes by methylmercury

N.V.C. Ralston* & L.J. Raymond
University of North Dakota, Grand Forks, USA

1 INTRODUCTION

Selenium (Se) is a nutritionally essential trace element that is required by enzymes that prevent and reverse oxidative damage, particularly in the brain (Ralston & Raymond, 2010). Because CH_3Hg irreversibly inhibits selenoenzyme activities (Prohaska et al., 1977; Sepannen et al., 2004; Carvalho et al., 2008) the effects of CH_3Hg exposures cannot be assessed without concomitant assessments of Se. The Selenium (Se) Health Benefit Value (HBV_{Se}) is a seafood safety index that considers both CH_3Hg and Se concentrations in fish. This criteria predicts child neurological health outcomes of epidemiological (Ralston, 2008) and laboratory animal studies with more reliable accuracy than predictions based on CH_3Hg alone (Ralston et al., 2008). It is calculated: $HBV_{Se} = [(Se - Hg)/Se] \times (Se + Hg)$. During high CH_3Hg exposures, increased dietary Se is expected to improve brain Se availability and offset losses in brain glutathione peroxidase (GPx) and thioredoxin reductase (TRx), thus diminishing oxidative damage as indicated by the formation and/or accumulation of F2-isoprostanes (F2-IsoP). This study compares risk predictions based on CH_3Hg exposures alone vs. dietary HBV_{Se}.

2 FEEDING STUDY

Weanling male Long Evans rats were fed 0.2 μmol Se/kg diets for 5 weeks to deplete their Se body stores, making them highly dependent on dietary Se to support growth and brain selenoenzyme synthesis. Following the depletion period, rats were weighed and pseudo-randomly assigned to diets that contained either 0.2, 1.0, or 10 μmol Se/kg prepared with the addition of 0 or 40 μmol CH_3Hg/kg (5 rats per diet treatment group) fed ab libitum for 5 weeks. The calculated HBV_{Se} of these diets were: 0.2, 1.0, 10, -152, $-1,599$, and $-7,999$.

3 RESULTS

Brain Se and CH_3Hg levels are reported in μmol/kg concentrations. The results of enzyme assays and body weights were normalized in relation to the group mean observed for rats fed diets containing 1.0 μmol Se/kg with no added CH_3Hg.

Neurotoxicity was only noted in rats fed diets containing 0.2 μmol Se with 40 μmol CH_3Hg. Results were compared with each other and relative to diet Se, CH_3Hg, HBV_{Se}, as well as Brain Se, CH_3Hg, and Se availability. Selenium availability relates the Se and Hg contents observed in individual rats and was calculated using the same equation as HBV_{Se}.

Statistical comparisons were performed prior to normalization. The p-values in Table 1 indicate the results of one-sided t-test comparisons of CH_3Hg effects. Values with different superscripts in the table differ significantly ($p < 0.05$) from one another.

3.1 Regression analysis

Brain Se was significantly related to dietary Se ($F = 42.0$, $p < 0.0001$) and HBV_{Se} ($F = 10.9$, $p < 0.01$), but not to CH_3Hg in the diet. *Brain Hg* was strongly related to diet Hg ($F = 248.9$, $p < 0.0001$) compared to HBV_{Se} ($F = 7.3$, $p = 0.01$), but not dietary Se. *Brain Se availability* was strongly related to dietary CH_3Hg ($F = 111.6$, $p < 0.0001$) and also to dietary HBV_{Se} ($F = 70.5$, $p < 0.0001$), but not to diet Se. *Brain GPx* activity was significantly associated with diet HBV_{Se} ($F = 27.2$, $p < 0.001$) but not with dietary Se or CH_3Hg. Brain GPx activity was significantly associated with brain Se availability ($F = 13.8$, $p < 0.001$) and brain Se contents ($F = 7.4$, $p = 0.01$), but not to brain Hg. *Brain TRx* activity was significantly associated with diet HBV_{Se} ($F = 17.0$, $p < 0.01$) and CH_3Hg ($F = 26.5$, $p < 0.001$) but not to Se concentrations. Brain TRx activity was significantly associated with brain Se availability ($F = 48.2$, $p < 0.00001$) and brain Hg contents ($F = 25.8$; $p < 0.0001$), but not to brain Se levels. *Brain F2-IsoP* was significantly associated with diet HBV_{Se} ($F = 44.2$, $p < 0.0001$) and CH_3Hg ($F = 9.8$, $p < 0.01$) but not to Se. Isoprostane concentrations were significantly associated with brain CH_3Hg ($F = 7.0$, $p < 0.001$) and brain Se availability ($F = 39.4$; $p < 0.0001$), but not with brain Se. *Body weights* were associated with diet HBV_{Se} ($F = 61.5$, $p < 0.0001$) and CH_3Hg ($F = 4.92$, $p = 0.03$) but not dietary Se. Weights were significantly associated with brain Se availability ($F = 26.3$, $p < 0.0001$) and brain Se contents ($F = 7.8$, $p < 0.01$), but not with brain Hg.

Table 1. Elemental and enzyme concentrations and statistical analysis.

Dietary Se (μmol/kg)	MeHg Treatment in Diet (μmol/kg)		P value
	0	40	
	Brain Se (μmol/kg)		
0.2	1.13 ± 0.22^a	0.78 ± 0.02^c	< 0.01
1.0	1.39 ± 0.03^b	1.19 ± 0.14^b	ns
10.0	1.60 ± 0.09^b	2.94 ± 0.35^b	ns
	Brain CH$_3$Hg (μmol/kg)		
0.2	0.00 ± 0.01^a	15.88 ± 1.28^c	< 0.01
1.0	0.03 ± 0.04^b	14.35 ± 1.59^b	ns
10.0	0.02 ± 0.02^b	23.77 ± 2.05^b	ns
	Brain Se Availability (calculated)		
0.2	1.13 ± 0.20^a	-325.2 ± 50.4^d	< 0.001
1.0	1.39 ± 0.03^b	-172.3 ± 25.2^e	< 0.001
10.0	1.60 ± 0.09^c	-190.1 ± 20.8^e	< 0.001
	Brain GPx (normalized)		
0.2	0.68 ± 0.21a	0.26 ± 0.09c	< 0.01
1.0	1.00 ± 0.23^b	0.90 ± 0.14^b	ns
10.0	0.94 ± 0.11^b	0.88 ± 0.09^b	ns
	Brain TRx (normalized)		
0.2	0.74 ± 0.08^a	0.41 ± 0.12^c	< 0.01
1.0	1.00 ± 0.20^b	0.68 ± 0.14^a	(0.09)
10.0	0.97 ± 0.13^b	0.56 ± 0.15^a	< 0.01
	Brain F2-IsoP (normalized)		
0.2	1.01 ± 0.12^a	4.15 ± 1.96^b	< 0.01
1.0	1.00 ± 0.13^a	1.53 ± 0.17^c	< 0.01
10.0	0.97 ± 0.12^a	1.55 ± 0.30^c	< 0.01
	Body Weight (normalized)		
0.2	0.94 ± 0.03^a	0.73 ± 0.03^b	< 0.001
1.0	1.00 ± 0.05^a	0.97 ± 0.03^a	ns
10.0	0.97 ± 0.05^a	0.97 ± 0.04^a	ns

4 CONCLUSIONS

Dietary Se influences CH$_3$Hg retention in brain tissues, but more importantly, high CH$_3$Hg exposures can diminish brain Se availability and thus impair brain selenoenzyme activities. When dietary Se intakes were low, brain GPx activities diminished, but the loss of GPx activity was severe if the rats were exposed to toxic amounts of CH$_3$Hg in their diets. In contrast, when diets provided sufficient amounts of Se, brain GPx activities were protected against the effects of high CH$_3$Hg exposures. Similarly, brain TRx also diminished when low Se diets were fed, but the loss of activity became severe when challenged by toxic CH$_3$Hg exposures. Normal or rich amounts of dietary Se partially restored brain TRx in brains of rats challenged with high CH$_3$Hg diets, but not to the same degree as for GPx. Brain F2-Isoprostane contents were inversely related to brain TRx ($F = 23.1, p < 0.00001$) and GPx ($F = 18.45, p < 0.001$) activities, indicating loss of GPx and TRx contributes to the increase in oxidative damage in CH$_3$Hg exposed rat brains.

In this study, dietary HBV$_{Se}$ was highly consistent in its associations with significant diminishments in brain GPx and TRx activities, increases in oxidative damage as measured by F2-Isoprostane, and growth inhibition due to toxic CH$_3$Hg exposures. Brain Se availability was a better index of selenonenzyme activities and oxidative damage that occurs in their absence. This study indicates that the HBV$_{Se}$ provides more reliably accurate risk assessments than seafood safety criteria based on CH$_3$Hg alone.

Fish and background diets consumed by human populations can vary widely in Se content, and thus influence health outcomes observed in association with CH$_3$Hg exposures. Epidemiological studies are likely to obtain more consistently reliable results if they consider dietary Se intakes as well as CH$_3$Hg exposures in relation to health outcomes.

REFERENCES

Anestål, K. & Arnér, E.S. 2003. Rapid induction of cell death by selenium-compromised thioredoxin reductase 1 but not by the fully active enzyme containing selenocysteine. *Journal of Biological Chemistry.* 278: 15966–72.

Behne, D., Pfeifer, H., Rothlein, D. & Kyriakopoulos, A. 2000. Cellular and subcellular distribution of selenium and selenium–containing proteins in the rat. In: *Trace Elements in Man and Animals 10.* Kluwer Academic/Plenum Publishers, NY. pp. 29–34.

Grandjean, P., Weihe, P., White, R. F., Debes, F., Araki, S., Yokoyama, K., et. al. 1997. Cognitive deficit in 7-year-old children with prenatal exposure to methylmercury. *Neurotoxicology and Teratology.* 19: 417–28.

Ralston, N.V.C. 2008. Selenium health benefit values as seafood safety criteria. *EcoHealth.* 5: 442–55.

Ralston, N.V.C., Ralston, C.R., Blackwell, J.L., & Raymond, L.J. 2008. Dietary and tissue selenium in relation to methylmercury toxicity. *Neurotoxicology.* 29: 802–11.

Ralston, N.V.C. & Raymond, L.J. 2010. Dietary selenium's protective effects against methylmercury toxicity. *Toxicology.* 278: 112–123.

Seale, L., Ralston, N.V.C. & Berry, M.J. 2011. The Role of Selenium in Mitigating Mercury Toxicity. In; *Biochemistry Research Trends.* Nova Science Publishers Inc., New York.

The "SOS" mechanisms of methylmercury toxicity

N.V.C. Ralston* & L.J. Raymond
University of North Dakota, Grand Forks, ND, USA

1 INTRODUCTION

Selenium (Se) incorporated as selenocysteine (Sec), the 21st proteinogenic amino acid, is the defining characteristic of the ~25 genetically unique selenoproteins expressed in humans. Several catalytically active members of this family (selenoenzymes) have pivotal roles in maintaining intracellular reducing conditions, particularly in brain tissues. Certain selenoenzymes prevent damage from reactive oxygen species (ROS) and free radicals, while others reverse oxidative damage (Ralston & Raymond, 2010).

Oxygen consumption in the brain is ~10 fold higher than in other tissues, making it vulnerable to oxidative damage. Furthermore, the distal compartments of dendrites and axons are remote from the soma of the neuron, making it difficult for the cell to repair damage to their cellular components. Because selenoenzyme-dependent prevention and reversal of oxidative damage is so vital in these tissues, homeostatic mechanisms have evolved to ensure their expression and activities are without interruption (Behn et al., 2000). The only insult known to impair their activities is high mercury (Hg) exposures, particularly during fetal development (Watanabe et al., 1999).

2 MECHANISMS OF MERCURY TOXICITY

Methylmercury (CH_3Hg) toxicity is characterized by a long latency of onset of symptoms following toxic exposure, neuroendocrine tissue specificity of pathological effects, oxidative damage in the affected tissues, accentuated fetal vulnerability, and effects accompanying supplemental dietary Se.

The involvement of Se-physiology in mechanisms of CH_3Hg toxicity were not initially recognized because selenoenzyme metabolism and its unique roles in brain and endocrine tissues were generally unknown. However, in recent years, the Se-dependent aspects of their physiology are becoming better understood. As a result, the characteristic features of CH_3Hg toxicity have become much easier to understand. High CH_3Hg exposures occur through a sequence of biochemical reactions referred to here as "SOS" mechanisms. These disruptions have consequences that increase in severity as tissue CH_3Hg concentrations approach, and especially as they exceed equimolar stoichiometry with tissue Se.

2.1 Synergies of sequestration (SOS-1)

Enzyme binding sites capture substrates and orient them for proper introduction into their active sites, thus accelerating the rates of catalyzed reactions. Selenoenzymes such as glutathione (GSH) and thioredoxin ($T[SH]_2$) are designed to bring the thiol moieties of sulfhydryls (-SH) of the thiomolecules they act upon into close proximity with the Se atom of the Sec in the enzyme's active site. Since CH_3Hg readily binds to thiols, but steadily exchanges from one to another, its association as CH_3Hg-SG or $T[S-S-Hg-CH_3]$ is in equilibrium with its binding to other cellular thiols. This mechanism expedites CH_3Hg introduction into selenoenzyme active sites.

2.2 Silencing of selenoenzymes (SOS-2)

The vulnerability of selenoenzymes to CH_3Hg was proposed by Ganther et al., (1972), demonstrated by Prohaska et al., (1977), characterized by Seppanen et al. (2004), and defined by Carvalho et al., (2008). Furthermore, Ralston and Raymond have shown that supplemental Se prevents interruption of selenoenzyme activities in brains of lab animals (Ralston et al., 2008, Ralston and Raymond 2010). These studies are supported by the findings of Watanabe et al., (1999) and Stringari et al., (2008) that indicate selenoenzyme activities of the fetal brain are far more vulnerable to high maternal CH_3Hg exposures than maternal brains, but are protected by increased maternal Se in the diet. The CH_3Hg that becomes bound to thiols of the substrates for these enzymes act as suicide inhibitors that transfer CH_3Hg directly to the Sec of the enzyme's active site, thus forming the CH_3Hg-Sec inhibitor-enzyme complex. Thus, by biochemical definition, CH_3Hg is a highly selective irreversible inhibitor of selenoenzymes.

2.3 Sequestration of selenium (SOS-3)

Mercury affinities for Se molecules are ~10^6 times higher than those of analogous sulfur molecules (Dyrssen & Wedborg, 1991). Neurological consequences of high CH_3Hg exposures in humans are consistently associated with accumulation of mercury selenide (HgSe) in brain tissues (Korbas et al., 2010) and appears to arise as the breakdown product of CH_3Hg-Sec in lysosomes. Since laboratory studies indicate that HgSe can only be decomposed by aqua regia, or through heating to temperatures greater than 300°C, it is biologically refractory. Various animal and human studies (e.g. Korbas et al., 2010) demonstrate that as tissue Hg levels approach a 1:1 stoichiometry with Se, the Se contents of brain and endocrine tissues increase to maintain free Se to support selenoenzyme activities.

2.4 Suicide of selenium-deprived cells (SOS-4)

Selenium-deprived cells that cannot synthesize the Sec that is needed for thioredoxin reductase will instead produce truncated molecules that are known as GRIM-12 – a potent apoptosis (cell suicide) initiator (Anestål & Arnér, 2003) This observation implies that sequestration of cellular Se by CH_3Hg will not only deprive cells of the selenoenzymes needed to counteract oxidative damage, but can also induce apoptosis. This mechanism remains untested.

2.5 Sustained oblivion of Sec synthesis (SOS-5)

Selenophosphate synthetase, the enzyme that makes the selenophosphate that is critical for Sec production, is itself a selenoenzyme. In the absence of its activities, selenophosphate can no longer be produced, thus abolishing Sec synthesis, and preventing creation of selenophosphate synthetase itself. Therefore, cells which lose their reserves of bioavailable Se due to CH_3Hg exposures high enough to abolish selenophosphate synthetase, are unlikely to recover. There is growing evidence for this mechanism. For example, Stringari et al. (2008) recently found that high CH_3Hg exposures during fetal growth exerted a sustained effect on brain selenoenzyme activities that showed no signs of recovery, thus providing evidence in support of this hypothesized mechanism.

3 CONCLUSIONS

The effects of "SOS" mechanisms on Se physiology and biochemistry coincide with features of CH_3Hg toxicity which were previously difficult to explain. These consequences appear to arise as intracellular Se is bound to CH_3Hg, thus inhibiting selenoenzyme activities and preventing recycling of bound Se.

REFERENCES

Anestål, K. & Arnér, E.S. 2003. Rapid induction of cell death by selenium-compromised thioredoxin reductase 1 but not by the fully active enzyme containing selenocysteine. *Journal of Biological Chemistry*. 278: 15966–72.

Carvalho, C.M.L., Chew, E.–H., Hashemy, S.I., Lu J., & Holmgren, A. 2008. Inhibition of the human thioredoxin system: A molecular mechanism of mercury toxicity. *Journal of Biological Chemistry*. 283: 11913–11923.

Dyrssen, D. & Wedborg, M. 1991. The Sulfur-mercury(II) system in natural waters. *Water Air Soil Pollut.*, 56: 507–519.

Ganther, H.E., Goudie, C., Sunde, M.L., Kopecky, M.J., Wagner, P., et al. 1972. Selenium: Relation to decreased toxicity of methylmercury added to diets containing tuna, *Science*. 175: 1122.

Korbas, M., O'Donoghue, J.L., Watson, G.E., Pickering, I.J., Singh, S.P., Myers, G.J., Clarkson, T.W. & George, G.N. 2010. The chemical nature of mercury in human brain following poisoning or environmental exposure. *ACS Chemical Neuroscience* 1(12): 810–818.

Prohaska, J.R., & Ganther, H.E. 1977. Interactions between selenium and methylmercury in rat brain. *Chemico-Biological Interactions*. 16: 155–67.

Ralston, N.V.C. 2008. Selenium health benefit values as seafood safety criteria. *EcoHealth*. 5: 442–55.

Ralston, N.V.C., Ralston, C.R., Blackwell, J.L., & Raymond, L.J. 2008. Dietary and tissue selenium in relation to methylmercury toxicity. *Neurotoxicology*. 29: 802–11.

Ralston, N.V.C. & Raymond, L.J. 2010. Dietary selenium's protective effects against methylmercury toxicity. *Toxicology*. 278: 112–123.

Ralston, N.V.C. & Raymond, L.J. 2010. Selenium Health Benefit Values as Seafood Safety Criteria. Final Project Report to NOAA 2010: 1–15.

Seppanen, K.P., Soininen, J.T., Salonen, S., Lotjonen & Laatikainen. R. 2004. Does mercury promote lipid peroxidation? An in vitro study concerning mercury, copper, and iron in peroxidation of low-density lipoprotein. *Biological Trace Element Research*.101: 117–32.

Stringari, J., Nunes, A.K.C., Franco, J.L., Bohrer, D., Garcia, S.C., et al. 2008. Prenatal methylmercury exposure hampers glutathione antioxidant system ontogenesis and causes long-lasting oxidative stress in the mouse brain. *Toxicology and Applied Pharmacology*. 227: 147–154.

Watanabe, C.K., Yin, Y., Kasanuma, Y. & Satoh, H. 1999. In utero exposure to methylmercury and selenium deficiency converge on the neurobehavioral outcome in mice. *Neurotoxicology and Teratology*. 21: 83–88.

Effects of selenium on animal, human and plant health

Role of the selenium in articular cartilage metabolism, growth, and maturation

Caroline Bissardon* & Laurent Charlet
ISTerre (Institut des Sciences de la Terre) – Université Joseph Fourier, Grenoble, France

Sylvain Bohic
Inserm GIN U836, & Nanoimaging ESRF Beamline ID16, Grenoble, France

Ilyas Khan
Regenerative Medicine Group, Swansea University Medical School, Swansea University, Wales, UK

1 INTRODUCTION

Selenium (Se) is an essential trace element for mammals and a key player in the cellular metabolism. Nevertheless, it is toxic at a concentration slightly higher than required for normal physiological function. This "double-edged sword" element accumulates in organism in minute quantities, which limits our ability to fully understand the Se-distribution and function within living organisms. However, several studies attest that Se plays an important role in the tissue development such as for articular cartilage (AC), especially via selenoproteins directly involved in cartilage growth and homeostasis (Yan et al., 2013). AC is formed by chondrocytes that produce an extracellular matrix (ECM) composed of diverse proteoglycans that are highly hydrated and entrapped in a collagen fiber meshwork. AC exhibits significant viscoelastic behaviors and biomechanical properties in order to distribute forces across the joint during activities of daily motion (Mow et al., 2008). Studies have shown that Se is indirectly involved in normal cartilage growth, as exemplified by Kashin-Beck disease (KBD) – a little-known endemic ailment that has crippled and stunted the growth of hundreds of thousands of children and adolescents in China's heartland (Stone, 2009). In this area, soils and thus cereals, the main source of Se in human diet, are Se depleted. In KBD patients, as well as in the tens of millions of people suffering osteoarthritis and rheumatoid arthritis, replacement of matrix components is inadequate; the ECM collapses, increasing biomechanical stresses on chondrocytes. As KBD progresses, catabolism of the ECM outstrips anabolic repair mechanisms, the collagen network frays, fibrils break under mechanical stress and collagen degrades. Some chondrocytes undergo cell death and cannot maintain the remaining cartilage, which progressively degenerate. The discovery that Se is a pathogenic factor in KBD has increased interest in selenoproteins (glutathione peroxidases, thioredoxin reductases), involved in antioxidative defense and in thyroid hormone activation (Roman et al., 2014). Studies have shown that low Se-levels combined with mycotoxins, found in poorly stored cereals, exacerbate AC degradation, an effect that seems to be reversed by Se-supplementation (Yao et al., 2012). The limited ability of AC to regenerate is still far from being overcome by current research techniques proposed to repair or regenerate this tissue with a similar quality and stability compared to the cartilage tissue. One approach appears to slow or stop degeneration of AC by understanding more about the role of Se in normal and abnormal situations. By using an in vitro model of AC development, that covers a vital period of growth and adaptation of this tissue to mechanical forces, we hope to understand more about the role of Se in AC homeostasis.

2 MATERIALS AND METHODS

During this study, in vitro culture of 6-mm explants excised under sterile conditions from the lateral aspect of the medial condyle of the metacarpophalangeal joints of immature male (7 days-old) bovine calves is used. A thin layer of the subchondral bone is present on the basal aspect of each explant. To induce maturation in cartilage explants, cartilage in serum-free medium (Dulbecco's modified Eagles medium. Insulin-transferrin-Se, HEPES buffer, and antibiotic and antimycotics) were supplemented with 100 ng/mL fibroblast growth factor 2 (FGF-2) and 10 ng/mL transforming growth factor $\beta 1$ (TGF$\beta 1$). The combination of growth factors induces profound morphologic changes in immature articular cartilage consistent with a highly accelerated maturational response within 21 days. In addition, explants were also cultured in culture medium depleted of Se. Growth factor stimulation induced apoptosis and resorption from the basal aspect and cellular proliferation in surface chondrocytes (Khan et al., 2013). After 3 weeks in culture, 10 µm-cryosections explants were used for synchrotron FTIR microscopy. This technique provide biochemical information on the ECM according to tabulated indices given in the literature (Camacho et al., 2001). Safranin-O staining was also applied on μm-wax sections. qPCR analyses allow to study the up and down regulations of diverse genes related to selenoproteins expressed in bovine AC upon diverse maturation treatments.

3 RESULTS

Understanding how Se affects tissue organization and function during cartilage maturation was the main objective of this study. A precise analysis of the structural, biochemical, morphological and biomechanical properties changes during maturation were explored through the use of complementary techniques. Explants placed in Se-depleted medium seem to have properties close to those of KBD patients. From absorption maps obtained via FTIR microscopy, we showed that explants placed in Se-depleted medium display a reduced collagen content. This observation is accompanied by disruption of growth plate organization and architecture that can be related to cellular changes (cell number & morphology). It can reveal the impact of Se on collagen remodeling to build the stratified architecture of the complex collagen fibre meshwork. Furthermore, tissue sections of explants cultured in the absence of Se and stained with Safranin-O, showed that chondrocytes in the surface zones had an abnormal organization. They were organized as cell clusters similar to patterns observed in early stage of osteoarthritis. Further investigations have to be performed to confirm and fully understand the biological processes acting behind cell and ECM changes observed in Se-depleted cartilage. The importance of Se in AC maturation is further supported by qPCR studies that reveal a number of selenoproteins such as DIO2, GPX1 or TXNRD1 to be modulated during AC maturation and also during Se-depletion. It appears that the expression of TXNRD1 is correlated to the presence of Se in culture medium: a decrease of its expression is observed. In parallel, GPX1 and DIO2 are up-regulated in presence of Se for AC explants treated with maturation cocktail. Control groups without growth factor treatments possess a similar expression of these genes.

4 CONCLUSIONS

These experiments have shown interesting evidence that Se deficiency can induce an osteoarthritis-like phenotype in articular cartilage. Further ongoing investigations will provide insights in the biological and biomechanical reasons for the occurrence of abnormal cartilage following Se-depletion. qPCR, and immunohistochemical tests for a large range of selenoproteins will be considered. Meanwhile, ongoing complementary analyses such as X-ray fluorescence, pSHG, DUV microscopies should provide a deeper understanding of the Se influence in the tissue organisation. Whether these abnormal changes are related to AC degeneration observed in people suffering from Kashin-Beck disease remains to be tested, but we have shown that critical enzymes such as glutathione peroxidase has her expression modulated in a similar way both in our in vitro model or in diseased tissues. The multidisciplinary approach we have implemented, through the molecular and cellular analysis of the role of Se in cartilage homeostasis, could result in a new strategy for OA-treatments through metabolic rebalancing to relieve pain generated by the joint cartilage degeneration.

REFERENCES

Camacho, N.P., West, P., Torzilli, P.A. & Mendelsohn, R. 2001. FTIR microscopic imaging of collagen and proteoglycan in bovine cartilage. *Biopolymers* 62:1–8.

Khan, I.M., Francis, L., Theobald, P.S., Perni, S., Young, R.D., Prokopovich, P. et al. 2013. In vitro growth factor-induced bio engineering of mature articular cartilage. *Biomaterials*. 34(5):1478–1487.

Roman, M., Jitaru, P. & Barbante, C. 2014. Selenium biochemistry and its role for human health. *Metallomics* 6: 25–54.

Stone, R. 2009. A medical mystery in middle China. AAAS 324.5933:1378–1381.

Xiong, Y.M., Mo, X.Y., Zou, X.Z., Song, R.X., Sun, W.Y., Lu, W. et al. 2010 Association study between polymorphisms in selenoprotein genes and susceptibility to Kashin-Beck disease. *Osteoarthritis and Cartilage* 18:817–824.

Yan, J., Zheng, Y., Min, Z., Ning, Q. & Lu, S. 2013. Selenium effect on selenoprotein transcriptome in chondrocytes. Biometals. 2-2:285–296.

Yao, Y., Pei, F., Li, X., Yang, J., Shen, B., Zhou, Z. et al. 2012. Preventive effects of supplemental selenium and selenium plus iodine on bone and cartilage development in rats fed with diet from Kashin-Beck Disease endemic area. *Biol. Trace Elem. Res.* 146:199–206.

Zhang, A., Cao, J.L., Chan, J.H., Zhang, Z.T., Li, S.Y., Fu, Q. et al. 2010. Effects of moniliformin and selenium on human articular cartilage metabolism and their potential relationships to the pathogenesis of Kashin-Beck disease. *Biomedicine & Biotechnology* 11(3):200–208.

Intestinal bioaccessibility and bioavailability of selenium in elemental selenium nano-/microparticles

G. Du Laing*, R.V.S. Lavu, B. Hosseinkhani, V.L. Pratti, F.M.G. Tack & T. Van de Wiele
Ghent University, Ghent, Belgium

1 INTRODUCTION

In selenium (Se) deficient populations, Se-enriched food crops and food supplements are often recommended to overcome Se deficiency. Moreover, Se nano- and microparticles have recently been cited as potential ingredients of functional food or feed components. It is well-known that the effective bioavailability of Se in Se-enriched products may depend on the chemical form in which Se occurs, as well as the matrix within which Se is embedded in the different products (Lavu et al., 2014). The term "bioaccessibility" has been defined as the fraction of a compound that is released from its matrix in the gastrointestinal tract and thus becomes available for intestinal absorption. Bioaccessibility data for pure Se standards, a yoghurt-based food supplement containing elemental Se particles, a Se+vitACE supplement, and Se-enriched crops (leek and kenaf) were previously reported by Lavu et al. (2014). In the current study, we investigated the bioaccessibility of Se contained in chemically and microbiologically synthesized nano- and microparticles in vitro and compared it with the bioaccessibility of Se in SelenoPrecise tablets and two different types of selenized yeast.

2 METHODS

The chemically synthesized particles were prepared based on a procedure described by Chiou and Hsu (2011). In brief, H_2SeO_3 is reduced using $NaHB_4$. The amount of reducing agent varied to obtain Se nanoparticles. Carboxymethylcellulose (CMC) is used during synthesis to stabilize particles in suspension. The particle size distribution of two types of synthesized particles is presented in Figure 1. *Pediococcus acidilactici* was grown during 2 days under anoxic conditions in MRS medium, harvested and washed with phosphate buffer solution (PBS). Afterwards, the cells were incubated during 2 days in presence of Na_2SeO_3.

To assess the bioaccessibility in vitro, thirty mL of simulated gastric juice (10 g/L pepsin adjusted to pH 2.0 with 2 M HCl) were added to the samples and the suspensions were shaken in an incubator (37°C) for 1 h. After 1 h of incubation a 5 mL sample was taken, considered to represent the gastric phase. Afterwards,

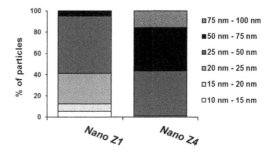

Figure 1. Particle-size distribution of the chemically synthesized Se nanoparticles.

12.5 mL of small intestine fluid (containing bile powder, pancreatin and sodium bicarbonate) were added and the mixture was shaken in the incubator for 2 h again. After 2 h, 5 mL of sample, representing the small intestine, was taken. Subsequently, 30 mL of colon suspension, sampled from the colon compartments of the SHIME® system (Simulator of the Human Intestinal Microbial Ecosystem, Van de Wiele et al., 2004) was added. The bottles were capped and flushed with nitrogen gas. The bottles were shaken in the incubator and sampled after 0 (T0), 2 (T2), 24 (T24) and 48 hours (T48). Collected samples were centrifuged at 10,000 g for 10 minutes and the supernatant was filtered (0.45 μm). Filtrates and residues were analyzed for total Se using ICP-MS.

3 RESULTS AND DISCUSSION

The evolution of Se bioaccessibility in the different intestinal phases after introducing the Se nanoparticles is presented in Figure 2.

The relative bioaccessibility of Se in the chemically synthesized particles was very low in the gastric phase, which should probably be attributed to coagulation of the particles and/or their association with pepsin at the pH prevailing in the stomach, or very slow disaggregation after adding the powdered nanoparticles to the gastric solution. Subsequently, the Se bioaccessibility increased in the small intestine, which may be due to the pH shift inducing disaggregation or dissolution of the particles, or association or reaction of the

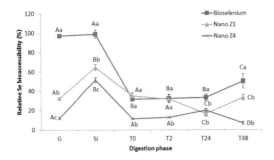
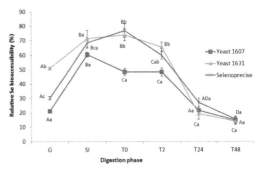

Figure 2. Relative bioaccessibility of Se (%) for (A) microbiologically synthesized nanoparticles (Bioselenium) and (B) two types of chemically synthesized nanoparticles (Z1 and Z2) in different steps of an in vitro simulation of gastrointestinal digestion (G refers to stomach; SI refers to small intestine).

Figure 3. Relative bioaccessibility of Se (%) for (A) crushed Selenoprecise tablets and (B) two types of selenized yeast (1607 and 1631) in different steps of an in vitro simulation of gastrointestinal digestion.

particles with other compounds present in this environment, which may result in a mobilization of the particles or release of Se from the particles.

In the stomach and small intestine the bioaccessibility was much higher for Bioselenium compared to the chemically synthesized nanoparticles. This observation may be attributed to the fact that a significant amount of Se in the Bioselenium suspension was still present in its original form prior to reduction, i.e., selenite. Around 45% of the Se in the Bioselenium suspension could be recovered in the filtrate after centrifugal ultrafiltration, which is actually expected to remove Se associated with the nano- and microparticles from the solution.

For all sample types, the bioaccessibility decreases significantly when moving from small intestine to colon. This occurrence may be attributed to external adsorption of the particles to microbial surfaces or active uptake of the particles by colon microorganisms. The bioaccessibility slightly increased again in two of the three samples after 48 h of colon incubation. The evolution of Se bioaccessibility in the different intestinal phases after introducing the selenized yeast samples and Selenoprecise tablets is presented in Figure 3. In all yeast samples, the bioaccessibility was low in the stomach but it increased significantly in the small intestine. Yeast sample 1607 had the lowest Se bioaccessibility and showed the largest decrease in bioaccessibility when moving from the small intestine to the colon. This yeast sample had a reddish-brownish color instead of the typical white color, which indicates that elemental Se particles had been formed during its production.

4 CONCLUSIONS

The bioaccessibility of Se is lower when Se is present in the form of elemental nano- or microparticles, compared to when it is present in selenized yeast or selenium-enriched crops. Whereas elemental Se nanoparticles are probably removed from the liquid phase in the colon through coagulation and precipitation or adsorption processes, the other Se species are mainly removed through active uptake by the micro-organisms present in this environment.

REFERENCES

Chiou, Y.D. & Hsu, Y.J. 2011. Room-temperature synthesis of single-crystalline Se nanorods with remarkable photocatalytic properties. *Applied Catalysis B-Environmental*, 105: 211–219.

Lavu, R.V.S., Van de Wiele, T., Pratti, V.L., et al. 2014. Bioaccessibility and transformations of selenium in the human intestine: selenium-enriched crops versus food supplements. In: G.S. Banuelos, Z.-Q. Lin, X.Y. Yin (eds). Selenium in the Environment and Human Health, Taylor and Francis Group, London, pp. 44–47.

Van de Wiele, T., Boon, N., Possemiers, S., Jacobs, H., Verstraete, W. 2004. Prebiotic effects of chicory inulin in the simulator of the human intestinal microbial ecosystem. *FEMS Microbiology Ecology* 51: 143–153.

Autoimmunity against selenium transport in human sera

Waldemar B. Minich*, Andrea Schuette, Christian Schwiebert, Tim Welsink,
Kostja Renko & Lutz Schomburg*
Institute for Experimental Endocrinology, Charité – University Medicine Berlin, Berlin, Germany

1 INTRODUCTION

Incidence and prevalence of autoimmune diseases and allergies are constantly increasing, especially in young children and adolescents (Brune & Hochberg, 2013). Besides differences in the microbiome, also deficiencies in certain micronutrients, vitamins and trace elements, e.g. polyunsaturated fatty acids, vitamin E, vitamin D, zinc or selenium (Se), have been implicated as underlying reasons for this trend in modern societies (Allan & Devereux, 2011). Most Europeans have certain micronutrient deficiencies, especially with respect to vitamin D, iodine and Se. Vitamin D deficiency is mainly due to lack of UV exposure, especially in winter. Iodine deficiency is largely due to a minimal intake of seafood like fish, algae or other marine food items. However supplementation of table salt with iodide considerably improved the situation.

In contrast, the wide-spread deficiency in Se is not adequately addressed by respective recommendations or dietary programs, except for in Finland, where fertilizers have been supplemented with sodium selenate for about three decades (Alftahn et al., 2015). Other European countries have a general Se intake below the amount needed for a full expression of selenoproteins and must be considered as Se deficient, which has a negative impact on immune system and contributes to a higher risk for autoimmune diseases and allergies (Huang et al., 2012).

We reasoned that also the inverse interaction may be of physiological relevance, i.e., that autoimmunity is the central factor controlling mammalian Se transport and distribution and may induce target cell Se deficiency by interfering with SePP transport and uptake. To this end, we developed a novel assay that is capable of detecting SePP-binding autoantibodies (SePP-aAb) from human blood. Surprisingly, we identified SePP-aAb in more than 5% of apparently healthy subjects in our first cross sectional analyses.

We speculated that a Se deficiency may both predispose humans to autoimmune diseases and develop an autoimmunity against SePP. This occurrence potentially closes a feed-forward cycle with unknown but potentially severe clinical consequences and disease relevance.

2 ASSAY DEVELOPMENT

Expression vectors encoding human recombinant selenocysteine (Sec)-free SePP were generated and used to construct baculovirus particles for infecting insect cells and producing recombinant protein. High efficient production of secreted Sec-free SePP (up to 20 mg/L in cell medium) was achieved after replacing all of the Sec residues in the SePP open reading frame by cysteine. The recombinant protein was isolated from culture supernatants using immobilized metal affinity chromatography and labeled with MACN-acridinium-NHS-ester.

Human serum samples were then mixed with labeled SePP, allowing the endogenous autoantibodies to bind. Immune complexes were precipitated by protein A–sepharose and gentle centrifugation. Amount of precipitated MACN-labeled SePP was determined by luminescence as a measure of SePP-aAb concentration. Results are recorded as relative light units (RLU) and used to calculate a Binding Index (BI). BI is calculated in relation to the average signal obtained from the lower 50% of samples. The calculation assumes that less than 50% of tested samples are positive for SePP-aAb, a very prudent and safe guess, as there have been no reports up to now about such autoantibodies. In addition, to test for the potential biological activity and function of SePP-aAb, the complete set of IgG was isolated from both positive and negative human serum aliquots by chromatography on protein A-sepharose, and then used in cell culture experiments. That was to test whether SePP-aAb interfere with SePP and SePP-receptor interaction.

3 RESULTS

Using well-characterized monoclonal antibodies, we verified reproducibility and detection limit of the SePP-aAb assay. SePP-specific antibodies but not unrelated antibodies were able to detect the recombinant Sec-free SePP both by Western Blot analysis and by the precipitation assay of MACN-labeled recombinant SePP protein. Next, we analyzed prevalence of SePP-aAb in human serum samples. To this end,

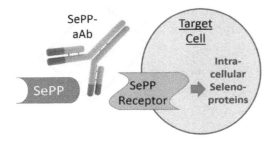

Figure 1. Interference of endogenous autoantibodies against selenoprotein P (SePP-aAb) with SePP and SePP-receptor interaction and Se delivery to target cells. The presence of SePP-aAb in human serum may interfere with the interaction of SePP with its receptors on target cells. Hereby, SePP uptake and intracellular Se liberation from SePP may become severly impaired, negatively affecting intracellular selenoprotein biosynthesis and tissue Se status. Notably, this target cell Se deficiency may proceed unnoted by commonly used Se status biomarkers from serum or blood.

a cross sectional collection of approx. 200 human serum samples was analyzed with the labeled protein. Assuming the detection of SePP-aAb is rare, we were surprised to observe that >5% of sera tested turned out to be positive for SePP-aAB. Autoantibody concentrations varied strongly as evident from the range of BI values obtained. The signals obtained from human sera proved to be stable upon dilution and repeated freeze-thawing cycles. Similar results were obtained when whole serum or purified immunoglobulin (IgG) preparations were analyzed, indicating that IgG was responsible for the signals detected.

We also tested whether SePP-aAb may be of functional relevance for SePP-dependent Se transport and metabolism. For this purpose, we purified IgG from SePP-aAb positive and negative sera. The interaction between SePP and its receptors on the eukaryotic cell membrane was selectively inhibited in the in vitro experiments when IgG preparations from SePP-aAb positive sera and not SePP-aAb negative sera were tested. These experiments indicated that SePP-aAb may be of clinical relevance as they affect SePP-dependent Se transport and Se uptake into human target cells (Fig. 1). Despite normal Se or SePP concentrations, such probands may experience target cell Se deficiency and develop similar albeit less severe symptoms than known from the few examples of inherited defects in selenoprotein expression or function (Schomburg, 2010; Schweizer et al., 2011).

4 CONCLUSIONS

We conclude that Se deficiency is not only a potential risk factor for the development of autoimmune diseases but inversely, autoimmunity may affect Se organification, transport and metabolism. The clinical significance of our finding that >5% of regular unselected probands are positive for SePP-aAb remains to be established. Especially in light of the tight interaction of Se status and autoimmune thyroid disorders, it may be worthwhile to conduct a large study on thyroid disease patients to test whether SePP-aAb impose an increased risk for this widespread and steadily increasing disease (Schomburg, 2012). A potential feedback circle of Se-dependent autoimmunity negatively impacting the Se transport via interfering with SePP and SePP-receptor interaction can be envisaged that may aggravate Se-dependent disease risks without becoming easily obvious by established biomarkers of Se status. This hypothesis remains to be tested in large epidemiological studies.

REFERENCES

Alfthan, G., Eurola, M., Ekholm, P., Venalainen, E. R., Root, T., Korkalainen, K., et al. 2015. Effects of nationwide addition of selenium to fertilizers on foods, and animal and human health in Finland: From deficiency to optimal selenium status of the population. *J Trace Elem Med Biol.* 31: 142–147.

Allan, K., & Devereux, G. 2011. Diet and asthma: nutrition implications from prevention to treatment. *J Am Diet Assoc.* 111(2): 258–268.

Brune, M., & Hochberg, Z. 2013. Secular trends in new childhood epidemics: insights from evolutionary medicine. *BMC Med.* 11: 226.

Huang, Z., Rose, A.H., & Hoffmann, P.R. 2012. The role of selenium in inflammation and immunity: from molecular mechanisms to therapeutic opportunities. *Antioxid Redox Signal* 16(7): 705–743.

Schomburg, L. 2010. Genetics and phenomics of selenoenzymes – how to identify an impaired biosynthesis? *Mol Cell Endocrinol.* 322(1-2): 114–124.

Schomburg, L. 2012. Selenium, selenoproteins and the thyroid gland: interactions in health and disease. *Nat Rev Endocrinol* 8(3): 160–171.

Schweizer, U., Dehina, N., & Schomburg, L. 2011. Disorders of selenium metabolism and selenoprotein function. *Curr Opin Pediatr.* 23(4): 429–435.

… *Global Advances in Selenium Research from Theory to Application – Bañuelos et al (Eds)*

Natural small molecules in protection of environment and health: Sulfur-selenium deficiency and risk factor for man, animal and environment

R.C. Gupta
SASRD Nagaland University, Medziphema, India

1 INTRODUCTION

Historically both selenium (Se) and sulfur (S) have common geochemical behaviors, and appear later in periodic classification due to similar physicochemical characteristics. Physicochemical nature is also index to physiological behavior and in practice, both share many common physiological activities. Looking at the Agro-ecosystem, due to continuous cutting or burning of forests, sulfur is becoming increasingly mobile. Sulfur loss in soil is also accelerating as fires convert soil sulfur to highly volatile sulfur dioxide (SO_2). A similar fate can also be observed for Se, in which soil Se can be convert into selenium dioxide (SeO_2). Both of these volatile processes can make a soil Se-S deficient and any crops grown in such soils will also be Se-S deficient. Subsequently, consumers of these crops may eventually be prone to Se-S deficiencies, inviting diseases and disorders (Moraes et al., 2009).

Selenium is critical for the health of living organisms but the majority of the population in some countries, e.g. China, only have a suboptimal intake of Se, resulting in life threatening diseases related to cancer, the heart, and increased oxidative stress. Furthermore, the amount of Se in soil and rate of cancer at that location are inversely proportional (Reference?). Agricultural crops have low Se concentrations either because of a low Se concentration in soil or because of a poor availability of soil Se for plant uptake (Haug et al., 2007; Lyons et al., 2003).

In biochemical processes, sulfur amino acids also metabolize to elemental sulfur; necessary for growth and development. Sulfur amino acids and Se deficiency can both occur, especially where food intake is of much more botanical in nature. One of the most promising sulfur amino acids is taurine (2-amino ethane sulfonic acid). Selenium and taurine share many biological actions from anti-oxidation to 'host defense' (Gupta & Kim, 2003; Gupta et al., 2005; Rasmussen, 2006; Reilly, 1996).

To reduce the occurrence of Se-related diseases, an improved intake of Se and S amino acids is the best option and most critical for optimizing functions of the immune system. Selenium concentrations in plants can be increased with the help of seleno compounds (Traulsen et al., 2004), and Se as a soil amendment in fertilizers can also be used. The recent data provide a clue that a Se and taurine deficient population is prone to viral attack thus requiring optimal protection. There is a vast volume of literature reporting on the protective role of Se and taurine against tissue damage caused by ischemia/reperfusion in several vital organs like the brain, heart, kidney, liver, lungs and eyes (Ueno et al., 2007; Gupta, 2011). Human dietary intake of Se can also be increased by adding Se to animal feed and thus enhancing the human Se intake through livestock products. Growing plants in soil naturally high in Se or soil fertilized with Se can produce Se rich products that may be regarded as 'functional food'. Functional foods are now part of "New Treatment Modalities". It is dishearten to know that 1/7 of the world population depends on a diet that is Se deficient, mostly in the under develop & developing areas of the world.

One good source of Se is nuts but green leafy vegetable, mushrooms and whole grains are also high in Se (Christopheren & Haug, 2005; Boucher et al., 2008: Kiremidjian-Schumacher et al., 1996). Similarly, taurine is a constituent of functional food and taurine enriched; eggs, fishes, meat, as well as nutraceuticals are available in the market (Cotter et al., 2009; Shapira, 2009).

2 MATERIALS AND METHODS

Pub med and other data for the past twenty five years was utilized for this study.

3 RESULTS AND DISCUSSION

"We are what we eat." Technically food also falls under category "new treatment modalities," giving the concept of functional food. Due to their common beneficial properties, Se and taurine are now classifying as components of "functional food." Selenium concentrations in the human diet can be increased by consuming food with higher Se concentrations. Besides human being, Se is involved with several physiological actions of animals too. In a parallel manner, taurine deficiency has a significant effect on retinal degeneration, cardiac dysfunction, cardiomyopathy, platelet hyper aggregation, growth and survival of mammalian cells, fetus development and development of new born. It is now established that taurine deficiency is linked to blindness in cats and dogs. The majority of world population is nurtured on cow's milk, which lacks taurine. Mother's milk has sufficient amount of taurine and thus a taurine deficiency during fetus development may linked to 'disease of generation.' The better prophylaxis of several important diseases will not be possible unless the nutritional quality of the diet can be made improved. Unless we bring the concept of "functional food" into reality, a dream like "good quality of life", will remain a dream.

4 CONCLUSIONS

Elemental sulfur as well as S in sulfur amino acids and Se are imperative for growth, development, and maintenance of good health. It will be best to have both in one food product, preferably though functional food, nutraceutical, designer or tailor made food. Several examples exist where Se- taurine joint therapy has produced desire effects and thus we may conclude that a combination of both dietary and pharmacological intervention rather than pharmacological intervention alone is needed for sound health.

REFERENCES

Boucher, F.R., Jouan MGMoro, C., Rakotovao, A.N., Tanguy, S., & De Leiris, J. 2008. Does selenium exert cardioprotective effects against oxidative stress in mycocardial ischemia? Acta Physiol Hung, 95:187–94

Christopheren, O.A. & Haug, A. 2005. Possible roles of oxidative stress, local circulatory failure and nutrition factors in the pathogenesis of hyper virulent influenza: implications for therapy and global emergency preparedness. Microb Ecol Health Dis, 17:189–199.

Cotter, P.A., McLean, E. & Craig, S.R. 2009. Designing fish for improved human health status. Nutr Health, 20:1–9.

Gupta, R.C. 2013. Selenium and Taurine: Components of functional food for sound health. In: G.S. Banuelos, Z.O. Lin, X. Yin, and N. Duan (eds.), Selenium: Global perspectives of Impacts on Human, Animals and the Environment. University of Science and Technology of China Press. Hefei, China.

Gupta, R.C. & Kim, S.J. 2003. Role of taurine in organs dysfunction and in their alleviation. Critical Care and Shock, 6:171–175.

Gupta, R.C., Win, T. & Bittner, S. 2005. Taurine analogues; a new class of therapeutics Retrospect and prospects. Current Medicinal Chemistry, 12:2021–2039.

Haug, A, Graham, R.D. Christophersen, O.A. & Lyons, G.H. 2007. How to use the world's scarce selenium resources efficiently to increase the selenium concentration in food. Microb Ecol Health Dis, 19:209–228

Kiremidjian-Schumacher, L., Roy, M., Wishe, H.I., Cohen, M.W., & Stozky, G. 1996 Supplementation with selenium augments the functions of natural killer and lymphokine-activated killer cells Free Tadic Res, 38:123–128.

Lyons, G., Stangoulis, J., & Graham, R. 2003. High-Se wheat biofortification for better health. Nutr Res Rev., 16:45–60.

Moraes, M.F., Welch, R.M., Nutti, M.R., Carrvalho, J.L.V. & Watanabe, E. 2009. Evidence of selenium deficiency in Brazil: from soil to human nutrition. pp. 73–74. In: G.S. Banuelos, Z.-Q. Lin, and X.B. Yin (eds.), Selenium: Deficiency, Toxicity and Biofortification for Human Health. 116p. Hefei: University of Science and Technology of Chinal Press.

Rasmussen, L.B., Mejborn, H., Andersen, N.L., Dragsted, L.O., Krogholm, K.S., & Larsen, E.H. 2006. Selenium and health. Ugeskt laeger, 168:3311–3313.

Reilly, C. 1996. Selenium in Food and Health. London: Blackie

Shapira, N. 2009. Modified eggs as nutritional supplement during brain development, a new target for fortification. Nutr Health, 20:107–108.

Traulsen, H., Steinbrenner, H., Buchczyk, D.P., Klotz, L.O. & Sies, H. 2004. Selenoprotein P protects low-density lipoprotein against oxidation. Biol Trace Elem Res, 52:227–239.

Ueno, T., Iguro, Y., Yotsumoto, G., Fukumoto, Y., Nakamura, K., Miyamoto, T.A. et al. 2007, Taurine at early repersusion significantly reduces myocardial damage and preserves cardic function in the isolated rat heart. Resuscitation, 73:287–295.

Incubation and crude protein separation from selenium-enriched earthworms

Yuhui Qiao*, Xiaofei Sun, Shizhong Yue, Zhenjun Sun & Shengnan Li
College of Resource and Environment, China Agricultural University, Beijing, China

1 INTRODUCTION

Selenium (Se) is one of the essential trace elements for organisms. Compared to inorganic Se, organic Se has advantages with low toxicity, more effective absorption and can be safely used as supplements. Currently, selenium-enriched animals, plants and microorganisms are produced by biotransformation, such as selenium-enriched silkworms, rice, soybeans, yeast and lucid ganoderma (Gallego-Gallegos et al., 2012; Gladyshev, 2012; Sánchez-Martínez et al., 2012; Schmidt et al., 2013). Earthworm is a source of high-quality animal protein with 70% of its dry-weight potentially useful as a Se selenium bio-carrier (Mester et al., 2006). The separation of proteins from earthworms is mainly focused on fibrinolytic enzymes, antimicrobial peptides, and metallothioneins, etc. (Edwards & Niederer, 2010; Papp et al., 2007; Procházková et al., 2011; Sanchez-Hernandez et al., 2014; Whanger, 2002). However, there are few studies on the separation of selenoproteins from earthworms. In this study, incubation conditions will be screened out for selenium-enriched earthworms. Meanwhile, selenium-enriched proteins from earthworms will be separated on the basis of different solubility levels with different methods.

2 MATERIALS AND METHODS

The effect of five Se concentrations (20, 40, 60, 80, 100 mg/kg) and four incubation times (15, 30, 45, 60 days in earthworm were investigated) on earthworm viability and Se concentrations. After the harvest of earthworms, selenium-enriched protein in earthworm was sequentially separated and extraction with ammonium sulfate precipitation method and Q Sepharose Fast Flow (QFF) ion exchanging chromatography. According to four different extraction methods (water extraction, salt extraction, ethanol extraction and alkali extraction), three percentages of ammonium sulfate saturation (30, 60, 90%) and five groups of elution fractions, a crude separation method will be screened out with highest purification and high Se content in protein as an optimizing index.

3 RESULTS AND DISCUSSION

Compared to the control treatment, survival rates remained the same under the Se concentration of 20,

Table 1. Concentrations ($\mu g/g$) of selenium in earthworm.

Se Treat. (mg/kg)	Selenium concentration in earthworm ($\mu g/g$)			
	Day 15	Day 30	Day 45	Day 60
0	5.59 ± 0.74a[†]	5.73 ± 1.61a	5.56 ± 2.44a	5.80 ± 2.20a
20	12.17 ± 3.14b	12.93 ± 2.06bc	13.99 ± 4.18b	16.42 ± 4.12b
40	14.62 ± 4.54b	21.51 ± 1.30b	26.92 ± 5.08c	30.01 ± 4.74c
60	17.05 ± 3.54bc	21.45 ± 5.40c	30.13 ± 1.61c	36.38 ± 1.51cd
80	17.47 ± 5.57bc	18.18 ± 5.35bc	33.25 ± 1.61cd	41.99 ± 5.39d
100	22.85 ± 0.76c	28.34 ± 2.49d	39.14 ± 4.85d	42.42 ± 4.23d

[†]Mean followed by SD. Different letters represent level of significance at the $p < 0.05$ level.

40, 60, 80 mg/kg, while the survival rate significantly decreased when Se concentration reached 100 mg/kg. The survival rate in day 30, 45 and 60 was 88.89%, 83.33% and 78.89% respectively. Selenium content in earthworm increased with the Se concentration and the extension of incubation days. When the Se content in the incubation medium is 80 mg/kg and the incubation time was 45 days, the Se content in earthworm was 33.25 $\mu g/g$ (Table 1).

Selenium content of water-soluble protein was significantly greater than salt-soluble protein and alkali-soluble protein, and there was no ethanol-soluble protein. The Se content of Se-enriched protein in 30% ammonium sulfate was significantly greater than 60 and 90% ammonium sulfate for water-soluble protein. Meanwhile, crude separation of water soluble selenoprotein in the NaCl gradient elution chromatography at 17, 23, 30, 65, 80 min was eluted into 5 distinct components (Fig. 1). The second peak had the greatest protein content (1142 $\mu g/ml$), followed by the fourth peak at 516 $\mu g/ml$, whereas the protein contents of peak 1, 3 and 5 were lower as shown in Table 2. Compared to other concentrations of Se, the fourth peak reached the greatest level (228 $\mu g/l$), followed by the fifth peak (144 $\mu g/l$). Selenium concentrations of peak 1, 2 and 3 were less than 100 $\mu g/ml$ and thus were significantly lower than peak 4 and 5. Comparing the Se concentrations of each seleno-protein, the fourth peak reached the greatest level at 399.66 $\mu g/g$. Its concentration was 76.06% higher than the water-soluble protein. After the third step by QFF, protein content, Se concentration and purification folds have been changed to different degrees. The Se concentration in Se-enriched earthworm eventually increased

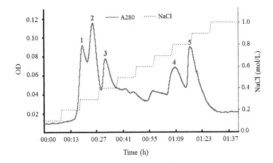

Figure 1. Elution profile of crude water protein on QFF column chromatography.

Table 2. Protein content and Se concentration in different protein from Q Sepharose Fast Flow (QFF).

Peak	Protein content (μg/ml)	Se in protein (μg/l)	Se in Se-enriched protein (μg/g)
Water soluble protein	1267.86 ± 40[†f]	288.07 ± 24.68[d]	227.21 ± 17.86[c]
1	172.54 ± 7.07[a]	20.15 ± 0.44[a]	116.87 ± 4.02[b]
2	1142.00 ± 25.53[e]	72.10 ± 0.92[c]	63.15 ± 1.15[a]
3	274.00 ± 5.57[b]	62.26 ± 1.06[b]	227.35 ± 8.40[c]
4	516.33 ± 13.23[d]	227.72 ± 5.94[f]	399.66 ± 7.75[e]
5	396.33 ± 3.00[c]	143.57 ± 8.51[e]	309.88 ± 15.81[d]

[†] Different letters represent significant differences ($p < 0.05$).

Table 3. Selenium content and purified folds of Se-enriched crude protein in earthworms.

Separation step	Protein content (μg/ml)	Yield (%)	Se concentration in Se-protein (μg/g)	Purified folds
Water Extraction	7710	100	72.32	1
30% ammonium sulphate	1267.86	16.44	227.21	3.14
QFF ion exchange column chromatography	516.33	6.70	399.66	5.53

from 72 μg/g to 399.66 μg/g. The purification fold increased by 5.53 times (Table 3).

4 CONCLUSIONS

Based on the effects of Se and incubation time on viability and Se concentration in earthworms, the optimized incubation condition for selenium-enriched earthworms is 45–60 days when Se concentration is 80 mg/kg in the incubation medium. This study shows that aqueous extraction is the first step for separation of seleno-protein from Se-enriched earthworm, followed with 30% ammonium sulphate. After QFF ion exchange column chromatography, the fourth eluting peak is selected to obtain seleno-proteins due to the highest Se content.

REFERENCES

Edwards, C. and Niederer, A. 2010. The production of earthworm protein for animal feed from organic wastes. In: C.A. Edwards, N.Q. Arancon & R. Sherman (eds.), Vermiculture Tecnology, pp. 323–334.

Gallego-Gallegos, M., Doig, L.E., Tse, J.J., Pickering, I.J., Liber, K. 2012. Bioavailability, toxicity and biotransformation of selenium in midge (Chironomus dilutus) larvae exposed via water or diet to elemental selenium particles, selenite, or selenized algae. Environ. Sci. Technol., 47: 584–592.

Gladyshev, V.N., 2012. Selenoproteins and Selenoproteomes, Selenium. Springer, pp. 109–123.

Mester, Z., Willie, S., Yang, L., Sturgeon, R., Caruso, J.A., Fernández, M.L., et al. 2006. Certification of a new selenized yeast reference material (SELM-1) for methionine, selenomethinone and total selenium content and its use in an intercomparison exercise for quantifying these analytes. Analytical and Bioanalytical Chemistry 385: 168–180.

Papp, L.V., Lu, J., Holmgren, A., and Khanna, K.K. 2007. From selenium to selenoproteins: synthesis, identity, and their role in human health. Antioxidants & Redox signaling 9: 775–806.

Procházková, P., Dvořák, J., Šilerová, M., Roubalová, R., Škanta, F., Halada, P., et al. 2011. Molecular characterization of the iron binding protein ferritin in Eisenia andrei earthworms. Gene, 485: 73–80.

Sanchez-Hernandez, J.C., Martínez Morcillo, S., Notario del Pino, J., Ruiz, P., 2014. Earthworm activity increases pesticide-sensitive esterases in soil. Soil Biology and Biochemistry 75, 186–196.

Sánchez-Martínez, M., da Silva, E.G.P., Pérez-Corona, T., Cámara, C., Ferreira, S.L., Madrid, Y., 2012. Selenite biotransformation during brewing. Evaluation by HPLC–ICP-MS. Talanta, 88: 272–276.

Schmidt, R., Tantoyotai, P., Fakra, S.C., Marcus, M.A., Yang, S.I., Pickering, I.J., et al. 2013. Selenium biotransformations in an engineered aquatic ecosystem for bioremediation of agricultural wastewater via brine shrimp production. Environ. Sci. & Technol., 47: 5057–5065.

Whanger, P. 2002. Selenocompounds in plants and animals and their biological significance. Journal of the American College of Nutrition, 21: 223–232.

Health impact of dietary selenium nanoparticles on mahseer fish

Kifayat Ullah Khan*
Fisheries and Aquaculture Laboratory, Department of Animal Sciences, Faculty of Biological Sciences, Quaid-i-Azam University, Islamabad, Pakistan; Centro de Aquicultura da UNESP, Jaboticabal, SP, Brazil

Amina Zuberi, Zeenat Jamil & Huda Sarwar
Fisheries and Aquaculture Laboratory, Department of Animal Sciences, Faculty of Biological Sciences, Quaid-i-Azam University, Islamabad, Pakistan

Samina Nazir
National Centre for Physics, Quaid-i-Azam University, Islamabad, Pakistan

João Batista Kochenborger Fernandes
Centro de Aquicultura da UNESP, Jaboticabal, SP, Brazil

1 INTRODUCTION

Selenium (Se) is an important dietary micronutrient (Dare et al., 2000) that is required for normal body functions and metabolism of all living organisms (Hamilton, 2004). Its supplementation in fish feed prevents cell damage and plays an important role in the growth, hematology, fertility, and immune functions of all vertebrates including fish (El-Hammady et al., 2007; Hoffmann & Berry, 2008). However, a deficiency of Se produces many disorders like calf pneumonia, white muscle disease in cattle, exudative diathesis and infertility in fish and other vertebrates (Muller et al., 2002).

As fish accept and eat feed containing nano-materials, nanotechnology strategies can be used to promote the growth, hematology, and immunity of fish through the transfer of micronutrients and constituent in aqua feeds (Handy, 2012). Mahseer (*Tor putitora*) is an important fish species of Pakistan and is currently facing overfishing and pollution-like threats (Islam & Tanaka, 2004). Thus, this species needs to be physiologically strong enough to resist all threats and in this regard. The present study was designed to determine the effects of dietary Se nanoparticles (Se-NPs) on physio-biochemical aspects of *Tor putitora*.

2 MATERIALS AND METHODS

A 70 day feeding trial was conducted in six fiber tanks in a semi-static condition in the Fisheries and Aquaculture Lab, Department of Animal Sciences, QAU, Islamabad, Pakistan. Se-NPs average size was 95 nm, prepared through precipitation method in NCP, QAU Islamabad, Pakistan. Juvenile mahseer (*Tor putitora*) were obtained from Hattian Nursery unit, District Attock, Pakistan, and were divided into two groups.

Group 1: fish fed 40% protein basal diet. Group 2: fish fed the diet supplemented with Se-NPs at a rate of 0.68 mg Se/kg diet (Jamil, 2013) for mahseer (*Tor putitora*).

At the end of feeding trial, all fish from each tank were immediately anesthetized with MS222 (60 mg/L) and blood was collected in *Eppendorf* tubes by tail ablation. Some of the fresh blood was used to analyze fish hematology directly through hematology analyzer. The remaining blood was centrifuged and used for the determination of lysozyme activity, and serum growth hormone levels. After taking blood, the fish were decapitated on an ice box and liver and muscle tissues of each fish were removed. These samples were then flash frozen in liquid nitrogen and stored at $-20°C$ for further biochemical analysis.

3 RESULTS AND DISCUSSION

Several research studies indicated that Se has a profound effect on hematological indices and immunity of fish (Molnar et al., 2011; El-Hammady et al., 2007; Kumar et al., 2008). In the present study, the diet of *Tor putitora* supplemented with Se-NPs significantly ($p < 0.05$) increased the RBCs count ($2.61 \pm 0.11 \ 10^6/\mu L$), Hb ($7.8 \pm 0.08$ g/dL), Hct values ($33.46 \pm 0.03\%$), lysozyme activity ($1.73 \pm 0.08 \ \mu g/mL$), and WBCs count ($5.7 \pm 0.14 /10$ mm), as compared to the control diet. Liver GSH-Px activity ($2.20 \pm 0.01 \ \mu mol/min/mg$ proteins) also showed a significant increase.

Selenium is associated with protein synthesis in all vertebrates including fish (Burk & Hill, 1993). Protein content in liver (14.92 ± 0.20 mg/g) and muscle (27.91 ± 0.38 mg/g) tissue was significantly ($p < 0.05$) increased with Se-NPs supplementation as compared to control group. Collipp et al. (1984)

observed a significant relationship between growth hormone and Se nutritional status. Dietary Se-NPs supplementation elevated significantly ($p < 0.05$) the growth hormone levels (0.098 ± 0.04 ng/L) of *Tor putitora*.

4 CONCLUSIONS

Physical and biochemical status of *T. putitora* showed close relationship with dietary Se-NPs supplementation. These results clearly indicated the positive effects of dietary Se-NPs and suggested that 0.68 mg/kg dose of Se-NPs is adequate for the improvement of physio-biochemical aspects of juvenile mahseer (*Tor putitora*).

REFERENCES

Burk, R.F. & Hill, K.E. 1993. Regulation of selenoproteins. *Annual Review of Nutrition* 13: 65–81.

Collipp, P. J., Kelemen, J., Chen, S. Y., Castro-magana, M., Angulo, M. & Derenoncourt, A. 1984. Growth hormone inhibition causes increased selenium levels in Duchenne muscular dystrophy: a possible new approach to therapy. *Journal of Medical Genetics* 21: 254–256.

Dare, C., Eisler, I., Russell, G., Treasure, J. & Doge, L. 2000. Psychological therapies for adults with anorexia nervosa: Randomized controlled trial of out-patient treatments. *British Journal of Pharmacology* 178: 216–221.

El-Hammady, A. K. I., Ibrahim, S. A. & El-Kasheif, M. A. 2007. Synergistic reactions between vitamin E and selenium in diets of hybrid Tilapia (*Oreochromis niloticus x Oreochromis aureus*) and their effect on the growth and liver histological structure. *Egypt Journal of Aquatic Biology and Fish* Vol. 11, No. 1: 53–58.

Hamilton, S.J. 2004. Review of selenium toxicity in the aquatic food chain. *Science of the Total Environment* 326: 1–31.

Handy, R.D. 2012. FSBI briefing paper: Nanotechnology in Fisheries and Aquaculture. Fisheries Society of the British Isles. 1–29. Available at: www.fsbi.org.uk/assets/brief-nanotechnology-fisheriesaquaculture.pdf

Hoffmann, P.R. & Berry, M.J. 2008. The influence of selenium on immune responses. *Molecular Nutrition and Food Research* 52: 1273–1280.

Islam, M.S. & Tanaka, M. 2004. Optimization of dietary protein requirement for pond reared mahseer (*Tor putitora*) Hamilton (Cypriniformes: Cyprinidae). *Aquaculture Research* 35: 1270–1276.

Jamil, Z. 2013. Effects of Inorganic and Nanoform of Selenium on Growth Performance and Biochemical Indices of Mahseer (*Tor putitora*). Department of Animal Sciences, Quaid-i-Azam University, Islamabad, Pakistan.

Kumar, N., Garg A.K. & Mudgal, V. 2008. Effect of different levels of selenium supplementation on growth rate, nutrient utilization, blood metabolic profile, and immune response in lambs. *Biological Trace Elements Research* 126: 44–566.

Molnár, T., Biró, J., Balogh, k., Mézes, M. & Hancz, C. 2011. Improving the Nutritional Value of Nile Tilapia Fillet by Dietary Selenium Supplementation. *The Israeli Journal of Aquaculture Bamidgeh* 64: 744–750.

Muller, A.S., Pallauf, J. & Most, E. 2002. Parameters of dietary selenium and Vitamin E deficiency in growing rabbits. *Journal of Trace Elements in Medicine and Biology* 16: 47–55.

Pathways of human selenium exposure and poisoning in Enshi, China

Jian-Ming Zhu*
State Key Laboratory of Geological Processes and Mineral Resources, China University of Geosciences, Beijing, China; State Key Laboratory of Environmental Geochemistry, Institute of Geochemistry Chinese Academy of Sciences, Guiyang, China

Hai-Bo Qin
State Key Laboratory of Environmental Geochemistry, Institute of Geochemistry Chinese Academy of Sciences, Guiyang, China

Zhi-Qing Lin
Environmental Sciences Program, Southern Illinois University, Edwardsville, Illinois, USA

Thomas M. Johnson
Department of Geology, University of Illinois at Urbana-Champaign, Urbana, IL, USA

Bao-Shan Zheng
State Key Laboratory of Environmental Geochemistry, Institute of Geochemistry Chinese Academy of Sciences, Guiyang, China

1 INTRODUCTION

Enshi is located in the southwest region of Hubei Province of China. It has high selenium (Se) areas where several incidences of human Se poisoning were observed in the early 1960s. All cases of human Se poisoning were found in the high-Se areas, but not all people living in high-Se areas suffered from Se toxicity. The pathways of human Se toxicity were ascribed to be coal burning, weathering of carbonaceous rocks, climate change, and human activities, etc. However, the reasons for this distribution of human Se poisoning and main pathways are still debated. In recent studies, Se concentrations in soil were comprehensively investigated in Yutangba, Enshi and soil Se concentrations were reported at 4.75 ± 7.43 mg/kg. There were 11 soil samples that contained extremely high Se concentration, ranging from 346 to 2118 mg/kg with an average of 899 ± 548 mg/kg were also found at different cropland sites and discarded stone-coal spoils. Although the micro-topographic features such as slope and lithological variations may result in considerable ranges in Se concentration in local rock and soil samples, although this finding cannot explain such extremely high Se concentration in soils. In this study, Se fractionation and speciation in extremely high Se concentration were studied to evaluate bioavailable and insoluble Se species in these samples, and to better understand the potential pathways of human Se poisoning in Enshi, China.

2 MATERIALS AND METHODS

Top soil samples (0–30 cm depth) were collected from Yutangba village in the southwestern Enshi. The geology and topography has been well described in a previous study by Zhu et al. (2008). All collected samples were transported to the laboratory, then sealed in polyester plastic bags after freeze dry.

Selenium fractions in soils were determined mainly by modifying a sequential extraction procedure (SEP) reported by Kulp and Pratt (2004): water soluble (MQ water) (F1), ligand exchangeable (0.1 M K_2HPO_4-KH_2PO_4) (F2), base soluble (0.1 M NaOH) (F3), elemental (1 M Na_2SO_3) (F4), acetic acid soluble (15% CH_3COOH) (F5), sulfide/selenide (0.5 M $CrCl_2$) (F6), and residual (HNO_3+HF+H_2O_2) (F7).

The elemental Se mineral was characterized using Scanning Electron Microscope (KY1010B-AMRAY and JEOL JSM-840A, equipped with an energy disperse X-ray analyzer (EDX). Selenium k-edge XANES spectra in selected samples were obtained at 1W1B beam line at the Beijing Synchrotron Radiation Facility (BSRF), while others at BL01B1 beam line at SPring-8 (Hyogo, Japan). The XAS data were analyzed using REX2000 software (Rigaku Co. Ltd.) and FEFF 7.02.

3 RESULTS AND DISCUSSION

Results of SEP for soil samples are shown in Figure 1. The organic matter associated Se (F3) is the main form identified in common Se-rich soil samples, while elemental Se (F4) is the dominant form in soils with extremely high Se content (accounting for up to 95% of total Se). Selenium in baked soils mainly exist in the form of organic matter associated Se (F3) and elemental Se (F4).

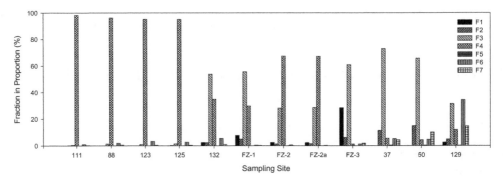

Figure 1. Fractionation of selenium in soils in Enshi, China. F1: water soluble (Milli-Q water); F2: ligand exchangeable (0.1 M K_2HPO_4-KH_2PO_4); F3: base soluble (0.1 M NaOH); F4: elemental Se (1 M Na_2SO_3); F5: acetic acid soluble (15% CH_3COOH): F6: sulfide/selenide (0.5 M $CrCl_2$); F7: residual Se (HNO_3+HF+H_2O_2).

The SEM observations showed that native Se crystals occurred in these samples, are very similar to those reported for Se crystals derived from natural combustion of stone coal and found in abandoned stone coal spoils.

The XANES analysis further indicated that the sample spectrum is very similar with one of elemental Se standard, with a white line at 12655 eV. Thus, the soil Se mineral likely occurs in the form of orthorhombic rather than monoclinic elemental Se in the soils containing high Se concentrations.

Previous studies indicated that elemental Se can be formed via three approaches of microbial reduction, abiotic reduction, and natural burning of stone coal. Elemental Se can be present in the form of nanoparticle, tubular or amorphous, and/or large particles with a good crystal structure. To the best of our knowledge, this study was the first to report that orthorhombic elemental Se observed in agricultural soils in Enshi. The chemical structure of orthorhombic elemental Se is similar with the elemental Se resulted from burning stone coals that were commonly used to make lime or bake local agricultural soils. During the burning process, volatile Se was concentrated and dispensed in the baked soils to improve soil quality. Thus, more Se was introduced into cropping soils through burning stone coals by local farmers. When abandoned stone coal spoils and high Se soils were reclaimed for cropland, elemental Se was oxidized and transformed into water soluble selenate or selenite, which can be further accumulated in local food chain. Therefore, Se-rich carbonaceous rocks provide the necessary condition for inducing local Se poisoning incidences. Mining stone coals for fuel or fertilizer will also lead to chemical oxidation and the transport of elemental Se in the environment.

4 CONCLUSIONS

This study provides new direct evidence and better understanding on effective environmental management of Se pollution that were resulted from inappropriate farming practices, such as burning stone coals for crop fertilizer to avoid possible Se poisoning in local residents in Enshi.

REFERENCES

Kulp, T.R. & Pratt, L.M. 2004. Speciation and weathering of selenium in Upper Cretaceous chalk and shale from South Dakota and Wyoming, USA. *Geochim. Cosmochim. Acta*, 68: 3687–3701.

Lenz, M. & Lens, P.N.L. 2009. The essential toxin: The changing perception of selenium in environmental sciences. *Sci. Total. Environ.* 407: 3620–3633.

Matamoros-Veloza, A., Newton, R.J. & Benning, L.G. 2011. What controls selenium release during shale weathering. *Appl. Geochem.* 26: S222–S226.

Pickering, I.J., Brown, G.E. & Tokunaga, T.K. 1995. Quantitative speciation of selenium in soil using x-ray absorption spectroscopy. *Environ. Sci. Technol.* 29: 2456–2459.

Qin, H.B., Zhu, J.M. & Su, H. 2012. Selenium fractions in organic matter from Se-rich soils and weathered stone coal in selenosis areas of China. *Chemosphere* 86: 626–633.

Wright, M.T., Parker, D.R. & Amrhein, C. 2003. Critical evaluation of the ability of sequential extraction procedures to quantify discrete forms of selenium in sediments and soils. *Environ. Sci. Technol.* 37: 4709–4716.

Zhu, J.M., Zuo, W., Liang, X. B., Li, S.H. & Zheng, B.S. 2004, Occurrence of native selenium in Yutangba and its environmental implications. Applied Geochemistry, 19(3): 461–467.

Zhu, J.M., Wang, N., Li, S.H., Li, L. & Su, H.C. 2008. Distribution and transport of selenium in Yutangba, China: Impact of human activities. *Sci. Total. Environ.* 392: 252–261.

Zhu, J.M., Johnson, T.M., Finkelman, R.B., Zheng, B.S., Sykorova, I. & Pesek, J. 2012. The occurrence and origin of selenium minerals in Se-rich stone coals, spoils and their adjacent soils in Yutangba, China. Chemical Geology, 330-331: 27–38

Selenium status in Iran: A soil and human health point of view

B. Atarodi* & A. Fotovat
Department of Soil Science, Ferdowsi University of Mashhad, Mashhad, Iran

1 INTRODUCTION

Selenium (Se) has been accepted as an essential micronutrient for animals and humans. Since insufficient and excess Se have detrimental consequences, this element is often defined as a "two-edged sword." Its biomagnification in the food chain results in toxicity, while its deficiency results from low concentrations in food sources (Farzin et al., 2009). In Iran, researchers have concluded that excessive Se in soil and food might be responsible for the spread of some kinds of cancers based upon soil Se concentrations (Semnani et al., 2010). Selenium deficiency might play a role in the high incidence of gastric cancer in Ardabil province (Nouraie, 2004).

2 SELENIUM IN SOIL AND CROP

Soils with Se concentrations of 0.1 to 0.6 mg/kg are generally classified as Se-deficient soils (Yläranta, 1983). The average total Se-soil concentration in the Center of Iran was 0.23 mg/kg (Nazemi et al., 2012). This value was greater than that in the North and South of Iran where soil Se concentrations were considered Se-deficient. In some countries, such as Canada and some European countries, Se fertilization was recommended as soil or foliar application (Lyons et al., 2003), while there are few published reports that suggest the same of Iran. Consequently, the dense cultivation and the continuous harvest of crops for many years without any Se applications has reduced the level of Se in soils.

Wheat is an important source of bioavailable Se and wheat bread is the staple food of the people of Iran. Other dietary sources like fish and meat contribute only small amount of Se to the total intake of Se for the Iranian people (Farzin et al., 2009). Thus, the wheat Se concentration may be a good guide to know the Se intake by people of Iran. Concentrations of Se for most of the world's wheat gains has been reported to be between 0.02–0.60 mg/kg (Alfthan & Neve, 1996), although, this value was reported to be as high as 2.0 mg/kg for wheat produced in North or South Dakota, USA (Combs, 2001; Lyons, 2003).

In Iran, Se concentrations of wheat grain in Kerman and Rafsanjan were about 0.36 and 0.3 mg/kg, respectively (Nazemi et al., 2012). The average Se concentration could be considered to be 0.5 mg/kg. We note that a substantial amount of nutrients including Se are removed during milling of wheat to white flour and baking into bread. For example, in the process of transforming wheat into white flour, an average reduction of 27% in Se concentration has been reported (Ahmad et al., 1994). Moreover, other studies found that baking procedures can result in Se losses by about 15% or more (Olson & Palmer, 1984). Additionally, the bioavailability of wheat Se is less than 100%. Lyons et al. (2004) believed that to increase daily Se intake, it is necessary to have breakfast cereal, flour, pasta, or bread containing Se concentrations of at least 1 mg/kg, which is more than Iran's 0.5 mg/kg. Hence, Se deficiency is more likely to be observed.

3 SELENIUM DAILY DIETARY INTAKE

Data released by the Ministry of Agriculture of Iran reported that Iranian's consumed on average about half a kilogram of wheat per person per day. Considering this quantity and Se concentration of wheat produced in Iran, as well as the loss of Se during milling and baking, the Se intake in Iran is estimated to be about 120 μg per person per day. However, other studies calculated the Iranian daily Se intake to be 62 μg in female and 67 μg in male (Safaralizade et al., 2005). Clearly, these values are above the recommended daily dietary Se intake of 55 μg per day per adult.

Some data on cancer prevention would suggest an intake rate of 200–300 μg Se per day per person (Clark et al., 1998) and the World Health Organization (1996) recommended an upper safe limit of 400 μg Se per day per adult. Hence, it is unlikely that the rate of daily Se intake in Iran can be a serious concern for Se toxicity and promoting the occurrence of cancer.

4 BLOOD SERUM SELENIUM CONCENTRATION

Blood serum or plasma Se concentration is another important key factor for determining Se intake and its status in the human body. According to some research studies, serum Se concentration of a healthy adult in Australia, Canada, and western states of USA are 92, 146, and 198 μg/L, respectively (Combs,

2001). In Iran, this value was reported to be 82, 119, and 123 µg/L in Ardabil, Kerman, and Mazandaran provinces, respectively (Nouraie, 2004). Also this value was reported 84.3 µg/L for healthy children (age 1–16) living in Tehran (Safaralizade et al., 2005). Since the baseline level of blood serum Se concentration is approximately 120 µg/L (Combs, 2001), the comparison of Iranian blood serum Se with the baseline level and with its status in other countries indicates that incidence of cancer might not be related to the excessive amount of Se in Iranian blood serum or plasma. On the contrary, it seems that Se deficiency is an issue of concern that should be addressed. In this regard, Keshteli (2009) reported that Se deficiency is among the contributors of goiter in Isfahan. In the meantime, findings of Hosseini Nezhad et al. (2015) suggested that lower serum Se might have some association with the risk of gastric non-cardia cancer (GNCC) in Iranian patients

5 CONCLUSIONS

The intake of Se by Iranian people is not high enough to raise concern for toxicity. In contrast, we concluded that the issue of Se deficiency may be a potential concern. Therefore, further studies on Se status in soils, crops, and blood serum are recommended in Iran.

REFERENCES

Ahmad, S., Waheed, S., Mannan, A., Fatima, I. & Qureshi, I. 1994. Evaluation of trace elements in wheat and wheat by-products. *J. of Association of Official Analytical Chemist International* 77: 11–18.

Alfthan, G. & Neve, J. 1996. Selenium intakes and plasma selenium levels in various populations. In: Kumpulainen, J. & Salonen, J. (eds.), *Natural Antioxidants and Food Quality in Atherosclerosis and Cancer Prevention*. Cambridge: Royal Society of Chemistry. pp. 161–167.

Clark, L., Dalkin, B., Krongrad, A., Combs G., Turnbull, B., Slate, E., et al. 1998. Decreased incidence of prostate cancer with selenium supplementation: Results of a double-blind cancer prevention trial. *British Journal of Urology* 81: 730–734.

Combs, G. 2001. Selenium in global food systems. *British J. of Nutrition*. 85: 517–547.

Farzin, L., Moassesi, M., Sajadi, F., Amiri, M. & Shams, H. 2009. Serum levels of antioxidants (Zn, Cu, Se) in healthy volunteers living in Tehran. *Biol Trace Elem Res* 129:36–45.

Hosseini Nezhad, Z., Darvish Moghaddam, S., Zahedi, M., Hayatbakhsh, M., Sharififar, F., Ebrahimi Meimand, F., et al. 2015. Serum selenium level in patients with gastric non-cardia cancer and functional dyspepsia. *Iran J. Med. Sci.* 40 (3): 214–218.

Keshteli, A., Hashemipour, M., Siavash, M. & Amini, M. 2009. Selenium deficiency as a possible contributor of goiter in schoolchildren of Isfahan, Iran. *Biol Trace Elem Res*. 129: 70–77

Lyons, G., Lewis, J., Lorimer, M., Holloway, R.E., Brace, D.M., Stangoulis, J.C.R., et al. 2004. High-selenium wheat: agronomic biofortification strategies to improve human nutrition. *Food, Agriculture & Environment* 2 (1): 171–178.

Lyons, G., Stangoulis, J. & Graham, R. 2003. High-selenium wheat: biofortification for better health. *Nutrition Research Reviews*. 16: 45–60.

Nazemi, l., Nazmara, S., Eshraghyan, M., Nasseri, S., Djafarian, K., Younesian, M., et al. 2012. Selenium status in soil, water and essential crops of Iran. *Iranian Journal of Environmental Health Sciences & Engineering* 9(11): 1–8

Nouraie, M., Pourshams, A., Kamangar, F., Sotoudeh, M., Derakhshan, M., Akbari, M., et al. 2004. Ecologic study of serum selenium and upper gastrointestinal cancers in Iran. *World J. Gastroenterol*. 10 (17): 2544–2546.

Olson, O. & Palmer, I. 1984. Selenium in foods purchased or produced in South Dakota. *J. Food Sci.* 49: 446–452.

Safaralizadeh, R., Kardar, G., Pourpak, Z., Moin, M., Zare, A. & Teimourian, S. 2005. Serum concentration of Selenium in healthy individuals living in Tehran. N*utrition Journal* **4**(32): 1–4.

Semnani, S., Roshandel, G., Keshtkar, A., Zendehbad, A., Rahimzadeh, H., Besharat, S., et al. 2010. Relationship between soil selenium level and esophageal cancer: An ecological study in Golestan province of Iran. *J. of Gorgan University of Medical Sciences* 12 (3): 51–56.

World Health Organization 1996. Selenium. pp. 105–122. In: *Trace Elements in Human Nutrition and Health*, Geneva WHO.

Yläranta, T. 1983. Effect of added selenite and selenate on the selenium content of Italian rye grass (*Lolium multiflorum*) in different soils. *Ann. Agric. Fenniae* 22: 139–151.

Impact of high selenium exposure on organ function & biochemical profile of the rural population living in seleniferous soils in Punjab, India

Rajinder Chawla*, Rinchu Loomba, Rohit J. Chaudhary & Shavinder Singh
Christian Medical College & Hospital, Ludhiana (Pb), India

K.S. Dhillon
Punjab Agricultural University, Ludhiana (Pb), India

1 INTRODUCTION

Rural areas in the three districts of Punjab in India (lying in the Shivalik foothills) have been found to have soil rich in selenium (Se). About 2160 hectare area is seleniferous and is populated by about 10,000 inhabitants. A number of studies have been conducted in the area to identify the high Se pockets and to study the effect of Se toxicity in plants and animals. Selenium concentrations in these pockets have been reported to be as high as 65 times over non-seleniferous areas. However, few reports are available on selenium's impact on human health from these areas, except one by Hira et al. (2003). The present study was, therefore, undertaken to assess the impact of high Se exposure to the rural population living in these areas.

2 MATERIALS AND METHODS

The tissue samples viz. blood, hair and nail clippings were collected from the human subjects and analyzed for complete blood cell counts, liver function, pancreatic function, renal function and thyroid function tests. Selenium concentration was estimated in the hair, nail clippings and serum samples by atomic absorption spectrophotometry at National Dairy Research Institute (NDRI), Karnal, India. Data thus generated was analyzed using epi-info 7.0 statistical software.

Control group: Fifty healthy volunteers from a village in non-seleniferous areas of Punjab served as control. The same samples (as already described) were collected, processed and analyzed similarly.

3 RESULTS AND DISCUSSION

A total of 690 human subjects and 50 healthy controls were recruited for this study. The laborers and farmers were equally distributed in the study group and 60% were females. Overt symptoms of Se toxicity were observed in 43% of the subjects, 42.2% (290) had dystrophic changes in nails, 40% (279) hair loss and 4.22% (29) had garlicky odor in their breath. Similar clinical presentations have been first reported in China by Yang et al. (1983).

Concentrations of Se (mean ± SD) in the hair samples were $50.9 \pm 58.0\,\mu g/g$ (range, 8.7–583.9) in the study group, compared to $22.5 \pm 10.7\,\mu g/g$ (range, 8.4–58.5) in the control group, showing a significant difference ($p < 0.01$). The corresponding Se concentrations in the nail clippings in the study and control groups were $154.0 \pm 91.5\,\mu g/g$ (range, 21.5–819.6) and $117.4 \pm 49.8\,\mu g/g$ (range, 51.8–267.5), respectively, which is also significantly different ($p = 0.013$) (Table 1). Huang et al. (2013) reported higher intake of Se in Enshi region of China and suggested that the high Se intake could place the local residents to potential risk of selenosis.

The impaired organ function tests were observed in significantly higher proportion of subjects in the study group compared to that in the control group (Table 2). Available epidemiologic evidence suggests increased Se intake in healthy individuals possibly increased the risk of diseases and disorders (Vinceti et al., 2013). Nuttall (2006) reviewed the Se poisoning and Se levels in human tissues, including liver, kidney and heart in poisoning cases. Liver and kidney are the target organs for Se accumulation and could lead to compromised organ function.

Table 1. Selenium concentrations in hair and nail samples in study and control groups.

Location	Hair Se (μg/g)	Nail Se (μg/g)
Simbly	49.7 ± 27.7	136.5 ± 100.9
Barwa	54.1 ± 55.2	192.3 ± 101.9
Mahindpr	38.1 ± 7.3	191.3 ± 53.5
Naajarpur	106.9 ± 133.4	153.0 ± 99.4
Rakran	28.1 ± 9.9	129.8 ± 50.0
Bhaguran	29.4 ± 29.1	120.7 ± 36.9
Jaadli	33.1 ± 14.4	117.7 ± 36.4
All Cases	50.9 ± 58.0	154.0 ± 91.5†
Control	22.5 ± 10.7	117.4 ± 49.8†

†$p < 0.01$

Table 2. Proportion of study subjects with deranged organ function.

Location	Abnormal organ function tests (proportion of subjects)		
	Liver	Kidneys	Pancreas
Simbly	8 (5.6%)	35 (24.6%)	23 (16.2%)
Barwa	20 (12.7%)	27 (17.1%)	34 (21.5%)
Mahindpr	10 (8.0%)	4 (3.2%)	14 (11.1%)
Naajarpur	5 (11.6%)	14 (32.5%)	8 (18.6%)
Rakran	7 (12.5%)	15 (26.8%)	7 (12.5%)
Bhaguran	1 (1.8%)	2 (3.6%)	9 (16.4%)
Jaadli	7 (6.9%)	3 (3.0%)	12 (11.9%)
All Cases	58 (8.5%)	100 (14.7%)	107 (15.7%)
Control	3 (6.0%)	5 (10.0%)	2 (4.0%)

High Se exposure might also be imparting some beneficial effects to the exposed population. The total antioxidant capacity (TAC) in the study group was higher than the control group. The very high TAC ($940.8 \pm 202.1\,\mu M/L$) might contribute to the resistant to infectious diseases, as well as susceptibility to oncogenesis. The serum TAC correlated with the Se concentrations in hair and nail samples. The incidence of cancer appears to be lower in the study population since we did not notice a single case of malignancy during our visits to the area. The incidence of cancer in the study population shall be analysed after completion of the study at the end of 2015. However, Vinceti et al. (2013) reviewed the results from randomized control trials on Se intake and concluded that the Se might not play an important role in reducing the risk of cancer.

The study population has been aware of the high Se content of their soil and has, over the past decade, learned to minimize Se intake by crop diversification (Dhillon & Dhillon, 2009b). In spite of the awareness, high Se in human tissues and prevalence of organ dysfunction indicate chronic effects of Se accumulation in the target tissues.

4 CONCLUSIONS

Chronic exposure to high Se through the soil-plant-water continuum can place the human population at risk of developing impaired organ function. The increased Se levels in hair and nails and impaired organ function, emphasize the need to educate the affected population on the appropriate cultivation practices in the areas rich in Se.

ACKNOWLEDGEMENT

The study was supported by the research grant from Department of Science and Technology (DST), India.

REFERENCES

Dhillon, S.K. & Dhillon, K.S. 2009a. Characterization and management of seleniferous soils of Punjab. Research Bulletin No. 1/2009. Punjab Agricultural University, Ludhiana, INDIA

Dhillon, S.K. & Dhillon, K.S. 2009b. Phytoremediation of selenium contaminated soils: the efficiency of different cropping systems. Soil Use and Management 25: 441–453

Hira, C.K., Partal, K. & Dhillon, K.S. (2003). Dietary selenium intake by men and women in high and low selenium areas of Punjab. Public Health Nutr. 7: 39–43

Huang, Y., Wang, Q., Gao, J., Lin, Z.Q., Bañuelos G.S., Yuan L. et al. 2013. Daily dietary selenium intake in a high selenium area of Enshi, China. Nutrients, 5: 700–710

Nuttall, K.L. 2006. Evaluating selenium poisoning. Annals of Clinical & Laboratory Science 36 (4): 409–420

Vinceti, M., Crespi, C.M., Malagoli, C., Giovane, C.D. & Krogh, V. 2013. Friend or foe? The current epidemiologic evidence on selenium and human cancer risk. Journal of Environmental Science and Health, Part C, 31:305–341

Yang, G., Wang S., Zhou R. & Sun S. 1983. Endemic selenium intoxication of humans in China. Am J Clin Nutr., 37: 872–881

Keshan disease and Kaschin-Beck disease in China: Is there still selenium deficiency?

Dacheng Wang*, Yan Liu & Dan Liu
Chemical Synthesis and Pollution Control Key Laboratory of Sichuan Province, College of Chemistry and Chemical Engineering, China West Normal University, Nanchong, China

1 INTRODUCTION

Keshan disease and Kaschin-Beck disease were previously found in about 15% of 2000 counties in China, where selenium (Se) was deficient. Due to rapid economic growth in China, patients affected by these two diseases are now only sparsely distributed among a few locations in former disease-affected areas. What caused the reduction of the two Se-deficiency-related diseases? Responsible factors may include improved Se intake from food growth in soil amended with Se, increased food diversity due to increased income, or enhanced Se-supplementation to the population. The Chuxiong area in Yunnan Province, and the Yongshou area in Shaanxi Province are a few areas where residents were severely affected by Keshan disease and Kaschin-Beck disease in the past. Many studies were conducted in the early 1980s. Thus, a comparative study at these two areas provided a clear picture of the evolution of Se deficiency through food chains in the last 30 years.

2 MATERIALS AND METHODS

Maize kernels and soil samples (0–20 cm) were collected in duplicate in 12 Keshan/Kaschin-Beck disease-affected villages in each study area. Children's hair samples were collected, including 15 boys and 15 girls, along with rice (Yunnan) or wheat (Shaanxi) from each child's house. Air-dried soil samples were analyzed for total Se, water soluble Se, selenite, selenate, and humic acid-bound Se (Wang et al., 1994). Concentrations of Se in acid digestion solutions were analyzed using 2,3-diaminonaphthalene (DAN) fluorometry.

3 RESULTS AND DISCUSSION

3.1 *The bioavailability of Se in soil*

In Chuxiong area soils only one of the 12 study villages was considered to be Se deficiency. The soil Se concentration was 87.6 µg/kg, increased from 64 µg/kg in 1984. However, water soluble Se concentration were lower than that reported in 1984. Similar results were also observed in Yongshou where shows Se deficient soils.

Water soluble Se in soil has a better correlation with plant uptake than with total Se. Only selenite and selenate in the water soluble fraction in soil are available for plant uptake. Both selenite and seleniate accounted for 28.8% of the water soluble Se (table 2), and 0.19% of the total Se in the soil in Chuxiong, while 43.9% of the water soluble and 0.81% of the total Se in soil in Yongshou.

3.2 *Daily intake of Se from cereals*

The Se concentration in rice was 19.6 µg/kg (15.1–28.6 µg/kg) Amongst 179 rice samples 95% of them were Se deficient, even though the rice Se concentrations increased from 15.0 µg/kg in 1984. The Se content in maize was nearly the same as 28 years ago, and was still Se deficient. In Yongshou of Shaanxi province, the staple food is wheat. There was a clear increase of Se concentrations compared to 1981 level for both wheat and maize, but their Se concentrations are deficient. Estimated Se daily intake for local population from cereals in these two studied areas was

Table 1. Concentrations of Se in soils (µg/kg).

Location	Year	Total Se	Water soluble
Chuxiong	2012	87.6	0.583
	1984[†]	64.0	0.795
Yongshou	2013	86.5	1.60
	1981[‡]	80.0	1.60

[†] Ecological Environment Group (1988)
[‡] Ecological Environmental Research on Kaschin-Beck's Disease in Yongshou County (1985)

Table 2. Soil selenium speciation (µg/kg).

Location	Selenite	Selenate	Humic Se
Chuxiong	0.082	0.086	0.133
Yongshou	0.816	0.240	0.462

Table 3. Concentrations of Se in cereals (μg/kg).

Location	Year	Rice	Wheat	Maize
Chuxiong	2012	19.6		15.8
	1984[†]	15.0		17.0
Yongshou	2013		20.8	16.0
	1981[‡]		4.1	3.6

[†]Ecological Environment Group (1988)
[‡]Ecological Environmental Research on Kaschin-Beck's Disease in Yongshou County (1985)

Table 4. Concentrations of Se in hair (mg/kg).

Location	Year	Mean	Boy	Girl
Chuxiong	2012	0.274	0.286	0.261
	1984[†]	0.160		
Yongshou	2013	0.299	0.296	0.301
	1981[‡]	0.074		

[†]Ecological Environment Group (1988)
[‡]Ecological Environmental Research on Kaschin-Beck's Disease in Yongshou County (1985)

<20 μg, which makes the population susceptible to Keshan or Kaschin-Beck's disease.

3.3 Se nutritional status of the local population

The Se concentration in children's hair was 0.274 mg/kg on average (0.208–0.827 mg/kg, n = 179, see Table 4) in Chuxiong, and 0.299 mg/kg (0.044–0.635 mg/kg $n = 149$) in Yongshou. Both increased significantly from the results reported in 1980s. In chuxiong and Yongshou, 70.6% and 96.6% of the children were adequate on Se nutrition, respectively. The increase of Se nutrition in Chuxiong could only be attributed to more diverse food sources other than locally-produced products, while the improvement of Se status in population of Yongshou was mainly due to the Se supplemented table salt enforced in Shaanxi province from December 1997 to July 2012 (3–5 mg/kg, or equivalent to 20 μg a day for adult).

3.4 Selenium concentrations in the Keshan disease environment in Chuxiong

The level of Se in soil, rice, maize, and children's hair decreased from none affected to severely affected by Keshan disease in four villages along a tributary downstream in Chuxiong (Table 5), which is similar to the results reported in 1984. This observation shows that there was no change on Se deficiency in the environment but the human Se nutritional status improved to adequate level. This is likely the result from the availability of more diverse food sources other than local agricultural products. Due to the fact that 1/3 children are in marginal Se deficient nutritional status, a monitoring program and efficient measures to increase Se intake should be enforced to prevent Keshan disease and other Se related diseases in Chuxiong. Luopingguan was not affected; Xindi was lightly affected; Hongdoushu & Longwumiao were severely affected.

Table 5. Selenium concentrations (μg/kg) of different environmental indicators in relation to the occurrence of Keshan disease.

	Not affected	Lightly affected	Severely affected	Severely affected
Soil Se	85.0	74.5	62.7	52.6
Soil Se(VI)	0.165	0.098	0.089	0.079
Rice	22.3	17.6	16.4	16.5
Maize	16.2	15.7	14.6	16.4
Hair	297	264	263	261

4 CONCLUSIONS

The total Se and water soluble Se in soil in Keshan disease area in Chuxiong, Yunnan, and in Kaschin-Beck's disease area in Yongshou, Shaanxi remained deficient compared to the 1980s. Less than 1% of the soil total Se was plant bioavailable. Most of rice samples (95%) and all maize samples were Se deficient in Chuxiong, but only 27.8% of 180 children were marginal deficient, suggesting additional Se was provided from other food production regions. All the wheat and maize samples from Yongshou were Se deficient (<25 μg/kg). The adequate Se nutritional status in the area was a resulted from Se-fortified salt. A significant decrease on the Se intake in Yongshou can be expected if Se fortified salt is not used by the population.

REFERENCES

Ecological Environmental Research on Kaschin-Beck Disease in Yongshou County. 1985. Yongshou Scientific Survey Group of Kaschin-Beck's Disease. Acta Scientiae Circumstantiae. 5(1): 1–19.

Wang, D.C., Alfthan, G. & Aro, A. 1994. Determination of total and dissolved selenium species in natural water samples using fluorometry. Environmental Science & Technology, 28: 383–387.

Ecological Environment Group. 1988. The ecological environment study on Keshan disease area in Chuxiong. In: Collected Works on Scientific Survey of Keshan Disease in Chuxiong (1984–1986). People's Health Press, pp. 48–83.

Tan, J.A., Li, R.B., Zheng, D.X., Zhu, Z.Y., Hou, S.F., Wang, W.Y., et al. 1987. Selenium ecological chemicogeography and endemic Keshan disease and Kaschin-Beck disease in China. In G.F. Combs, J.E. Spajjholz, O.A. Levander, J.E. Oldfield (eds), Selenium in Biology and Medicine; Proceedings of the Third International Symposium on Selenium in Biology and Medicine. Beijing, May 27–June 1, 1984. New York: Van Nostrand Reinhold Company.

Influence of canola oil, vitamin E and selenium on cattle meat quality and its effects on nutrition and health of humans

M.A. Zanetti*, L.B. Correa, A. Saran Netto, J.A. Cunha & R.S.S. Santana
Department of Animal Science, Faculty of Animal Science & Food Engineering, University of São Paulo, Pirassununga, Brazil

S.M.F. Cozzolino
Department of Food Sciences, Faculty of Pharmaceutical Sciences, University of São Paulo, Sao Paulo, Brazil

1 INTRODUCTION

Healthy eating is a challenge for modern humans and becomes a main issue for the majority of world's population. Beef contains important nutrients for human health and nutrition. High fat levels in human diet is often associated with increased blood cholesterol, cardiovascular disease and atherosclerosis, due to the saturated fatty acids and high cholesterol contents. However, some studies have demonstrated that the use of different fat sources in the diet of ruminants can change the fatty acid profile of meat.

Canola oil is considered as one of the most healthy vegetable oils because it has high contents of Omega-3 (reduces triglycerides and controls atherosclerosis), vitamin E (an antioxidant that reduces free radicals), mono-unsaturated fats (reduces LDL), and lower saturated fat, which helps control the level of cholesterol. Therefore, decreasing saturated fat levels in meat for human consumption would be beneficial for both marketing and public health. Thus, the objectives of this research were to determine the effect of canola oil, organic selenium (Se), and vitamin E in feedlot steers ration on meat quality and on human serum Se and cholesterol levels in the people who consumed the Se-biofortified meat.

2 MATERIALS AND METHODS

Forty eight Nellore steers were divided into four groups and assigned for four treatments (i.e., twelve animals/replicates per treatment), and placed in individual pens. Treatments were as follows: CK (control); CK + antioxidants [2.5 mg of organic Se per kg of dry matter (DM) + 1000 IU of vitamin E/day]; Oil (3% of canola oil in DM diet); Oil+antioxidants (3% of canola in DM diet + 2.5 mg of organic Se per kg of DM diet + 1000 IU of vitamin E/day). The diets were formulated in accordance with guidelines established by NRC (1999). The experiment lasted for 82 days. The steer blood was sampled for vitamin, Se, and cholesterol analyses in 0, 4, 8, and 12 weeks, respectively.

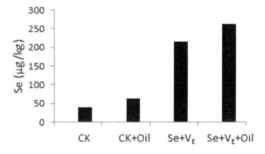

Figure 1. Concentrations (mg/kg) of selenium in meat. CK: control; VE: Vitamin E; Oil: Canola oil.

At the end of the trial, concentrations of vitamin E and Se in the meat were determined prior to the meat was provided to humans for a 90-day experimental time period. The effects of Se and vitamin E-enriched meat on Se accumulation in human blood serum were also explored. Statistical analysis was conducted based on a completely randomized design using the mixed procedure (SAS, 2000).

3 RESULTS AND DISCUSSION

The Se supplementation in steers ration significantly ($p < 0.01$) increased the Se concentrations in serum and meat ($p < 0.001$). In the supplemented steer meat, Se increased from 39.3 µg/kg in the control group to 262 µg/kg in the group/treatment that received Se, vitamin E, and canola oil in the ration (Fig. 1). Organic Se increased the Se level in animal products (Suray, 2006). At the end of the trial, the blood Se concentration in the steers were 42 µg/mL for control and the canola oil treatment, 105 µg/mL for the antioxidant treatment, and 103 µg/mL for the antioxidant plus canola oil treatment. The meat cholesterol decreased significantly ($p < 0.5$) in the animals treated with Se (Fig. 2). With high Se levels in the Bulls diet, the cholesterol concentration decreased in meat, but not in serum (Saran Netto et al., 2014). High Se contents

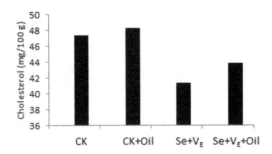

Figure 2. Cholesterol levels in meat (mg/100 g). CK: control; VE: Vitamin E; Oil: Canola oil.

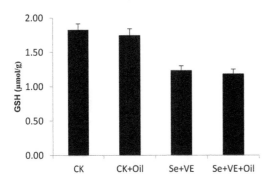

Figure 3. GSH levels in liver (μmol/g). CK: control; VE: Vitamin E; Oil: Canola oil.

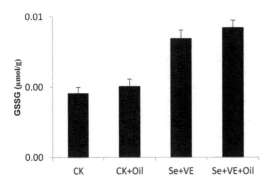

Figure 4. GSSG levels in liver (μmol/g). CK: control; VE: Vitamin E; Oil: Canola oil.

in the meat increased the human serum Se concentration significantly ($p < 0.05$), but decreased the level of serum cholesterol (data not shown).

In liver tissues concentrations of GSH decreased but GSSG increased significantly ($p < 0.5$) in the Se treatments, compared to the control (Figs. 3 & 4). When this occurred, there was negative effects on the enzyme HMG-CoA reductase that affects cholesterol synthesis (Correa et al., 2012). The HMG-CoA reductase needs thiols like GSH to become biochemically active (Armstrong et al., 2001).

4 CONCLUSIONS

We concluded that Se supplementation during an 82-day experiment effectively and significantly increased the Se concentration in the steers serum and meat, but decreased the concentrations of cholesterol in the serum and meat. People who consumed Se-biofortified meat had higher concentrations of serum Se and lower cholesterol levels.

REFERENCES

Armstrong, T.A., Spears, J.W., Engle See, M.T.E., & See, M.T. 2001. Effect of pharmacological concentrations of dietary copper on lipid and cholesterol metabolism in pigs. *Nutrition Research*, 2: 1299–1308.

Correa, L.B., Zanetti, M.A., Del Claro, G.R., Melo, M.P., Rosa, A.F., & Saran Netto, A. 2012. Effects of supplementation of two sources and two levels of copper on lipid metabolism in Nellore beef cattle. *Meat Science*, 91: 466–471.

National Research Council. 1996. Nutrient Requirements of Beef Cattle, 7th edition. *National Academy Press*, Washington, DC.

Saran Netto, A., Zanetti, M.A., Del Claro, G.R., Melo, M.P., Vilela, F.G., & Correa, L.B. 2014. Effects of copper and selenium supplementation on performance and lipid metabolism in confined Brangus bulls. *Asian Australas. J Anim. Sci.*, 27(4): 488–494.

SAS. 2000. Institute SAS/STAT. Guide for Personal Computers. Cary: SAS Institute Inc.

Suray, P.F. 2006. *Selenium in Nutrition and Health*. 1st edition. United Kingdom: Nottingham University Press.

Selenium content of food and estimation of dietary intake in Xi'an, China

Z.W. Cui, R. Wang, J. Huang, D.L. Liang* & Z.H. Wang
College of Natural Resources and environment, Northwest A&F University, Yangling, Shaanxi, China
Key Laboratory of Plant Nutrition and the Agri-environment in Northwest China, Ministry of Agriculture, Yangling, Shaanxi, China

1 INTRODUCTION

Selenium (Se) is an essential micronutrient with a narrow range between toxicity and deficiency for humans and other mammals, which has attracted people's attention to explore the harmful and benefits facets of Se. When Se intake is lower than 10 μg/day or exceeds 600 μg/day, it can cause negative effects on human health (Mahapatra et al., 2001). An Se-dependent enzyme by which human tissues can be protected against oxidative damage is called glutathione peroxidase (GPx); the first identified selenoprotein with Se being the part of its active center (Rotruck et al., 1973; Thiry et al., 2012).

Diet Se content is dependent on the local soil Se concentrations which can influence the Se concentration of plants and animals (Navarro-Alarcon & Cabrera-Vique, 2008). Selenium concentrations in humans largely depend on the Se content of food and beverages human consumed (Thiry et al., 2012). The recommended dietary Se allowance for both males and females is 55 μg per day based on the maximization of glutathione enzyme activity and the tolerable upper intake of 400 μg per day in the U.S. (Food and Nutrition Board, 2000). In some Se-rich countries such as Canada and the United States, daily Se intake was estimated to be as high as 224 μg/day (Gissel-Nielsen, 1998) and 160 μg/day (Longnecker et al., 1991), respectively, which is much higher than data collected in the Keshan area of China (3–11 μg/day) and Turkey (30 μg/day) (Dumont et al., 2006).

In previous studies, the investigation of Se content at Se rich and deficient areas has been determined, however, there is a lack of data based on a systemic investigation of a Chinese city. In this study, we measured the Se concentrations of food products from Xi'an for the following purposes: (1) to estimate daily Se dietary intake of residents in Xi'an and investigate the key sources of Se in selected food products; (2) to suggest strategies for city inhabitants to maintain and improve Se intake at healthy levels.

2 MATERIALS AND METHODS

Xi'an, the largest city in the northwest of China, has a population of 8.5 million. This region belong to the zone of the continental monsoon climate with a precipitation of 522 to 720 mm. The difference in altitude of this area is the greatest in China, widely ranged from 345 m to 2800 m.

Food samples were collected on 2015 from framers' markets and supermarkets in a wide range of geographical locations within the urban area of Xi'an. This included 48 kinds of foodstuffs containing grain ($n = 68$), vegetable ($n = 78$), fruit ($n = 15$), meat ($n = 15$), egg ($n = 16$), and milk ($n = 23$). After collecting these food products, gains were washed using deionized water, oven-dried at 60°C to a constant weight, and ground into fine powders using a stainless steel mill. Vegetables, fruit, and eggs were freshly shredded using a broke beater. Meat samples were rinsed three times with deionized water and blended using a pulp refiner. Concentrations of Se were measured based on dry weight for grains and fresh weight for vegetables, fruit, eggs and meat. All samples were acid-digested with HNO_3-$HClO_4$ (4:1, v/v), and prepared samples were analyzed using AFS-930.

The daily Se dietary of intake was calculated from the formula below:

$$PDI = \sum (C_{Se}^i \times IR)$$

where PDI = probable daily intake of Se; C_{Se}^i = Se concentration in food; IR = intake rate of each food.

3 RESULTS AND DISCUSSION

3.1 Selenium in samples

Concentrations of Se in each kind of food obtained from Xi'an is shown in Table 1 with the range and standard deviation. The results are expressed in microgram Se per gram. The highest Se concentration was measured in the egg, with a mean Se concentration of 0.122 ± 0.006 μg/g and a maximum concentration of 0.171 μg/g and a minimum concentration of 0.086 μg/g, similar to previously reported values (Pappa et al., 2006). Elevated Se concentrations were found in grain and meat products when compared to the other dairy products that contained similar concentrations of Se: vegetable, fruit and milk. Milk was shown to be a poor source of Se with a mean concentration of 0.014 μg/g.

Table 1. Se concentrations of each kind of food.

Dietary (n)	Product	Max. (μg/g)	Min.	Mean ± SD
Grain (68)	Rice	0.2191	0.0019	0.0981 ± 0.0558
	Flour	1.4953	0.0027	0.1454 ± 0.3050
Vegetable (78)		0.0357	0.0008	0.0125 ± 0.0082
Fruit (14)		0.0486	0.0091	0.0213 ± 0.0145
Meat (15)		0.0699	0.0143	0.0454 ± 0.0284
Egg (16)		0.1265	0.1175	0.1120 ± 0.0064
Milk		0.0163	0.0122	0.0143 ± 0.0029

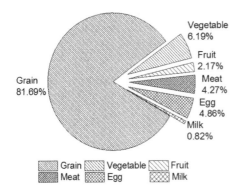

Figure 1. Food contribution to daily Se intake.

Due to the low protein fraction, Se concentrations in vegetable and fruit were very low, while meat and egg as protein-rich foods showed higher Se concentrations. (Navarro-Alarcon & Cabrera-Vique, 2008). There were four kinds of meat collected from supermarket; Se data for pork and chicken are significant higher than beef and mutton, which can be explained by the different Se concentrations in feeds consumed by animals.

3.2 Estimation of daily dietary intake

The annual consumption percentage of dairy products was provided by Zhang (2003). The most consumed food is grain (64.7%), followed by vegetable (21.7%), fruit (5.2%), meat (4.1%), egg (1.8%) and milk (2.5%). The calculated PDI is 26 μg/day. We noted that the greatest proportion of daily Se intake is through grain consumption (82%), while vegetable, fruit, meat, egg and milk contributed 6%, 2%, 4%, 5% and 1% to daily dietary Se intake for residents, respectively.

Considering the fact that PDI of habitants living in Xi'an is 26 μg/day, which is much lower than the Recommended Dietary Allowance (RDA) established by US Food and Nutrition Board, it is reasonable to conclude that Xi'an is a Se-deficient region and there is a risk of Se deficiency for the residents. More research will be needed to investigate the Se concentration and Se speciation in soil of Xi'an.

4 CONCLUSIONS

The results revealed that Se concentrations of food products consumed by inhabitants in Xi'an is relatively low, and the daily Se intake is 26 μg/day, which is lower than RDA. The study area belongs to a Se-deficient area and we should take measures to reduce the risk of getting Se deficiency disease by increasing consumption of Se-enriched food products.

REFERENCES

Dumont, E., Vanhaecke, F., & Cornelis, R. 2006. Selenium speciation from food source to metabolites: a critical review. *Analytical and bioanalytical chemistry*, 385(7): 1304–1323.

Food and Nutrition Board, Institute of Medicine. 2000. Dietary reference intakes for vitamin C, vitamin E, selenium, and carotenoids. *National Academy of Sciences*.

Gissel-Nielsen, G. 1998. Effects of selenium supplementation of field crops. *Environmental Chemistry of Selenium*: 99–112.

Longnecker, M.P., Taylor, P.R, Levander, O.A, Howe, M., Veillon, C., McAdam, P., et al. 1991. Selenium in diet, blood, and toenails in relation to human health in a seleniferous area. *The American Journal of Clinical Nutrition*, 53(5): 1288–1294.

Mahapatra, S., Tripathi, R., Raghunath, R., & Sadasivan, S. 2001. Daily intake of Se by adult population of Mumbai, India. *Science of the Total Environment*, 277(1): 217–223.

Navarro-Alarcon, M., & Cabrera-Vique, C. 2008. Selenium in food and the human body: a review. *Science of the Total Environment*, 400(1): 115–141.

Pappa, E.C., Pappas, A.C., & Surai, P.F. 2006. Selenium content in selected foods from the Greek market and estimation of the daily intake. *Science of the Total Environment*, 372(1): 100–108.

Rotruck, J., Pope, A., Ganther, H., Swanson, A., Hafeman, D.G., & Hoekstra, W. 1973. Selenium: biochemical role as a component of glutathione peroxidase. *Science*, 179(4073): 588–590.

Thiry, C., Ruttens, A., De Temmerman, L., Schneider, Y.-J., & Pussemier, L. 2012. Current knowledge in species-related bioavailability of selenium in food. *Food Chemistry*, 130(4): 767–784.

Zhang, X. 2003. Shaanxi Statistical Yearbook. China Statistics Press, Beijing. (in Chinese)

Assessment of selenium intake, status and influencing factors in Kenya

Peter Biu Ngigi* & Gijs Du Laing
Department of Applied Analytical and Physical Chemistry, Gent University, Gent, Belgium

Carl Lachat
Department of Food Technology & Nutrition, Gent University, Gent, Belgium

Peter Wafula Masinde
Department of Food Science, School of Agriculture & Food Science, Meru, Kenya

1 INTRODUCTION

Selenium (Se) is beneficial to human health due to its antioxidant capacity and therefore, its deficiency may result in clinical disorders (Nogales et al., 2013; Temmerman et al., 2014). In Africa, the risk of Se deficiency is high and it is greatest in the Eastern region (Joy et. al., 2014). However, Se has not received considerable research attention in Africa in the past because unlike other minerals, it has faced analytical limitations that probably have stimulated researchers to exclude it from their analysis (Bueno et al., 2007). As diet is the most important source of Se, collection of dietary intake data is of primary importance for the identification, treatment, and management of Se deficiency disorders (Chilimba et al., 2011; Hartikainen et al., 2000). Selenium interacts with other elements commonly known to be deficient in Sub-Saharan Africa (SSA) such as iron (Fe) and zinc (Zn) and therefore analysis of these elements will be included in this study (Eybl et al., 1999).

We are currently conducting a cross-sectional descriptive survey aimed at assessing the Se intake and status among 6–59 month old children and their mothers in Kenya. Dietary intake data combined with the Se content of local foods will represent the actual Se intake for selected 29 study areas in the agricultural production zones; 21 regions in the central Kenya highlands and 8 regions along the Lake Victoria basin. Selenium deficiency is expected in the highlands due to the geochemical characteristics of the soils and very low fish consumption as compared to high fish consumption, in the Lake Victoria basin. Moreover, soil, hair and nail samples will be taken from the selected households. The association between soil Se and crop Se, and between population Se status and Se intake will help to determine the soil characteristics that influence soil-to-plant Se transfer, and dietary factors that influence Se status among the population. Past nutrition interventions such as supplementation and food fortification have failed to reach the rural population. Therefore, we intend to also conduct Se, Fe, and Zn agronomic biofortification trials in a next phase, as a possible solution to the 'hidden hunger'.

2 MATERIALS AND METHODS

Study areas were identified first based on the various soil types used for agriculture in Kenya. A significant correlation between Se in plant and soil can be found if the relationship is investigated within various soils types. In this regard, the mobility and plant-availability of Se in soil is controlled by a number of soil's chemical and biochemical processes. The second criteria was based on fish consumption habits since fish is one of the best dietary sources of Se.

The target population are 6–59 month old children and their mothers because Se plays an important role in normal growth. Selenium deficiency depresses growth resulting in impaired intrauterine growth retardation. Hence, child growth can be used as an evaluation criteria for the impact of interventional measures beside the biochemical markers. Based on a Se deficiency proportion of 37% among Kenyan population, 1234 subjects for both target population groups are needed. From the randomly selected households in each study area, the following samples will be collected: a composite 100 g soil sample from the farm, 40 g of each food produced on the farm, and 1.5 to 2 g of hair samples and 10–50 mg of nail samples depending on availability. All samples will be transported to Belgium and analysed at the Laboratory of Analytical Chemistry and Applied Ecochemistry, Ghent University.

Laboratory results and dietary intake data will be analysed with STATA statistical package. Soil-plant and Se status-dietary intake regression models will be derived to determine the relationship between Se concentrations in crops and soil characteristics and the relationships between Se concentrations in hair and nail and dietary Se characteristics (concentration in hair and nail, dietary intake), respectively.

3 RESULTS AND DISCUSSION

Samples are currently still being collected in Kenya. Dietary intake data combined with the Se concentration of local foods plus soil Se concentration will reflect the Se exposure at the population level. Hair and nail samples will be used to assess Se status at the individual level. Selenium deficiency is expected in the highlands due to the geochemical characteristics of the soils and very low fish consumption, as compared to high fish consumption in the Lake Victoria basin. The fact that soil types were included as a selection criterion for the sampling sites allows us to study correlations between Se concentrations in plant and soil to assess the mobility and plant-availability of Se in the soil, which is controlled by a number of chemical and biochemical processes. The association between soil Se and crop Se concentration will help to determine the soil characteristics that influence soil-to-plant Se transfer, while the association between population/individual Se status and Se intake will help to determine dietary factors that influence Se status among the target population. Regression models will be used to determine the relationship between Se concentrations in crops and soil characteristics, and the relationship between Se concentration in hair/nail and dietary Se characteristics, respectively.

REFERENCES

Bueno, M., Pannier, F., & Potin-Gautie, M. 2007. Determination of organic and inorganic selenium species using HPLC-ICP-MS. Agilent Technologies, Inc., 5989–7073.

Chilimba, A.D., Young, S.D., Black, C.R., Rogerson, K.B., Ander, E.L., Watts, M.J. et al. 2011. Maize grain and soil surveys reveal suboptimal dietary selenium intake is widespread in Malawi. Scientific Reports, 1:72. doi:10.1038/srep00072.

De Temmerman, L., Waegeneers, N., Thiry, C., Du Laing, G., Tack, F. & Ruttens, A. 2014. Selenium content of Belgian cultivated soils and its uptake by field crops and vegetables. Science of the Total Environment, 468–469: 77–82.

Eybl, V., Kotyzova, D. & Koutensky, J. 1999. On the interaction of selenium with zinc in experiments with mice. Interdisciplinarni Cesko-Slovenska Toxikologicka Konference, Prednasky.

Hartikainen, H., Xue, T. & Piironen, V. 2000. Selenium as anti-oxidant and pro-oxidant in ryegrass. Plant and Soils, 225: 193–200.

Joy, E.J.M., Anderb, E.L., Younga, S.D., Black, C.R, Watts, M.J, Chilimba, A.D.C. et al. 2014. Dietary mineral supplies in Africa. Physiologia Plantarum, 151: 208–229.

Nogales, F., Ojeda, M.L., Fenutrıa, M., Murillo, M.L., & Carreras, O. 2013. Role of selenium and glutathione peroxidase on development, growth, and oxidative balance in rat offspring. Reproduction, 146: 659–667.

# Selenium improves the biocontrol activity of *Cryptococcus laurentii* against *Penicillium expansum* in tomato fruit

Jie Wu, Zhilin Wu & Miao Li*
Key Laboratory of Agri-Food Safety of Anhui Province, School of Plant Protection, Anhui Agriculture University, Hefei, Anhui, China

Gary S. Bañuelos
USDA-ARS, Parlier, California, USA

Zhi-Qing Lin
Environmental Sciences Program, Southern Illinois University Edwardsville, Edwardsville, Illinois, USA

1 INTRODUCTION

Selenium (Se) is an essential trace element for human and animal health, while high concentrations of Se can be toxic to plants (Mills et al., 2004). Our previous work has demonstrated that Se could be used for the control of postharvest diseases in fruits and vegetables caused by the fungus *Botrytis cinerea* and *Penicillium expansum* (Wu et al., 2014). The mode of action of Se against *B. cinerea* and *P. expansum* may be directly related to the disruptive effects of Se on cell membranes of the fungal pathogen. However, little is known about the efficacy of Se on the activity of biological control agents.

Tomato fruit are susceptible to postharvest diseases caused by various pathogenic fungi (Zhu et al., 2009). Blue mold rot, caused by the wound invading necrotrophic fungus *P. expansum*, is one of the most destructive postharvest diseases of tomato fruit (Qin et al., 2010). The objective of this study was to evaluate the synergistic effects of combining Se with the antagonistic yeast *Cryptococcus laurentii* against blue mold rot caused by *P. expansum* in tomato fruit. The possible mechanisms by which Se enhanced the efficacy of biological control agents were also discussed.

2 MATERIALS AND METHODS

Tomato fruit were harvested at commercial maturity. Fruit without physical injuries were surface sterilized with 2% (v/v) sodium hypochlorite for 2 min, washed with tap water, and air-dried prior to use. Antagonistic yeast *C. laurentii* and *P. expansum* were obtained from Key Laboratory of Plant Resources in North China, Institute of Botany, Chinese Academy of Sciences were cultured in nutrient yeast dextrose broth (NYDB; 8 g of nutrient broth, 5 g of yeast extract, and 10 g of dextrose in 1000 mL water) for 48 h at 25°C. Yeast cells were collected by centrifugation at 8000x g for 10 min. After washing twice with sterile distilled water, the yeast cells were suspended and adjusted to a concentration of 1×10^7 cells/mL with a haemocytometer. *P. expansum* was maintained on potato dextrose agar (PDA). Spores were obtained from PDA cultures incubated for two weeks at 23° C. The spores were scraped from the surface of the cultures and suspended in 5 mL of sterile distilled water containing 0.05% (v/v) Tween 80. Spore suspensions were filtered through four layers of sterile cheesecloth to remove any adhering mycelia, and spore concentration was determined with a hemacytometer adjusted to 5×10^4 spores/mL with sterile distilled water.

Fruit were wounded with a sterile nail at midpoint of the fruit equator and to each wound were added 20 µL of the treatment suspensions as follows: sodium selenite (0.1%, w/v), *C. laurentii* (1×10^7 cells/mL), (1×10^8 cells/mL), and *C. laurentii* (1×10^7 cells/mL) + sodium selenite (0.1%, w/v). Fruit treated with sterile distilled water served as controls. After wounds were air-dried, fruit were challenge-inoculated with 15 µL of a conidial suspension of *P. expansum* at 5×10^4 spores/mL. In another experiment, fruit were challenge-inoculated with 15 µL *P. expansum* conidial suspension at 5×10^4 spores/mL before the treatment suspensions as described above were added. Treated fruit were stored at 25°C with high humidity (RH about 95%) and blue mold rot was measured daily after treatment. There were three replicates of each treatment with 20 fruits per replicate, and the experiment was repeated twice.

In addition, the effects of Se was evaluated on the growth of *C. laurentii in vitro*. Firstly, *C. laurentii* was cultured in NYDB overnight. Then, 50 µL of *C. laurentii* culture was added to 5 mL NYDB containing sodium selenite (0.05%, 0.1%, w/v). After cultured at 25°C on a rotary shaker at 200 rpm for indicated time (0, 12, 24, 48, 72, 96, and 120 h), the OD600 of the yeasts was measured with a spectrophotometer (UV-160). Each treatment was replicated three times and the experiment was repeated twice.

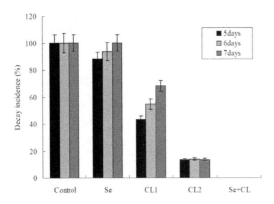

Figure 1. Effect of *C. laurentii* (CL), alone or in combination with Se (0.1%, w/v) on the efficacy against blue mold rot of tomato fruit caused by *P. expansum*.

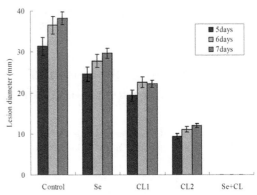

Figure 2. Fruit were treated with Se (0.1%, w/v), CL 1 (1×10^7 cells/mL), CL2 (1×0^8 cells/mL), and CL (1×10^7 cells/mL) + Se (0.1%, w/v), followed by challenge inoculation with *P. expansum*.

3 RESULTS AND DISCUSSION

A single application of Se was effective at controlling blue mold caused by *P. expansum*, but better control was achieved when Se was used in combination with the biocontrol agent *C. laurentii*. The timing of Se and *C. laurentii* application affected the efficacy of disease control. Application of Se alone before pathogen inoculation resulted in significantly higher incidence of *P. expansum* disease and lesion diameters than did Se treatment after pathogen inoculation (Fig. 1). In comparison, *C. laurentii* alone was effective for controlling *P. expansum* when applied before pathogen inoculation, but it showed little effect when applied after pathogen inoculation. Fruit decay was completely suppressed by a treatment of Se in combination with *C. laurentii* before *P. expansum* inoculation (Fig. 2) after 9 days of storage at 25°C. Although *C. laurentii* showed little effect on fruit decay when applied after pathogen inoculation, *C. laurentii* combined with Se significantly reduced the disease incidence and lesion diameter of blue mold rot.

Population dynamics of *C. laurentii* were determined in the wounds of tomato fruit with or without Se (0.1%, w/v). Se at 0.1% (w/v) has been shown to inhibit *P. expansum* in tomato fruit (Fig. 2). However, Se at 0.1% (w/v) did not significantly influence ($p < 0.05$) the growth of *C. laurentii* in fruit over the duration of the experiment. Populations of *C. laurentii* increased with or without Se during the first 48 h and then they stabilized. In addition, Se did not significantly inhibit yeast growth when the concentration was lower than 0.1% (w/v) (data not shown).

4 CONCLUSIONS

In conclusion, we have shown that Se improved the efficacy of the antagonistic yeast for control of blue mold rot on tomato fruit. The enhancement of disease control may be associated with the difference in sensitivity to Se between the fungal pathogen and the antagonistic yeast. Selenium exhibited inhibitory effects against *P. expansum* via targeting the mitochondria, leading to the disruption of mitochondrial function and the fungal death. Compared with other salts such as sodium carbonate, injury to tomato fruit was not observed in this study by using Se. These results suggested that Se may be promising as an environmentally friendly additive to enhance the performance of antagonistic yeasts against postharvest decay of fruits and vegetables.

REFERENCES

Mills, A.A.S., Platt, H.W. & Hurta, R.A.R. 2004. Effect of salt compounds on mycelial growth, sporulation and spore germination of various potato pathogens. *Postharvest Biology Technology*. 34: 341–350.

Qin, G.Z., Zong, Y.Y., Chen, Q.L., Hua, D.L. & Tian, S.P. 2010. Inhibitory effect of boron against *Botrytis cinerea* on table grapes and its possible mechanisms of action. *International Journal of Food Microbiology*, 138: 145–150.

Wu, Z.L., Yin, X.B., Lin, Z.Q., Bañuelos, G.S., Yuan, L.X., Liu, Y. & Li M. 2014. Inhibitory effect of selenium against *Penicillium expansum* and its possible mechanisms of action. *Current Microbiology* 69(2): 192–201.

Zhu, Y.G., Pilon-Smits, E.A., Zhao, F.J., Williams, P.N. & Meharg, A.A. 2009. Selenium in higher plants: understanding mechanisms for biofortification and phytoremediation. *Trends in Plant Science*. 14(8): 436–442.

Antifungal activity of selenium on two plant pathogens *Sclerotinia sclerotiorum* and *Colletotrichum gloeosporioides*

Yu Zhang, Han Jiang, Huawei Zang, Genjia Tan & Miao Li*
Key Laboratory of Agri-Food Safety of Anhui Province, School of Resources and Environment–School of Plant Protection, Anhui Agriculture University, Hefei, Anhui, China

Linxi Yuan & Xuebin Yin
Advanced Lab for Selenium and Human Health, Suzhou Institute for Advanced Study, University of Science and Technology of China, Suzhou, Jiangsu, China

1 INTRODUCTION

Various pesticides have been used to control plant diseases for a long time. However, widespread use of agrochemicals has contributed to the development of resistant pathogens, and may be lethal to beneficial microorganisms and useful insects and they may also enter the food chain and accumulate in agricultural products as pesticides residues. Consequently, many researchers have tried to develop new and effective antimicrobial reagents (Wu et al., 2014).

Selenium (Se) is an essential trace element for human and animal health. Low concentrations of Se may be a benefit for the growth of some plants, but large amounts of Se can also be toxic to biological organisms (Zhu et al., 2009). Our previous work has demonstrated that Se could be used for the control of postharvest diseases in fruits and vegetables caused by *Botrytis cinerea* and *Penicillium expansum* (Wu et al., 2014). However, only postharvest fungal pathogen *B. cinerea* and *P. expansum* were tested in the study, and little is known about the efficacy of Se on other plant pathogens. Therefore, many plant pathogenic fungal and bacterial species need to be tested to investigate if the antimicrobial effect of sodium selenite and/or sodium selenate at low concentrations has any effect against other fungal and/or bacterial species.

Sclerotinia sclerotiorum and *Colletotrichum gloeosporioides* cause highly destructive diseases in oilseed rape and pear, respectively, resulting in significant economic losses (Shi et al., 2011; Wang et al., 2012). In the present study, we have demonstrated the inhibitory action of Se against plant pathogens *S. sclerotiorum* and *C. gloeosporioides* under different concentrations in growth medium, as well as attempt to understand the control mechanisms of Se in selenate and selenite, respectively, on the two tested plant pathogens.

2 MATERIALS AND METHODS

The phytopathogenic fungi *Sclerotinia sclerotiorum* and *Colletotrichum gloeosporioides* used in this study was obtained from Laboratory of Plant Health, Anhui Agriculture University, China. The fungi were routinely grown on potato dextrose agar (PDA) plates for two weeks at $23 \pm 2°C$. The spores were obtained from the surface of the agar and suspended in 5 mL of sterile distilled water containing 0.1% v/v Tween 20. Spore suspensions were filtered through four layers of sterile cheesecloth to remove mycelia fragments. A hemocytometer was used to calculate the number of spores and the spore concentration was adjusted to 5.0×10^5 spores/mL.

Sodium selenite was used as a Se source in this study and was purchased from Sigma-Aldrich (St. Louis, MO). All other chemicals used in this study were of high analytical grade.

The antifungal activity of Se was measured on PDA media. It were supplemented with 0, 1, 5, 10 and 15 mg/L of sodium selenite and sodium selenate, respectively. An agar plug of 9 mm diameter containing the tested fungi was inoculated simultaneously at the center of each petri dish and incubated at $23 \pm 2°C$. Growth diameter was measured after 2, 4, and 6 days of incubation, and each treatment was triplicated.

3 RESULTS AND DISCUSSION

The inhibition effect of Se at different concentrations was estimated against *S. sclerotiorum* (Table 1) and *C. gloeosporioides* (Table 2) for different incubation time periods. In all cases, a higher inhibition of fungal growth was recorded at the Se concentration of 15 mg/L of two tested Se salts and the two fungi showed growth inhibition with increasing incubation time. Selenite treatment at 15 mg/L concentration resulted in about 69.2 and 72.6% inhibition percentage with *S. sclerotiorum* and *C. gloeosporioides*, respectively. The lowest level of inhibition against both fungi was observed with selenite at a concentration 15 mg/L. The results suggested that maximum inhibition was obtained in both fungal isolates treated with a Se concentration of 15 mg/L.

Management of fungal diseases on crops and fruits is economically important. Recently, more efforts have

Table 1. Effect of Se on fungal growth diameter (mm) of S. sclerotiorum.

Incubation time (d)	Control	Se Treatment (15 mg/L)	
		Selenite	Selenate
2	28.4 ± 0.5	16.3 ± 0.2	21.1 ± 0.5
4	45.7 ± 1.0	20.6 ± 1.0	25.7 ± 0.5
6	72.8 ± 0.5	31.2 ± 2.0	37.4 ± 1.0

Table 2. Effect of Se on fungal growth diameter (mm) of C. gloeosporioides.

Incubation time (d)	Control	Se Treatment (15 mg/L)	
		Selenite	Selenate
2	34.1 ± 1.0	21.2 ± 1.0	26.9 ± 2.0
4	56.6 ± 0.5	31.3 ± 2.0	38.6 ± 1.0
6	81.3 ± 0.5	39.8 ± 0.5	48.5 ± 0.5

been given to develop safe management methods with synthetic fungicides that pose less danger to humans and animals. The current investigation showed that Se was very effective for reducing plant diseases caused by spore producing phytopathogenic fungi. Results presented in this study showed the inhibitory effect of Se on growth of *S. sclerotiorum* and *C. gloeosporioides* were different. Growth of *S. sclerotiorum* and *C. gloeosporioides* were inhibited at different concentrations of selenite and selenite, especially at 15 mg Se/L. This study was the first to demonstrate that the inhibitory effect of Se against *S. sclerotiorum* and *C. gloeosporioides*.

Previous reports stated that, antimicrobial activity of Se was different depending on microbial species (Wu *et al.*, 2014). Selenium can significantly delay mycelial growth in a dose-dependent manner in vitro (Wu *et al.*, 2014). In this way, Se may directly attach to and penetrate the cell membrane to kill spores, although penetration of Se into microbial cell membranes is not completely understood (Wu *et al.*, 2014). The present study also suggests that selenite was more effective than selenate for the control of both plant pathogens *S. sclerotiorum* and *C. gloeosporioides*. Thus, Se may be a cost effective method with great promise as an antimicrobial agent and have a great potential for use in controlling phytopathogenic fungi.

4 CONCLUSIONS

The results of this study clearly demonstrate that Se can inhibit *S. sclerotiorum* and *C. gloeosporioides* growth and development. However, some parameters will require evaluation prior to practical application, including the evaluation of its phytotoxicity and antimicrobial effects in situ, to develop systems for delivering Se into host tissues that have been colonized by phytopathogens. The widespread use of Se agent needs to be further evaluated in the context of its potential ecological impacts at higher concentrations.

REFERENCES

Shi, X.Q., Li, B.Q., Qin, G.Z. & Tian, S.P. 2011. Antifungal activity of borate against *Colletotrichum gloeosporioides* and its possible mechanism. *Plant Diseases* 295: 63–69.
Wang, Z., Tan, X., Zhang, Z., Gu, S., Li, G., & Shi, H. 2012. Defense to *Sclerotinia sclerotiorum* in oilseed rape is associated with the sequential activations of salicylic acid signaling and jasmonic acid signaling. *Plant Science* 184: 75–82.
Wu, Z.L., Yin, X.B., Lin, Z.Q., Bañuelos, G.S., Yuan, L.X., Liu, Y. & Li M. 2014. Inhibitory effect of selenium against *Penicillium expansum* and its possible mechanisms of action. *Current Microbiology* 69(2): 192–201.
Zhu, Y.G., Pilon-Smits, E.A., Zhao, F.J., Williams, P.N. & Meharg, A.A. 2009. Selenium in higher plants: understanding mechanisms for biofortification and phytoremediation. *Trends in Plant Science* 14(8): 436–442.

Exogenous selenium application influences lettuce on bolting and tipburn

X.X. Wang*, Y.Y. Han & S.X. Fan
Beijing University of Agriculture, Beijing, China

1 INTRODUCTION

Lettuce (*Lactuca sativa* L.) is a popular vegetable throughout the world. Early bolting and tipburn are two major problems in lettuce production. Previous studies have showed that the HSP70 family played a crucial role in heat-shock response (Feige et al., 1994) in which tipburn is included (Sylvie et al., 2013). Recent research suggests that besides fighting against the heat, the expression of HSP genes affects bolting in different ways (Passam et al., 2008). We found using q-PCR that the expression of two LsHsp 70 genes are strongly relevant with bolting (Ting et al., 2015). In addition, studies on mice showed that selenium-rich diet hsp70 significantly increased their sperm cells (Kaur et al., 2003). Later research on male mice also observed an increase of hsp70 in the high Se content group (Naveen et al., 2009). In human research, we found that the HSP70 mRNA level in "high Se+F group" was significantly higher compared to "high F group" (Chen et al., 2009). Based on those former studies we assumed that exogenous selenium (Se) application should help lettuce raise its expression of LsHsp 70, and eventually, suppress its bolting and tipburn.

2 MATERIALS AND METHODS

In this study, we selected 3 well-performed lettuce main cultivars (Grand Rapids, Rapid, and Fang Ni) that have been widely-grown and are popular vegetables on the market. On the 30th and 45th day after germination, sodium selenite solution at a Se concentration of 33 mg/kg was applied to leaves. The application rate was equivalent 37.5 g Se per hectare (hm^2). Plants were harvested on the 60th day. The plant bolting rate, height, stem length, the stem length between the 5th and 15th blade attaching points, and the disease index of tipburn were measured. Concentrations of total Se in lettuce tissues were also determined using HG-AFS.

3 RESULTS AND DISCUSSION

3.1 *Selenium accumulation*

The applying of exogenous Se significantly increase the total Se content in all 3 varieties, among which Rapid has experienced the greatest Se concentration at 334 μg/kg (Table 1).

3.2 *Bolting characters*

There have been observable reductions in bolting rate, height, steam length, and length between 10 blade attachment points in all three varieties. Comparing the three varieties, we found that Fang Ni experienced the most reduction in bolting rate and steam length, which are the two major characteristics in bolting (Table 1).

3.3 *Tipburn*

The disease index of tipburn of different lettuce cultivars are shown in Table 1. The foliar Se application significantly reduced the disease index, and the Se treatment prevented all three lettuce cultivars from

Table 1. Effects of foliar Se application on lettuce growth and Se accumulation and tipburn. Different letters indicate a significant difference between the treatment and control ($p < 0.05$).

	Grand Rapids Control	Se treatment	Rapid Control	Treat-Se	Fang Ni Control	Se treatment
Se application (g/ha)		37.5	0	37.5	0	37.5
Total Se in plant (μg/kg)	<10 a	249 ± 2.0b	<10a	334 ± 1.0b	<10 a	255 ± 0.6b
Bolting rate (%)	27.8 ± 1.3b	19.4 ± 1.9 a	33.0 ± 0.9b	25.7 ± 1.0 a	2.2 ± 0.4b	0.3 ± 0.2 a
Height (mm)	440.17 ± 26.4b	353.1 ± 12.8 a	435.6 ± 15.9b	327.0 ± 14.1 a	271.3 ± 4.2b	219.3 ± 8.0 a
Stem length (mm)	236.3 ± 8.0b	195.7 ± 17.7 a	309.3 ± 22.3b	196.6 ± 29.1 a	133.4 ± 11.6b	83.5 ± 1.6 a
Distance between 5th and 15th leaves (mm)	79.8 ± 14.9b	67.7 ± 4.4 a	90.1 ± 4.9b	63.8 ± 6.3 a	41.0 ± 4.0b	29.3 ± 1.3 a
Disease index of tipburn	48 ± 1b	25 ± 1 a	50 ± 2b	21 ± 1a	2 ± 1b	0 ± 0 a

tipburn. In particular, there was no individual Fang Ni plants showing any tipburn symptoms.

4 CONCLUSIONS

The research showed that exogenous Se application not only suppressed bolting in lettuce but also restricted tipburn. We hypothesize that HSP 70, which was up-regulated by the exogenous Se, helped reduce bolting and tipburn in lettuce.

REFERENCES

Chen, Q., Wang, Z., Xiong, Y., Xue, W., Kao, X., Gao, Y., et al. 2009. Selenium increases expression of HSP70 and antioxidant enzymes to lessen oxidative damage in Fincoal-type fluorosis. *The Journal of Toxicological Science* 34(4):399–405.

Feige, U. & Polla, B.S. 1994. Heat shock proteins: the hsp70 family. *Cellular and Molecular Life Sciences*, 50(11-12):979–986.

Jenni, S., José Truco, M. & Michelmore, R.W. 2013. Quantitative trait loci associated with tipburn, heat stress-induced physiological disorders, and maturity traits in crisphead lettuce. *TAG Theoretical and Applied Genetics*, 126(12):3065–3079.

Kaushal, N. & Bansal, M.P. 2009. Diminished reproductive potential of male mice in response to selenium-induced oxidative stress: involvement of HSP70, HSP70-2, and MSJ-1. *J Biochem Molecular Toxicology* 23(2):125–136.

Parminder, K. & Bansal, M.P. 2003. Effect of oxidative stress on the spermatogenic process and hsp70 expressions in mice testes. *Indian Journal of Biochemistry and Biophysics* 40(04):246–251.

Passam, H.C., Koutri, A.C. & Karapanos, I.C. 2008. The effect of chlormequat chloride (CCC) application at the bolting stage on the flowering and seed production of lettuce plants previously treated with water or gibberellic acid (GA3). *Scientia Horticulturae*, 116(2):117–121.

Ting, L., Han, Y.Y., Fan, S.G., & Zhao, Z.Z. 2015. Correlation Analysis of Bolting and Two Hsp70 Genes in Lactuca sativa L. *Chinese Agricultural Science Bulletin* 31(10):58–62.

Biological uptake and accumulation of selenium

The genetics of selenium accumulation by plants

Philip J. White
The James Hutton Institute, UK
King Saud University, Riyadh, Kingdom of Saudi Arabia

1 INTRODUCTION

Selenium is an essential mineral element for human and animal nutrition but it is not required by plants (White & Broadley, 2009). Since most Se in human diets is derived from plant products, insufficient dietary Se is generally attributed to crop production on soils with low Se content or phytoavailability. At the opposite extreme, excess Se is harmful to humans and can be toxic to most plants (White et al., 2004). The concentration and chemical forms of Se in natural soils are primarily determined by geology (White et al., 2007b). Selenium concentrations in most soils lie between 0.01–2.0 mg/kg but soils associated with particular geological features can reach 1200 mg/kg. Selenate (SeO_4^{2-}) is the dominant Se species in solution in most cultivated soils, although selenite (SeO_3^{2-}) dominates under anaerobic and acidic conditions, such as in paddy soils. Selenium concentrations in crops can be increased simply by the application of Se-fertilisers. This article first provides an overview of Se uptake and metabolism in flowering plants (angiosperms) then describes genetic variation in Se uptake and accumulation between and within angiosperm species.

2 SELENIUM UPTAKE, TRANSLOCATION AND METABOLISM IN PLANTS

Plant roots take up SeO_4^{2-}, SeO_3^{2-} or organoselenium compounds, such as selenocysteine (SeCys) and selenomethionine (SeMet), but are unable to take up colloidal elemental Se or metal selenides (White & Broadley, 2009; Zhang et al., 2014). Selenate uptake is catalysed by high-affinity sulphate transporters homologous to arabidopsis AtSULTR1;1 and AtSULTR1;2. Phosphate transporters, such as rice OsPT2, catalyse the uptake of $HSeO_3^-$ and homologues of the rice aquaporin channel OsNIP2;1 catalyse the uptake of H_2SeO_3. Transporters for amino acids might catalyse the uptake of SeCys and SeMet. Selenite is rapidly converted to organoselenium compounds in the root, whereas selenate is delivered immediately to the xylem. Sulphate transporters homologous to AtSULTR1;3, AtSULTR2;1, AtSULTR2;2 and AtSULTR3;5 are thought to contribute to the long-distance transport of selenate.

Selenate is assimilated into organoselenium compounds in plastids (White et al., 2007b; Pilon-Smits & LeDuc, 2009). It is first activated by adenosine triphosphate sulphurylase to form adenosine 5′-phosphoselenate, which is reduced to selenite by adenosine 5′-phosphosulphate reductase. Selenite is then reduced to selenide by sulphite reductase. Selenocysteine is synthesised from serine and selenide by cysteine synthase. Selenomethionine is synthesised from SeCys through the sequential actions of cystathionine γ-synthase, which produces selenocystathionine, cystathionine β-lyase, which produces selenohomocysteine (SeHCys), and methionine synthase. Biofortification of some plants with Se results in the formation of selenohomolanthionine from SeHCys. Both SeCys and SeMet can be incorporated into proteins or methylated. The methylated products, Se-methylselenocysteine (SeMeSeCys) and Se-methylselenomethionine (SeMeSeMet), can be converted to volatile compounds, such as dimethyldiselenide or dimethylselenide, or conjugated with glutamate to form γ-glutamyl-SeMeSeCys or γ-glutamyl-SeMeSeMet. In some plants, γ-glutamyl-Secystionine has also been reported. Selenocysteine can be converted to alanine and elemental Se by SeCyslyase. Selenium toxicity has been attributed to the nonspecific replacement of cysteine and methionine in proteins by SeCys and SeMet. Thus, the conversion of these compounds to volatile or non-toxic Se-metabolites improves Se tolerance. Plant cells can accumulate selenate in their vacuoles. Sulphate transporters homologous to AtSULTR4;1 and AtSULTR4;2 are thought to catalyse the efflux of selenate from the vacuole. Selenium is redistributed within the plant as selenate and organoselenium compounds via the phloem.

3 SELENIUM ACCUMULATION VARIES BETWEEN AND WITHIN PLANT SPECIES

Considerable variation in shoot Se concentration has been observed among angiosperm species growing in the same environment (White et al., 2004, 2007a). However, little of this variation is attributed to systematic differences between angiosperm orders and is thought to reflect species-specific adaptations

(White et al., 2004). Angiosperms have been divided into three ecological types: non-accumulators, Se-indicators and Se-accumulators. Most angiosperms are non-accumulators. They cannot tolerate tissue Se concentrations exceeding 10–100 µg/g dry weight (DW) and are unable to colonise seleniferous soils. The Se-indicator plants, however, are able to tolerate tissue Se concentrations approaching 1 mg/g DW and colonise both non-seleniferous and seleniferous soils. Such plants avoid the accumulation of Se-amino acids and volatilise Se effectively. These traits can be improved by genetic manipulation of Se-assimilation pathways for phytoremediation of Se-contaminated soils (Pilon-Smits & LeDuc, 2009). The distribution of Se-accumulator plants is largely restricted to seleniferous soils. They include members of Fabales, Brassicales, Cornales, Gentianales, Lamiales and Asterales, which can accumulate shoot Se concentrations > 20–40 mg/g DW, and members of the Lecythidaceae family, which accumulate large Se concentrations in fruits and seeds. It is thought that the ability to accumulate Se arose by convergent evolution of appropriate Se transport and biochemical pathways in disparate angiosperm clades (White et al., 2007b). This probably occurred during geological periods when seleniferous soils were more widespread than they are today. A species is defined as a Se-hyperaccumulator if its leaves contain > 1 g/kg DW when sampled from its natural habitat. In addition to their ability to accumulate Se, Se-hyperaccumulators exhibit several other characteristics that distinguish them from Se-indicator and non-accumulator plants. When compared to other angiosperms, Se-hyperaccumulators (1) constitutively express genes encoding sulphate transporters, (2) have significantly greater leaf Se/S quotients, (3) exhibit reduced molybdate accumulation with increasing rhizosphere sulphate or selenate concentrations, (4) restrict the incorporation of Se-amino acids into proteins through greater expression of appropriate genes, and (5) accumulate Se in leaf trichomes and epidermal cells (White et al., 2007a,b; Cabannes et al., 2011; El Mehdawi & Pilon-Smits, 2012; Harris et al., 2014). Genotypes within non-accumulator and Se-accumulator species can differ markedly in shoot Se concentrations and chromosomal loci (QTL) affecting this trait have been identified in arabidopsis (Chao et al., 2014). Furthermore, there appears to be sufficient genetic variation to breed for edible crops with greater Se concentrations. Genetic variation in shoot Se concentration has been observed in alliums, brassicas and lettuce, in tomato fruit, and in seeds of cereals, soybean and lentil (White and Broadley, 2009). QTL affecting seed Se concentration have been identified in rice (Norton et al., 2010) and soybean (Ramamurthy et al., 2014).

REFERENCES

Cabannes, E., Buchner, P., Broadley, M.R. & Hawkesford, M.J. 2011. A comparison of sulfate and selenium accumulation in relation to the expression of sulfate transporter genes in *Astragalus* species. *Plant Physiol.* 157: 2227–2239.

Chao, D.-Y., Baraniecka, P., Danku, J., Koprivova, A., Lahner, B., Luo, H. et al. 2014. Variation in sulfur and selenium accumulation is controlled by naturally occurring isoforms of the key sulfur assimilation enzyme APR2 across the Arabidopsis species range. *Plant Physiol.* 166: 1593–1608.

El Mehdawi, A.F. & Pilon-Smits, E.A.H. 2012. Ecological aspects of plant selenium hyperaccumulation. *Plant Biol.* 14: 1–10.

Harris, J., Schneberg, K.A. & Pilon-Smits, E.A.H. 2014. Sulfur-selenium-molybdenum interactions distinguish selenium hyperaccumulator *Stanleya pinnata* from non-hyperaccumulator *Brassica juncea* (Brassicaceae). *Planta* 239: 479–491.

Norton, G.J., Deacon, C.M., Xiong, L., et al. 2010. Genetic mapping of the rice ionome in leaves and grain: identification of QTLs for 17 elements including arsenic, cadmium, iron and selenium. *Plant and Soil* 329: 139–153.

Pilon-Smits, E.A.H. & LeDuc, D.L. 2009. Phytoremediation of selenium using transgenic plants. *Curr. Opin. Biotech.* 20: 207–212.

Ramamurthy, R.K., Jedlicka, J., Graef, G.L. & Waters, B.M. 2014. Identification of new QTLs for seed mineral, cysteine, and methionine concentrations in soybean [*Glycine max* (L.) Merr.]. *Molec. Breed.* 34: 431–445.

White, P.J. & Broadley, M.R. 2009. Biofortification of crops with seven mineral elements often lacking in human diets – iron, zinc, copper, calcium, magnesium, selenium and iodine. *New Phytol.* 182: 49–84.

White, P.J., Bowen, H.C., Parmaguru, P., Fritz, M., Spracklen, W.P., Spiby, R.E. et al. 2004. Interactions between selenium and sulphur nutrition in *Arabidopsis thaliana*. *J. Exp. Bot.* 55: 1927–1937.

White, P.J., Bowen, H.C., Marshall, B. & Broadley, M.R. 2007a. Extraordinarily high leaf selenium to sulphur ratios define 'Se-accumulator' plants. *Ann. Bot.* 100: 111–118.

White, P.J., Broadley, M.R., Bowen, H.C. & Johnson, S.E. 2007b. Selenium and its relationship with sufur. In M.J. Hawkesford & L.J. de Kok (eds) *Sulfur in Plants – an Ecological Perspective*, pp. 225–252. Dordrecht: Springer.

Zhang, L., Hu, B., Li, W., Che, R., Deng, K., Li, H. et al. 2014. OsPT2, a phosphate transporter, is involved in the active uptake of selenite in rice. *New Phytol.* 201: 1183–1191.

The genetic loci associated with selenium accumulation in wheat grains under soil surface drenching and foliar spray fertilization methods

T. Li* & A. Wang
College of Agronomy, Yangzhou University, Yangzhou, China

G. Bai
USDA-ARS Hard Winter Wheat Genetics Research Unit, Manhattan, KS, USA

L. Yuan & X. Yin
Suzhou SeTek Ltd., Suzhou, China
Advanced Lab for Selenium and Human Health, Suzhou Institute of University of Science and Technology of China, Suzhou, China

1 INTRODUCTION

Selenium (Se) is one of the beneficial microelements in animals (Kato et al., 2010). Se deficiencies have been shown to be associated with more than 40 human diseases. Thus, improving human intake of Se can be beneficial for populations living in Se-deficient regions (Broadley et al., 2010). Wheat is one of the staple crops that has the capacity to accumulate Se in grains (Lyons et al., 2005). However, the potential for Se biofortification in wheat depends on many factors including wheat genotypes, genotype-environment interactions, fertilizer types, fertilization methods, and application time, etc. The objectives of this paper were to identify, using genome-wide association mapping strategy (GWAS) strategy, the genetic loci underlying Se accumulation in grains under foliar spray (FS) and soil surface drenching (SSD) fertilization methods. The results of this paper could be useful in improving the efficiency of Se biofortification by marker-assisted selection.

2 MATERIALS AND METHODS

2.1 Materials and Se fertilization

One hundred and twenty wheat varieties from domestic and overseas were fertilized with selenate - through SSD and FS methods. In the SSD experiment, the wheat was planted in plastic pots filled with 12 kg of soil, with four pots per cultivar and six plants per pot. Two pots (replicates) were for control (ck) and the other two pots for selenate application (trt). A 30 mg Se/L selenate solution was applied onto the surface of soil per pot at the booting stage. FS experiment was conducted in the field, with two replicates per cultivar and 15 plants per replicate (row). A 30 mg Se/L selenate solution was sprayed onto the leaves of wheat at the booting stage. Se concentrations in the seeds of both controls and Se treatments were determined by ICP-AES (IRIS Intrepid II XSP, Thermo-Fisher Scientific, St. Louis, USA).

2.2 Association mapping

Genome wide SNP markers were genotyped using Wheat-9000SNPInfinium method at Kansas State University, USA. Association of SNP markers with traits and estimation of their explained phenotypic variation were performed using Mixed Linear Model (MLM) in TASSEL 3.0 pipeline.

3 RESULTS AND DISCUSSION

3.1 SNP markers associated with Se accumulation in grain seeds in FS experiment

A total of 82 SNP markers were identified to be associated with Se accumulation in wheat grains under FS method ($p < 0.01$), and 58 of 82 markers were reproducible in two or more evaluations. These markers explained an average phenotypic variation of 9.1%, ranging from 6.0% (SNP604) to 16% (SNP4764). The majority of these markers were assigned to the wheat chromosomes 1A, 1B, 3B, 5B, 6B, and 7B, suggesting the corresponding chromosomal bins harboring these SNPs were responsible for Se uptake or accumulation.

3.2 SNP markers associated with selenium accumulation in SSD experiment

Thirty-four SNP markers were identified to be associated with Se accumulation in wheat grainsunder SSD method. These markers explained an average of phenotypic variation of 8.4%, ranging from 6.4% (SNP6119) to 15.3% (SNP6458). The majority of significant markers were on the chromosomes 1A, 2B, 3B, and 7B. Chromosome 2B had up to 17 significant

Table 1. SNP markers in common under the two methods and the phenotypic variation explained.

Markers	R^2 (FS)	R^2 (SSD)
SNP3089	0.075	0.067
SNP2619	0.067	0.075
SNP3901	0.071	0.068
SNP5646	0.065	0.072
SNP7125	0.065	0.072
SNP7391	0.063	0.072
SNP3686	0.107	0.09

markers, suggesting the QTL on 2B chromosome was stable and crucial for Se accumulation. No significant markers were found on the chromosomes 3D, 4B, 4D, and 5D.

3.3 *SNP markers associated with Se concentrations under both methods*

Seven SNP markers were in common both under FS and SSD methods (Table 1), indicating these loci might be crucial for Se translocation and distribution.

4 CONCLUSIONS

Although the genes for Se uptake and metabolism have been reported in model plants, the genetic loci responsible for Se accumulation in wheat under the two fertilization methods were probably reported here for the first time. Most of the genetic loci responsible for Se accumulation identified here were application method-unique. Those markers reproducible in two or more evaluations were reliable and the associated markers could be used in marker assisted improvement of Se uptake, translocation and accumulation.

ACKNOWLEDGEMENT

This work was supported by grants from the Priority Academic Program Development of Jiangsu Higher Education Institutions, and by the Innovative Research Team of Universities in Jiangsu Province.

REFERENCES

Broadley, M., Alcock, J., Alford, J., Cartwright, P., Foot, I., Fairweather-Tait, S., et al. 2010. Selenium biofortification of high-yielding winter wheat (*Triticum aestivum* L.) by liquid or granular Se fertilisation. *Plant and Soil,* 332: 5–18.

Kato, M., Finley, D., Lubitz, C., Zhu, B., Moo, T.A., Loeven, M.R., et al. 2010. Selenium decreases thyroid cancer cell growth by increasing expression of GADD153 and GADD34. *Nutr Cancer,* 62: 66–73.

Lyons, G.H., Judson, G.J., Ortiz-Monasterio, I., Genc, Y., Stangoulis, J.C., & Graham, R.D. 2005. Selenium in Australia: selenium status and biofortification of wheat for better health. *J Trace Elem Med Biol* 19: 75–82.

Molecular mechanisms of selenium hyperaccumulation in *Stanleya pinnata*: Potential key genes *SpSultr1;2* and *SpAPS2*

M. Pilon*, A.F. El Mehdawi, J.J. Cappa, J. Wang & E.A.H. Pilon-Smits
Biology Department, Colorado State University, Fort Collins, CO, USA

1 INTRODUCTION

Due to the chemical similarity of selenate to sulfate, all plants can take up selenate non-specifically via sulfate transporters and assimilate it into SeCys. The Se/S ratio in most plants reflects that of their environment, suggesting their transporters cannot distinguish between the two elements. However, Se hyperaccumulator species like *Stanleya pinnata* can distinguish between selenate and sulfate, and preferentially accumulates Se over S (Harris *et al.*, 2014). In nature, its plant Se levels on seleniferous soils can approach 0.5%, all accumulated in organic forms (Cappa *et al.*, 2014). Most other plant species, including the related species *S. elata*, do not accumulate more than 0.01% Se in nature, and accumulate more inorganic forms of Se (Cappa *et al.*, 2015). Our goal is to elucidate the mechanisms of Se hyperaccumulation and hypertolerance in *S. pinnata*. This information will provide important new insight into the mechanisms underlying the physiological processes of Se hyperaccumulation: selenate/sulfate discrimination, extraordinarily high selenate influx, assimilation and tolerance. The results will also be applicable for Se biofortification and phytoremediation, which are currently hindered by high S levels that impede plant Se uptake, and by suboptimal plant Se accumulation or Se phytotoxicity. The genes characterized here may be used to develop crops with selenate-specific uptake, and the capacity to efficiently convert inorganic selenate to less toxic, anticarcinogenic organic forms of Se.

2 MATERIALS AND METHODS

2.1 *RNA sequencing*

Seeds of *S. pinnata* and *S. elata* were surface-sterilized and grown for 30 days on 1/2 strength Murashige and Skoog agar medium containing 0 μM or 20 μM sodium selenate. At harvest, roots and shoots were separated (3 bioreplicates per treatment). A parallel experiment was run for elemental analysis and dry weights. Total RNA was extracted (RNA Mini Kit), mRNA purified and used to construct Illumina cDNA libraries using the TruSeq RNA Kit, then sequenced (Illumina HiSeq-2000). Pair-end 100 bp sequencing was performed for one biological replicate for each species, organ, and Se treatment combination (8 samples). The quality filtered data were assembled *de novo* using Trinity (Grabherr et al., 2011). Next, single-end 50 bp sequencing was performed on the remaining two biological replicates for each species, organ, and Se treatment combination (16 samples). Quality filtered reads for all three biological replicates were aligned to the de novo assemblies using NextGENe version 2.17. Assembled reads were annotated against the *A. thaliana* cDNA database (TAIR, http://www.arabidopsis.org/) using BLASTn and assigned homologs with an e-value threshold of 0.00005. Contigs annotated to the same ATID were associated with one gene and their RPKM values were summed.

2.2 *Molecular, physiological & biochemical assays*

The Illumina RNA seq information was used to design primers against the 5′ and 3′ ends of the *S. pinnata* and *S. elata* cDNAs for *Sultr1;2* and *APS2*. RNA was isolated from roots of the two species grown as described above (no Se), cDNA produced (Stratagene RT kit), and PCR was used to obtain the full-length cDNAs, which was subsequently sequenced and cloned into the TOPO®*E. coli* vector.

3 RESULTS AND DISCUSSION

3.1 *RNA sequencing of S. pinnata and S. elata*

To identify key genes for Se hyperaccumulation, the transcriptomes of *S. pinnata* and *S. elata* (both diploid accessions) were Illumina sequenced and analyzed (root and shoot, +/−20 μM selenate, in triplicate). Transcriptome comparison (Illumina RNA sequencing) between the physiologically divergent *S. pinnata* var. *pinnata* and *S. elata* revealed constitutive upregulation of sulfate/selenate transporters, the sulfate/selenate assimilation pathway, as well as genes involved in antioxidant functions and protein recycling and repair. These functions likely are related to upregulated production of the plant hormones jasmonic acid, salicylic acid, and ethylene. Several potential key genes will be subject of further study.

Table 1. Expression levels (in RPKM) of a putative sulfate/selenate transporter (homologue of AtSultr1;2) and ATP sulfurylase (homologue of AtAPS2) obtained from Illumina RNA sequencing of hyperaccumulator *S. pinnata* and non-accumulator *S. elata*. Values shown are means ($n = 3$).

	Root contr	Root SeO$_4$	Shoot contr	Shoot SeO$_4$
AtSultr1;2 homologue				
S. pinnata	732.035	783.272	17.229	14.698
S. elata	32.796	17.984	10.651	9.972
AtAPS2 homologue				
S. pinnata	6844.687	6886.744	374.704	163.390
S. elata	81.056	14.475	137.888	160.168

Two genes stood out because they were expressed at orders of magnitude higher levels in roots of *S. pinnata* than roots of *S. elata*, independent of Se supply (Table 1).

The first gene is a homologue of *A. thaliana* sulfate transporter *Sultr1;2*, i.e. the main transporter responsible for sulfate and selenate into plant roots. It is reasonable to hypothesize that this exceptionally high expression of *SpSultr1;2* is responsible for the hyperaccumulator Se levels in *S. pinnata*. Assuming *SpSultr1;2* is the main portal for selenate into the hyperaccumulator, it may also be hypothesized to be responsible for its capacity to distinguish selenate from sulfate and to preferentially take up selenate. The other gene that stood out is a homologue of *A. thaliana* ATP sulfurylase 2 (*AtAPS2*), the first enzyme of the sulfate/selenate assimilation pathway. Constitutive overexpression of ATP sulfurylase 1 (APS1) from *A. thaliana* in *B. juncea* was shown earlier to confer enhanced selenate tolerance, Se accumulation, and selenate assimilation to organic Se (Pilon-Smits et al., 1999). Since those early studies showed APS activity to be rate-limiting for selenate assimilation to organic Se, it is reasonable to hypothesize that the extra-ordinarily high expression of *SpAPS2* in roots is a key mechanism for *S. pinnata* Se tolerance via its capacity to assimilate Se to less toxic, organic forms.

3.2 *Characterization of Stanleya Sultr1;2 and APS2*

Our aims for further study are the following:

(1) To determine whether the elevated expression levels of *SpSultr1;2* and *SpAPS2* are associated with elevated uptake rates of selenate and/or sulfate into *S. pinnata* roots, and elevated levels of ATP sulfurylase activity.
(2) To functionally analyze the sulfate and selenate transport properties of the *S. pinnata* and *S. elata* SULTR1;2 transporters in baker's yeast *(Saccharomyces cerevisiae)*.
(3) To analyze the intracellular location of the *SpSULTR1;2* and *SpAPS2* proteins, as well as their tissue-specific expression patterns in the root.
(4) To determine the physiological effects of expression of *S. pinnata* and *S. elata Sultr1;2* and *APS2* in *B. juncea*.

We have currently cloned and sequenced the S. pinnata and S. elata Sultr1;2 and APS2 genes, and compared their expression. This revealed interesting structural differences will be further investigated. These include apparent differences in APS2 targeting (cytosolic vs. chloroplastic), as well as several aminoacid substitutions that may affect the kinetic properties of the proteins or their regulation. These will be further investigated by expression in yeast and Arabidopsis.

REFERENCES

Cappa, J.J., Yetter, C., Fakra, S.F., Cappa, P.J., DeTar, R., Landes, C. et al. 2014. Evolution of selenium hyperaccumulation in Stanleya (Brassicaceae) as inferred from phylogeny, physiology and x-ray microprobe analysis. *New Phytol.* 205: 583–595.

Cappa, J.J., Cappa, P.J., El Mehdawi, A.F., McAleer, J.M., Simmons, M.P., & Pilon-Smits E.A.H. 2015. Characterization of selenium and sulfur accumulation in Stanleya (Brassicaceae). A field survey and common-garden experiment. *Am. J. Bot.* 101: 830–839.

Harris, J., Schneberg, K.A., & Pilon-Smits, E.A.H. 2014. Sulfur – selenium – molybdenum interactions distinguish selenium hyperaccumulator *Stanleya pinnata* from non-hyperaccumulator *Brassica juncea* (Brassicaceae). *Planta* 239: 479–491.

Pilon-Smits, E.A.H., Hwang, S.B., Lytle, C.M., Zhu, Y.L., Tai, J.C., Bravo, R.C. et al. 1999. Overexpression of ATP sulfurylase in *Brassica juncea* leads to increased selenate uptake, reduction and tolerance. *Plant Physiol.* 119: 123–132.

Effect of soil pH on accumulation of native selenium by Maize (*Zea mays var. L*) grains grown in Uasin Gishu, Trans-Nzoia Kakamega and Kisii counties in Kenya

S.B. Otieno*
School of Public Health, Kenyatta University, Kenya; State Department for Livestock, Government of Kenya

T.S. Jayne & M. Muyanga
Department of Agricultural, Food, and Resource Economics, Michigan State University USA

1 INTRODUCTION

Selenium (Se) occurs naturally in metalloid form and it is essential for human health (Otieno et al., 2014b; Chilima et al., 2011; Longchamp et al., 2011; Fordyce, 2007) and is needed in traces for human health. Selenium plays important role in body antioxidation system (Jezek et al., 2012; Lien et al., 2008) and in the process protects the body cells from free radical activities, cardiovascular diseases, inhibits HIV virulence and reduces development of AIDS, protects body from complications associated with asthma and is important in thyroid hormone metabolism (Jezek et al., 2012). Selenium mainly enters the food chain through crops and vegetables growing in soil. Soil characteristics such as pH and porosity, affect Se uptake by plants (Diplock, 1993). Moreover, the decrease in redox potential (Eh) by water logging affects availability of Se, while an increase in clay and organic matter in the soil also reduces uptake of Se by crops (Boonstra et al., 2007).

In Kenya, maize is the main cereal consumed by 96% of the population. This crop is therefore an ideal and inexpensive source to provide adequate Se to target populations. The counties of Trans Nzoia, Kakamega, Uasin Gishu and Kisii comprise the maize belt in Kenya and are considered as the bread basket for the country. As a result our study evaluates the impact that soil pH may exert on Se uptake by maize.

2 MATERIALS AND METHODS

One hundred samples of maize grains (250 grams, respectively) were collected and put in separate labeled Ziplocs along with 100 soil samples from the same locations. The samples were transported at room temperature to laboratory for analysis. The maize Se level was analyzed at KEBS laboratory by a method reported by William et al. (2008) and Mitoko et al. (1979). In brief, concentrations of Se were determined by Atomic Absorption Spectrophotometry after making appropriate dilutions. Approximately 0.250 grams of maize samples were accurately weighed and put in a graduated tube and mixed with hydrogen peroxide, followed by digestion with perchloric acid. Nitric acid (0.75 ml) and 2.25 ml hydrochloric acid (HCl) were carefully added. The contents were thoroughly mixed with test tube shaker. The mixture was heated at $80°C$ for 1 hour on aluminum heated block, and allowed to cool. Approximately 11.5 ml of distilled water was added, mixed thoroughly and allowed to settle. A portion of the solution was centrifuged for analysis by AAS (Perkin Elmer Analyst 300, Germany).

The soil pH levels were measured in a soil water suspension at NAR Laboratory. Fifty ml of water was added to a glass beaker containing 10 grams of sediments. The suspension was stirred using a glass rode and then left to equilibrate for 16 hours. The pH was measured using a pH electrode 520ApH meter (Orion Research Inc., Boston, USA).

3 RESULTS AND DISCUSSION

The mean Se levels in *Zea mays* grains varied from 1.82 ± 0.76 mg/kg in Kakamega to 2.11 ± 0.86 mg/kg in Trans Nzoia, with overall mean being 1.938 mg/kg (Table 1). There was no significant difference in Se concentration in maize grains observed between the batches $\chi = 26.04$ ($p = 1.000$, df = 76). The mean soil pH varied from 5.43 ± 0.58 in Kisii to 5.85 ± 0.32 in Trans Nzoia. No significant difference in pH between the five batches were observed (p = 0.58). There was no significant positive correlation between the soil pH and the soil Se concentration (p = 0.162) with Pearson's correlation of -0.143.

The mean value of Se in the *Zea mays* grains in all batches were slightly higher than those earlier reported by Chilima et al. (2011) but lower than those reported by Funwe et al. (2012). However, the values were consistent with those reported by Otieno et al. (2014a). These current Se levels likely reflect the availability of the more soluble selenate (SeVI) species in the soil. The Se uptake by *Zea mays* roots depends on soil

Table 1. Mean selenium concentrations in *Zea mays* grains and soil pH ($n = 100$).

	Se (mg/kg)	Soil pH
Uasin Gishu 1	1.88 ± 0.98	5.5 ± 0.47
Uasin Gishu 2	1.90 ± 1.07	5.85 ± 0.32
Trans Nzoia	2.11 ± 0.86	5.77 ± 0.43
Kakamega	1.82 ± 0.76	5.71 ± 0.55
Kisii	1.97 ± 0.99	5.43 ± 0.58

conditions (Funwe, 2012; Zhang et al., 2010; Mayland et al., 2007), which includes the pH, CEC, and total carbon that in turn determines the Se speciation and the mobility in the soil (Jezek et al., 2012). Plants tend to absorb selenate via sulphate transporters. Lie et al. (2010) reported that lower levels of sulphur in the soil tend to upregulate the uptake of Se by stimulating manufacture of sulphate transporters in the roots, which increases the uptake of selenate. Since no significant correlation ($p > 0.05$) was found between Se content in grains and soil pH, it is likely that the soil was deficient of sulphur, which resulted in high levels of grain Se.

4 CONCLUSIONS

Using Se levels in *Zea mays* grains as proxy indicator of Se speciation, we conclude that soil pH levels did not significantly affect the uptake and accumulation of native Se in *Zea mays* grains at the study site.

REFERENCES

Chilima, A.D.C., Young, S.D., Black, C.R., Meachim, M.C., Lammel, J., Broadly, M.R. 2011. Agronomic biofortification of Maize with selenium in Malawi. pp. 79–80. In: G.S. Banuelos, Z.-Q. Lin, X.B. Yin, and D. Ning (eds.), Selenium: Global Perspectives of Impacts on Humans, Animals and Environment. University of Science and Technology of China Press, Hefei, China.

Fordyce, F. 2007. Selenium geochemistry and health. Ambio, 36(1): 94–97.

Losak, T., Hlusek, J., Juzl, M. Elzner, P. 2012. Selenium – An important antioxidant in crop biofortification. pp. 345–350. In: M.A. El-Missiry (ed.), Antioxidant Enzyme, Rijeka, Croatia.

Longchamp, M., Angeli, N., Castrec-Rouell, M. 2011. Uptake of selenate and/or selenite in hydroponically grown Maize plants and interaction with some essential elements. pp. 83–84. In: G.S. Banuelos, Z.-Q. Lin, X.B. Yin, and D. Ning (eds.), Selenium: Global Perspectives of Impacts on Humans, Animals and Environment. University of Science and Technology of China Press, Hefei, China.

Lien, A.I.P., Hua, H., Pham-Huy, C. 2008. Free radicals, antioxidants in disease and health. International Journal of Biomedical Science. 4(2): 89–96.

Mitoko-Ohayo, G.J.A. 1995. Concentrations of heavy metals, organochlorine pesticides, organic and microbial pollution in the Nairobi River and its Tributaries. Vol. 1: The Government of Netherlands; The Royal Netherlands Embassy, Nairobi.

Mayland, H.F. 1994. Selenium in plant and animal nutrition. 29–47. In: W.T. Frankenberger and S. Benson (eds.), Selenium in the Environment, Marcel-Decker, New York.

Otieno, S.B., Were, F., Kabiru, E.W., Waza, K. 2014a. The study of selenium content of foods in a high HIV prevalent community: A case study of Pala Bondo District Kenya. pp. 62–65. In: G.S. Banuelos, Z.-Q. Lin, and X.B. Yin (eds), Selenium in the Environment and Human Health. Francis and Taylor Group (CRC), London.

Otieno, S.B., Were, F., Kabiru, E.W., Waza, K. 2014b. The effect of selenium on CD4 T cell count of HIV1 positive orphan children at Orongo Widows and Orphans in Kisumu, Kenya. International Journal of Science and Technology 4(3): 233–241.

William, R.M. and Stephen, G.C. 2008. Elemental analysis manual: Section 4. 4A, Inductively-coupled plasma-atomic emission spectrometric determination of elements in food using microwave assisted digestion; appendix a supplemental information of in-house method validation version I (June 2008)

Zhang, L.H., Yu, F.Y., Li, Y.J. and Miao, Y.F. 2011. Effect of pH on physiological characteristic of selenite uptake by Maize Roots. 87–88. In: G.S. Banuelos, Z.-Q. Lin, X.B. Yin, and D. Ning (eds.), Selenium: Global Perspectives of Impacts on Humans, Animals and Environment. University of Science and Technology of China Press, Hefei, China.

Quantification, speciation and bioaccessibility of selenium from Se-rich cereals cultivated in seleniferous soils of India

N. Tejo Prakash
School of Energy and Environment, Thapar University, Patiala, India

1 BACKGROUND

The chemical speciation of selenium (Se) is a critical consideration in assessing its bioavailability and toxicity. In plants, selenate and selenite are the major inorganic forms of Se that are readily absorbed and assimilated to organic Se compounds. Consequently, the non-volatile organic Se forms are key to the bioavailability and transfer to animals and human.

For last few years the attention of our research group, in collaboration with ISS, Italy and University of Pau, France, has been towards the quantification, speciation and bioaccessibility of Se in cereal grains cultivated in Se-impacted soils. The focus of the work has been towards exploring potential use of Se rich dietary matrices as sources of supranutritional doses of Se. Supranutritional amounts of a nutrient are levels that are above the established needs of an individual but not toxic. Research, since recent past, indicates a significant role for Se in counteracting tumorigenesis, particularly at supranutritional levels (Rocourt & Cheng, 2013).

The observations presented are a compilation of studies carried out over five years (2009–2014) in independent and collaborative studies on the cereal grains cultivated in the Se-impacted region of Punjab, India.

2 QUANTIFICATION STUDIES

Samples of cereal grains viz., wheat, rice and maize, were collected at sites near the villages of Jainpur and Barwa geographically located at 32′46°N, 74′32°E, in the Nawanshahr-Hoshiarpur Region, Punjab, India. The samples were collected during the years 2009 and 2014.

Quantification of Se was carried out using neutron activation analysis (NAA) (Sharma et al., 2009) for samples collected in 2009; and using florescence spectrometry (FS) for the samples collected in 2014. The sample preparation protocol followed was similar in both the cases.

All the analyses were cross-validated using appropriate certified reference materials. The data presented in Table 1 summarizes the observations on Se levels in grains determined using various techniques.

Table 1. Selenium ($\mu g/g$) in Se rich cereal grains, collected during different periods, as analysed by various techniques

Whole grains	FS (2014)	NAA (2009)
Wheat	122.9 ± 0.6	115.1 ± 0.6
Maize	26.5 ± 0.2	13.0 ± 0.5
Rice	19.7 ± 0.2	16.2 ± 0.5

Samples of wheat and maize collected in 2009 were also validated using ICP-MS and the Se levels ($\mu g/g$) were 96.9 ± 0.2 and 37.2 ± 0.2, respectively.

These observations taken over time, and obtained using various techniques, are higher than those reported in the Se-impacted region of Enshi, China or from other Se-rich areas (Aureli et al., 2012). Cereals grown in soils with phytoavailable Se have a concentration ranging from $< 0.1 - 1.0 (\mu g/g)$ (Aureli et al., 2012).

3 SPECIATION STUDIES

Selenium species were determined by means of HPLC/HILIC-ICP-MS and obtained fractions were further characterized by SEC-ICP-MS and Orbitrap MS/MS. The detailed instrumentation and the experimental conditions are outlined in Cubadda et al. (2010) and Aureli et al. (2012). The speciation studies were carried in samples collected in 2009.

Initial studies on wheat using HPLC-ICP-MS in-dicated the 72% selenomethionine, 6% selenite and 0.2% methylselenocysteine. However, the total sum of species accounted to 89% of the Se extracted (Cubadda et al., 2010).

Detailed characterization carried out on wheat as well as other cereals using HILIC-ICP-MS and high resolution ESI-Orbitrap MS allowed a comprehensive insight into unaccounted metabolites in the 2010 studies, as selenosugars, in wheat as well as in rice and maize. The detection of selenosugars was a novel finding initiating interest on furthering the examination on role of these compounds in plants.

4 BIOACCESSIBILITY STUDIES

Bioaccessibility of Se in the cereal matrices were determined through simulated enzymatic gastric and

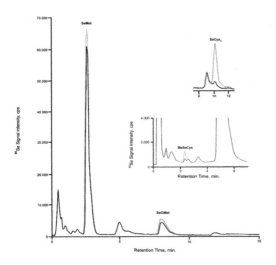

Figure 1. Chromatograms of wheat extracts with SeMet as a dominant fraction along with other minor fractions.

Table 2. Bioaccessibility of Se (%) in Se rich cereal grains across gastric and intestinal digestion.

Whole grains	Gastric	Intestinal
Wheat	75	98
Maize	60	95
Rice	51	82

intestinal digestion of these cereal grains. In general, the apparent absorption of the organic Se compounds in foods was appreciable (about 70–95%) (Combs & Combs, 1986). However, it can vary according to the digestibility of the various Se-containing food proteins and to the pattern of Se compounds present in the particular food. In the present study, the bioaccessibility varied with the variety of the cereals examined (Table 2).

The variation in the bioaccessibility is presumed to be due to varying concentrations of water soluble Se containing Osborne fractions and their bioaccessibility, an aspect that needs further investigations. A similar extent of bioaccessibility was observed in processed wheat (as wheat bread) and rice (as boiled rice) with SeMet being the dominant bioaccessible fraction in both the cereals.

5 CONCLUSIONS

The studies carried out during years 2009–2014 on the Se in cereal grains indicated that the levels of Se in food crops cultivated in the region contain significantly high concentrations hitherto less observed in other regions of the world. These cereals facilitate good bioaccessibility of Se and therefore can be potential dietary sources of this nutrient to Se deficient population.

ACKNOWLEDGMENTS

The author duly acknowledge the grants from BRNS, India; and collaborative support of Dr. F. Cubadda, ISS, Rome and Dr. L. Ouerdane, Univ. Pau, France, for extensive quantification and speciation studies.

REFERENCES

Aureli, F., Ouerdane, L., Bierla, K., Szpunar, J., Tejo Prakash, N., & Cubadda, F. 2012. Identification of selenosugars and other low-molecular weight selenium metabolites in high-selenium cereal crops. *Metallomics* 4: 968–978.

Combs, G.F. & Combs, S.B. 1986. Selenium in foods and feeds. In G.F. Combs and S.B. Combs (eds.). *The Role of Selenium in Nutrition*: 41–126. New York: Academic Press.

Cubadda, F., Aureli, A., Ciardullo, S., D'Amato, M., Raggi, A., Acharya, R., et al. 2010. Changes in selenium speciation associated with increasing tissue concentration of selenium in wheat grain. *Journal of Agricultural and Food Chemistry* 58: 2295–2301.

Rocourt, C.R.B. & Cheng, W.-H. 2013. Selenium supranutrition: Are the potential benefits of chemoprevention overweighed by promotion of diabetes and insulin resistance? *Nutrients* 5: 1349–1356.

Sharma, N., Prakash, R., Srivastava, A., Sadana, U.S., Acharya, R., Tejo Prakash, N., et al. 2009. Profile of selenium in soil and crops in seleniferous area of Punjab, India by neutron activation analysis. *Journal of Radioanalytical and Nuclear Chemistry* 281: 59–62.

Wautersiella enshiensis sp. nov. – selenite-reducing bacterium isolated from a selenium-mining area in Enshi, China

Zhihao Qu
Jiangsu Bio-Engineering Research Centre of Selenium, Suzhou, Jiangsu, China;
China Center for Type Culture Collection, College of Life Sciences, Wuhan University, Wuhan, Hubei, China

Linxi Yuan* & Xuebin Yin
Jiangsu Bio-Engineering Research Centre of Selenium, Suzhou, Jiangsu, China

Fang Peng*
China Center for Type Culture Collection, College of Life Sciences, Wuhan University, Wuhan, Hubei, China

1 INTRODUCTION

Selenium (Se) is an essential trace element for human beings and is required for the synthesis of the essential amino acid selenocysteine for bacteria but it can be poisonous at high concentrations (Jong et al., 2015). A previous study revealed that a Se-mine drainage area in Enshi, Hubei contained Se concentrations as high as 20–500 mg/kg (DW) in soil/sediment (Yuan et al., 2013), but until now no Se-tolerant bacteria were reported from the sampling sites. This study focused on identifying and isolating Se-tolerant bacteria in the Se-mining area in Enshi.

2 MATERIALS AND METHODS

Soil samples were collected from the Se-mine drainage area in Enshi, China (Yuan et al., 2013) and cultured in a medium containing Se of 200 μg/L. Strain YLX-1^T was selected to perform 16S rRNA gene analysis. Cell morphology was examined by phase-contrast (Olympus BX51) and transmission electron microscopy (Hitachi H-8100). The metabolism parameters and enzyme activities on YLX-1^T were also tested. The DNA G + C content was determined using HPLC following the method by Mesbah and Whitman (1989).The fatty acids methyl esters were obtained and tested according to Sherlock Microbial Identification System's protocol and analyzed by Agilent 6890 N, MIDI Sherlock TSBA6 (Sasser, 1990). Polar lipids were extracted and analyzed as described by Tindall (1990) and Ventosa et al. (1993) using two-dimensional TLC (silica gel 60 F254 plates, layer thickness 0.2 mm, Merck).

3 RESULTS AND DISCUSSION

The strain YLX-1^T cells were rod-shaped (0.4–0.7 × 0.8–2.0 μm), non-spore-forming, Gram-negative, facultatively anaerobic and non-motile. Its colonies were yellow, smooth, circular, and convex with entire margins on TSA medium after 1 day at 28°C. Growth occurred at 4-37°C (but optimum 28°C) and at pH 5.0–9.0 (but optimum at pH 7), but no growth in the presence of ≥ 1.0% NaCl. The strain was catalase- and oxidase-positive. Tween 40 was hydrolysed, but not Tween 20, 60, 80, starch, DNA and casein.

The strain YLX-1^T was identified as *Wautersiella enshiensis* based on 16S rRNA data (Fig. 1). The GenBank/EMBL/DDBJ accession number for the 16S rRNA gene sequence of strain YLX-1^T is KF923410.In API ZYM strips, cells were positive for the activities of alkaline phosphatase, esterase (C4), esterase lipase (C8), leucine arylamidase, trypsin, acid phosphatase, naphthol-AS-BI-phosphohydrolase, α-glucosidase and β-glucosidase, were weakly positive for the activities of valine arylamidase and negative for other reactions in the strip. In API 20NE and API 20E strips, the reduction of nitrate and the production of hydrogen sulfide did not occur. It is positive for urease, gelatinase, galactosidase, arginine dihydrolase, aesculin hydrolysis, citrate utilization and Voges-Proskauer reaction. Moreover, it could assimilate glucose, maltose, malic acid and sodium citrate.

In summary, this study showed that *Wautersiella enshiensis*was can grow in selenite-enriched medium, having a Se concentration of up to 6000 μg/mL, and it is able to reduce selenite into red elemental nano-Se.

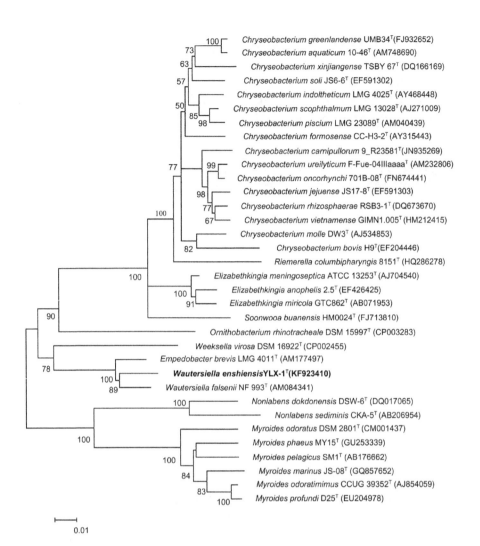

Figure 1. Neighbor-joining phylogenetic tree, based on 16S rRNA gene sequences, showing the phylogenetic position of strain YLX-1^T. Bootstrap values (1000 replications) are shown as percentage at each node only if they are 50% or greater. Bar, 1% sequence diverge.

ACKNOWLEDGEMENTS

This work was supported by National Natural Science Foundation of China (NNSFC31400091), Natural Science Foundation of Jiangsu Province (BK2012195, BK2012202), and Applied Basic Research Project of Suzhou (SYN201306).

REFERENCES

Jong, M.S., Reynolds, R.J.B., Richterova, K., Musilova, L., Staicu, L.C., Chocholata, I., et al. 2015. Selenium hyperaccumulators harbor a diverse endophytic bacterial community characterized by high selenium resistance and plant growth promoting properties. Frontiers in Plant Sciences, 6: 113.

Mesbah, M., Premachandran, U. & Whitman, W.B. 1989. Precise measurement of the G+C content of deoxyribonucleic acid by high-performance liquid chromatography. *Journal of Systematic and Evolutionary Microbiology*, 39: 159–167.

Sasser, M. 1990. Identification of bacteria by gas chromatography of cellular fatty acids, MIDI Technical Note 101. Newark, DE: MIDI.

Tindall, B.J. 1990. Lipid composition of *Halobacterium lacusprofundi*. *FEMS Microbiological Letter*, 66: 199–202.

Ventosa, A., Marquez, M.C., Kocur, M. & Tindall, B.J. 1993. Comparative study of "*Micrococcus* sp." strains CCM 168 and CCM1405 and members of the genus *Salinicoccus*. *International Journal of Systematic and Evolutionary Microbiology*, 43: 245–248.

Yuan, L.X., Zhu, Y.Y., Lin, Z.Q., Banuelos, G., Li, W., Yin, X.B. 2013. A novel selenocystine-accumulating plant in selenium-mine drainage area in Enshi, China. *PLoS ONE*, 8(6): e65615. doi:10.1371/journal.pone.0065615.

Selenium in Osborne fractions of Se-rich cereals and its bioaccessibility

Noorpreet I. Dhanjal* & Siddharth Sharma
Department of Biotechnology, Thapar University, Patiala, India

N. Tejo Prakash
School of Energy and Environment, Thapar University, Patiala, India

1 INTRODUCTION

Cereal grains contain upto 7–12% proteins, 65–75% carbohydrates, 2–6% lipids and 12–14% water. Osborne (1907) divided the cereal proteins into four fractions based on their solubility in different solutes i.e., albumins (water), globulins (saline solution), glutelins (alkaline solution) and prolamins (70–80% alcohol). The percentage protein concentration of these fractions is significantly different in different cereal grains.

Cereal grains contain significantly good Se concentration ranging between 29–185 μg/g (Cubadda et al., 2010) and can potentially be used as Se supplements in case of deficiency. Dietary sources rich in organic forms (SeMet and SeCys) are easily bioaccessible (≥90%) in intestine and efficient reservoir for long term storage (Behne et al., 2009). Dietary sources rich in organic forms of Se are easily bioaccessible (≥90%) in intestine, which bind with albumins along with β-proteins for the circulation in blood stream. In non-accumulators, selenium mainly accumulates in protein fractions as SeMet and SeCys.

Keeping in mind, growing importance of dietary sources of Se the objective of study was to determine the distribution and bioaccessability of selenium in Osborne fractions of cereal grains collected from seleniferous fields.

2 MATERIALS AND METHODS

Wheat (*Triticum aestivum*) rice (*Oryza sativa*) and maize (*Zea mays*) were collected in November 2014, near the villages of Jainpur and Barwa geographically located at 32°46′N, 74°32′E, in the Nawanshahr-Hoshiarpur districts of Punjab (India). Non-seleniferous samples were collected from agricultural fields of Patiala (30°32′N, 76°40′E), India.

Osborne fractions viz., albumin, globulin, glutelin and prolamin were extracted from cereal grains according to the Ju et al. (2001). The extracts were centrifuged, washed with acetone, cryo-dried and crushed to powder for further use.

Selenium was determined in defatted whole grain flour, starch residue and all protein fractions by using the procedure given by Levesque and Vendette (1971). Se quantification in each sample was carried out by relative method using NIST certified Selenium ICP standard solution (SRM-1349).

The *in vitro* gastrointestinal digestion was performed according to Bhatia et al. (2013). Samples were centrifuged at 5000 g for 10 min (4°C), supernatant was filtered through 0.45 μm pore size filter and subjected to selenium quantification.

3 RESULTS AND DISCUSSION

3.1 Selenium in soil and cereal grains

The mean selenium concentration in soil samples collected during different cropping seasons was 3.32 ± 0.15 μg/g ($n = 3$). Soil with ≥ 1 μg/g Se concentration is seleniferous in nature whereas with 0.4 μg/g concentration the soil are generally considered selenium deficit.

The concentrations of selenium (μg/g) in whole grains of cereals viz., wheat, maize and rice were 122.9 ± 0.6; 26.5 ± 0.2; and 19.7 ± 0.2 respectively, when compared to Se levels in grains (1.1 to 1.3 μg/g) from non-seleniferous area. Se concentration in most of the World's wheat ranges between 0.02–0.6 μg/g (Alfthan and Neve, 1996). Wheat from highly seleniferous areas of South Dakota, USA has Se to the extent of 30 μg/g. The present study, thus, reports selenium concentrations in wheat and maize at much higher levels than those reported elsewhere.

3.2 Osborne fractions and selenium content

With reference to the selenium levels in various proteins, glutelin fraction dominated over other proteins across all the grains tested. Prolamin was observed to contribute significant Se levels despite being the lowest amongst all protein fractions across all grains. Overall selenium content in prolamin and glutelin fraction was higher than other proteins. Multiple comparison test showed significant variation ($p < 0.05$) in Se levels across grains but not significant across proteins. Correspondingly, the selenium

Table 1. Selenium in whole grains and Osborne fractions of Se-rich cereal grains.

	Se content (µg/g)		
Sample	Wheat	Rice	Maize
Whole grain	122.86 ± 0.6	19.72 ± 0.2	26.45 ± 0.2
Albumin	401.04 ± 1.9	28.62 ± 1.5	280.23 ± 2.6
Globulin	263.56 ± 4.2	241.58 ± 1.6	191.97 ± 0.8
Glutelin	562.94 ± 2.1	177.83 ± 1.3	358.91 ± 2.3
Prolamin	628.46 ± 6.9	257.06 ± 1.3	338.53 ± 1.3

Table 2. Bioaccessibility of Se from Osborne fractions of Se rich cereal grains during gastric (G) and intestinal (I) digestion.

	% age bioaccessibility (G/I)		
Fraction	Wheat	Rice	Maize
Whole grain	75/98	51/82	59/95
Albumin	78/88	42/71	79/99
Globulin	74/79	91/95	75/98
Glutelin	79/85	71/82	60/95
Prolamin	78/82	94/97	45/80

levels in the residual starch was below detectable limits, whereas in the spent residue, the concentration was 46.5 ± 0.7; 6.4 ± 0.8; and 10.6 ± 0.1 µg/g in wheat, maize and rice respectively as residual percentage loss.

Abundance of methionine and cysteine in each fraction represents non-specific incorporation of selenomethionine and selenocysteine into proteins, respectively. Aureli et al. (2012) reported that the total Se in rice is present in organic form (SeMet) dominantly in glutelin fraction (31.3%) and only 2.85% as inorganic Se. In turn, Se-methyl-selenocysteine and selenocystathione are present in non-protein fractions of Se-accumulating plants. It would, therefore, be of further interest to understand the profile of Se moieties in these protein fractions.

3.3 *Selenium bioavailability*

The bioavailability of Se in different protein fractions is directly proportional to the concentration present in it and is maximum in cereal grains. Hakkarainen (1993) reported the Se bioavailability from wheat (83–100%), barley (78–85%), oats (41–45%), fish (64–80%) and meat meal (22–30%).

The bioaccessible fractions across the cereal grains showed significant variation with wheat providing higher accessibility of Se, followed by rice and maize (Table 2). Bioaccessibility of Se through intestinal digestion was higher as compared to gastric digestion in all the Se rich protein samples investigated.

This research finding proves that the intestinal fluids sufficiently breakdown the Se containing proteins which is then easily absorbed.

4 CONCLUSIONS

In this study, the Se fractionation of Osborne proteins indicates their trafficking and deposition in cereal grains. In most of the countries, food system does not provide enough selenium to support the expression of selenoenzymes. To overcome the major health issues due to its deficiency either Se-biofortification process is acquired in these countries or seleniferous areas can be considered as source of production of Se rich plants through food as supra-nutritional supplementation.

REFERENCES

Aureli, F., Ouerdane, L., Bierla, K., Szpunar, J., Tejo Prakash, N. & Cubadda, F. 2012. Identification of selenosugars and other low-molecular weight selenium metabolites in high-selenium cereal crops. *Metallomics* 4: 968–978.

Behne, D., Alber, D. & Kyriakopoulos, A. 2009. Effects of long-term selenium yeast supplementation on selenium status studied in the rat. *Journal of Trace Elements in Medicine and Biology* 23: 258–264.

Bhatia, P., Aureli, F., D'Amato, M., Prakash, R., Cameotra, S.S., Tejo Prakash, N., et al. 2013. Selenium bioaccessibility and speciation in biofortified Pleurotus mushrooms grown on selenium-rich agricultural residues. *Food Chemistry* 140: 225–30.

Cubadda, F., Aureli, F., Ciardullo, S., D'Amato, M., Raggi, A., Acharya, R., et al. 2010. Changes in selenium speciation associate with increasing concentrations of selenium in wheat grains. *Journal of Agricultural and Food Chemistry* 58: 2295–301.

Hakkarainen, J. 1993. Bioavailability of selenium. *Norwegian Journal of Agricultural Science* 11: 21–35.

Ju, Z.Y., Hettiarachchy, N.S. & Rath, N. 2001. Extraction, denaturation and hydrophobic properties of rice flour proteins. *Journal of Food Science* 66: 229–232.

Osborne, T.B. 1907. *The Proteins of the Wheat Kernel.* USA: Carnegie Institution of Washington Publ. 84. (http://archive.org/stream/proteinsofwheatk00osborich#page/, accessed on April 19, 2015)

New insights into the multifaceted ecological and evolutionary aspects of plant selenium hyperaccumulation

E.A.H. Pilon-Smits, A.F. El Mehdawi, J.J. Cappa, J. Wang, A.T. Cochran & R.J.B. Reynolds
Biology Department, Colorado State University, Fort Collins, CO, USA

M. Sura-de Jong
Department of Biochemistry and Microbiology, Institute of Chemical Technology, Prague, Czech Republic

1 INTRODUCTION

While most plant species do not accumulate more than 100 mg/kg DW (0.01%) selenium (Se) when growing in their native habitat, even on seleniferous soils, Se hyperaccumulators can concentrate Se to 0.1–1.5% of dry biomass. Hyperaccumulation probably evolved independently in the Brassicaceae (*Stanleya*), Fabaceae (*Astragalus*) and Asteraceae (*Oonopsis, Symphyotrichum, Xylorhiza*) (Cappa & Pilon-Smits, 2014).

Selenium hyperaccumulators distinguish themselves from other plants in that they preferentially take up Se over sulfur (S), efficiently assimilate selenate into organic forms, effectively translocate Se to the shoot via the xylem and remobilize Se to reproductive structures and young leaves via the phloem where they concentrate Se in specific locations (e.g., epidermis, pollen, and ovules).

Evolutionary questions related to this fascinating trait are: why and how did Se hyperaccumulation evolve? Which selection pressures favor the evolution of enhanced Se accumulation, and are there also evolutionary constraints? Which genes are under selection, and which mutations gave rise to Se hyperaccumulation and hypertolerance? How are the gene products affected, and how does this affect plant transport and metabolic functions? Further, ecological questions related to hyperaccumulation are: how does the hyperaccumulation affect the plant's ecological relationships, its ecological partners, and the local ecosystem as whole (e.g. Se distribution, Se cycling)?

Why study Se hyperaccumulation? Not only are hyperaccumulator plants intrinsically interesting because of their unique and extreme properties, but a better understanding of their mechanisms of Se hyperaccumulation and hypertolerance may make it possible to confer these traits to crop species via genetic engineering. Also, understanding the ecological effects of Se hyperaccumulation and characterizing their ecological partners (e.g. microbes) may lead to various applications in, e.g., pest management, crop production, nutritional value (biofortification), and phytoremediation.

2 MATERIALS AND METHODS

2.1 *Field surveys*

One approach we used to study the ecological effects of hyperaccumulators on the species composition of their surrounding vegetation and microbiome is field surveys. These surveys were carried out in several seleniferous areas near Fort Collins, Colorado, USA as well as a non-seleniferous control site nearby. The plant survey included linear transects comparing areas with and without hyperaccumulators (*Astragalus bisulcatus, Stanleya pinnata*) as well as circular plots surrounding one hyperaccumulator or comparable nonaccumulator plant. For microbiome comparison, several hyperaccumulator and nonaccumulator species from seleniferous or non-seleniferous sites were collected and culturable rhizosphere and endosphere bacteria isolated.

2.2 *Molecular, physiological & biochemical assays*

Microbiome-related methods are described by Sura-de Jong *et al.* (2015). In short, the bacterial isolates were analyzed for selenate and selenite resistance and some also for their capacity to promote plant growth. The species of these microbiome communities was also investigated using non-culture based methods (by Illumina sequencing, T-RFLP). Plant Se resistance and accumulation were analyzed on either agar medium or Turface®gravel spiked with different concentrations of selenate, or on naturally seleniferous soil (Cappa et al., 2015). Plant growth was determined as a measure of Se resistance. To determine Se accumulation, elemental analysis was carried out after nitric acid digestion of plant samples followed by analysis with inductively-coupled plasma optical emission spectrometry (ICP-OES). Chemical speciation and localization was analyzed using liquid chromatography mass spectrometry (LC-MS) or by X-ray microprobe analysis at the advanced light source (ALS) in Berkeley, California (USA), all as described by Freeman *et al.* (2010).

3 RESULTS AND DISCUSSION

A combined phylogenetic, physiological and biochemical analysis of all *Stanleya* taxa, using both field surveys and controlled laboratory experiments, revealed that Se hyperaccumulation likely is a derived trait that evolved within the *S. pinnata* clade (Cappa et al., 2015). Selenium tolerance probably evolved before Se hyperaccumulation, and is more prevalent in the genus. Transcriptome comparison (Illumina RNA sequencing) between the physiologically divergent *S. pinnata* var. *pinnata* and *S. elata* revealed constitutive upregulation of sulfate/selenate transporters, the sulfate/selenate assimilation pathway, as well as genes involved in antioxidant functions and protein recycling and repair. These functions likely are related to upregulated production of the plant hormones jasmonic acid, salicylic acid and ethylene. Several potential key genes will be subject of further study. Furthermore, a phylogenomics analysis of all Stanleya species was recently completed, where >1,500 loci were sequenced. This will give insight into which genes are under selection. Also, the genome of hyperaccumulator *S. pinnata* is currently being sequenced.

Selenium accumulation was shown to protect plants from a wide variety of herbivores, as well as from fungal pathogens, suggesting that hyperaccumulation has evolved as a protection mechanism from biotic stresses (El Mehdawi & Pilon-Smits, 2012). Hyperaccumulators also appear to enrich their surrounding soil with Se, creating a toxic zone for Se-sensitive plant neighbors. Thus, hyperaccumulation may also serve as a form of elemental allelopathy. Selenium also offers a physiological benefit: a hyperaccumulator's growth can be enhanced as much as 2-fold by supply with Se. The mechanisms underlying this effect remain to be determined.

By means of their high Se levels and that of the soil around them, hyperaccumulators tend to negatively affect Se-sensitive ecological partners (microbes, herbivores, pollinators, plants) through toxicity and deterrence. At the same time, they offer a niche for Se-resistant ecological partners, including herbivores, pollinators, neighboring plants and the hyperaccumulator microbiomes (El Mehdawi & Pilon-Smits, 2012). Our studies so far indicate that vegetative cover is lower around hyperaccumulators, and plant species richness (number of species per soil surface area) appears to be higher. Hyperaccumulator-dependent plant community structure and Se resistance is still being compared. Microbial species richness appears similar for hyperaccumulators, and their microbiomes are characterized by extreme Se tolerance (>200 mM selenate/selenite). By favoring some ecological partners over others, hyperaccumulators may be hypothesized to affect species composition at various trophic levels. They may also affect Se cycling in their local ecosystem.

4 CONCLUSIONS

Selenium hyperaccumulation likely has evolved independently in five genera from three families. The selection pressures that have driven the evolution of Se hyperaccumulation may include both ecological and physiological benefits. Selenium can confer biotic stress protection, elemental allelopathy, and enhance plant growth. Selenium hyperaccumulators may profoundly affect their local ecosystem by concentrating Se, assimilating it into organic forms and depositing it in their vicinity. The high Se levels in hyperaccumulators appear to negatively affect Se-sensitive ecological partners while offering a unique niche to Se-resistant partners. The high Se levels in hyperaccumulators likely also affect their microbiomes, which show exceptional Se resistance and are equally diverse to that of other plants. A comparative study is underway on the composition and Se-related properties of the microflora of hyperaccumulators and nonaccumulators. Another topic of study in progress is the evolution of Se hyperaccumulation in *Stanleya pinnata*, using a multidisciplinary approach including physiological, biochemical, molecular and x-ray microprobe approaches. RNA sequencing has pinpointed several putative key genes for Se hyperaccumulation that will be topic of further functional characterization.

REFERENCES

Cappa, J.J. & Pilon-Smits, E.A.H. 2014. Evolutionary Aspects of Hyperaccumulation. *Planta* 239: 267–275.

Cappa, J.J., Cappa, P.J., El Mehdawi, A.F., McAleer, J.M., Simmons, M.P. & Pilon-Smits, E.A.H. 2014. Characterization of selenium and sulfur accumulation in Stanleya (Brassicaceae). A field survey and common-garden experiment. *Am. J. Bot.* 101: 830–839.

El Mehdawi, A.F. & Pilon-Smits, E.A.H. 2012. Ecological Aspects of Plant Selenium Hyperaccumulation. *Plant Biol.* 14: 1–10.

Freeman, J.L., Tamaoki, M., Stushnoff, C., Quinn, C.F., Cappa, J.J., Devonshire, J. et al. 2010 Molecular mechanisms of selenium tolerance and hyperaccumulation in *Stanleya pinnata*. *Plant Physiol.* 153: 1630–1652.

Sura-de Jong, M., Reynolds, R.J., Richterova, K., Musilova, L., Hrochova, I., Frantik, T. et al. 2015. Selenium hyperaccumulators harbor a diverse endophytic bacterial community characterized by extreme selenium tolerance and plant growth promoting properties. *Front. Plant Sci.* 6: 113.

Comparative effects of selenite and selenate on growth and selenium uptake in hydroponically grown pakchoi (*Brassica chinensis* L.)

Q. Peng, Z. Li & D.L. Liang*
College of Natural Resources and Environment, Northwest A&F University, Yangling, Shaanxi, China
Key Laboratory of Plant Nutrition and the Agri-environment in Northwest China, Ministry of Agriculture, Yangling, Shaanxi, China

1 INTRODUCTION

Selenium (Se) is considered to be an essential trace element for humans as well as animals, and either deficient or excessive Se intake can be associated with several dramatic consequences for health. However, at present two-thirds of world regions and approximately 5 billion to 10 billion people lack Se. Especially in China, approximately 72% of land and soil in natural state are suffering from Se deficiency (<0.125 mg/kg) (Seppänen et al., 2010). Plants, although it remains to be controversial that whether Se is essential for plant growth, play an important role in the entrance of Se into food chains. Numerous studies have revealed that the Cruciferous vegetable family, like pakchoi, has great potential to provide as a Se-enriched diet. In general, the production of high Se cereals and vegetables requires agronomic Se-fertilized soil, and by supplementing inorganic water-soluble selenite or selenate in low Se soil areas (Liu et al., 2015).

Since the phytoavailability of Se in soil is relatively low, less than 30% of the total Se is commonly absorbed by plants. Furthermore, the difference in Se phytoavailability is considerable after adding different exogenous Se forms into the soil, for example selenite and selenate (Kamei-Ishikawa et al., 2007). Previous studies showed that soluble selenite added to the soil can be adsorbed and fixed strongly into soil components in a short time, hence reducing the mobility and phytoavailability of Se, for uptake by plants. In contrast, selenate is weakly absorbed to the soil, and exhibits relatively high mobility, therefore more bioavailable for plants (Fujita et al., 2005). In contrast, hydroponics is a closed system and can exclude the impact of soil adsorption and fixation of Se added by different Se fertilizers to the soil. Rios et al. (2013) demonstrated that selenite applications had a lower toxic critical concentration on lettuce plants than selenate in hydroponic experiments, suggesting that selenate was less toxic than selenite. Nevertheless, in other cases, the uptake of selenate by Indian mustard (*Brassica juncea*) was much faster than uptake of selenite for 7-day hydroponic experiments (de Souza et al., 1998).

This study proposes to explore the contrasting influences of selenite and selenate at the same concentration on the biomass and Se accumulation in pakchoi through short-term cultivation under hydroponic conditions.

2 MATERIALS AND METHODS

Three week-old pakchoi seedlings with 4 true leaves were transferred to 5 dm^3 plastic pot (three plants each) containing 0.25-strength Hoagland's nutrient solution. The hydroponic solution, in turn, was replaced with fresh 0.5-strength and full strength Hoagland's solution, respectively. After the transplanted seedlings were completely recovered and adapted themselves to the hydroponic environment for 9 days, the nutrition solution was replaced with full strength Hoagland's solution. The pH of the medium was adjusted to 6.0 ± 0.2. The nutrient solution was treated by supplementing with different forms and concentrations of Se: 0, 0.01, 0.1 and 1 mg/L applied as selenite (Na_2SeO_3) or selenate (Na_2SeO_4), respectively. The plants from each pot were harvested and separated into shoots and roots after 3 days from Se addition. They were dried at 50°C to a constant weight, weighed and ground. Then, the dried plant materials were digested by using HNO_3-$HClO_4$ (4:1, v/v), and Se concentrations were determined by atomic fluorescence spectrophotometer with hydride-generation according to the Standard Method (GB/T 5009.93-2003). One-way ANOVA analysis (SPSS, v. 20.0) was used to compare the means among different treatments at $p < 0.05$, and the means with different letters are significantly different.

3 RESULTS AND DISCUSSION

The results showed the greatest biomass growth occurred at a selenite concentration of 0.01 mg Se/L. Concentrations greater than this led to a growth reduction. All selenate treatments were contributed to pakchoi growth compared to control yields positively. The

Table 1. Effect of selenite or selenate on the biomass in shoots and roots of pakchoi.

Treatment (mg/L)	Selenite		Selenate	
	Shoots (g)	Roots (g)	Shoots (g)	Roots (g)
0	0.77 ± 0.05b	0.8 ± 0.00a	0.77 ± 0.05b	0.8 ± 0.00c
0.01	1.04 ± 0.9a	0.11 ± 0.03a	0.88 ± 0.09b	0.1 ± 0.01bc
0.1	0.83 ± 0.5b	0.08 ± 0.00a	1.10 ± 0.03a	0.13 ± 0.01a
1	0.73 ± 0.09b	0.08 ± 0.01a	0.84 ± 0.11b	0.12 ± 0.01ab

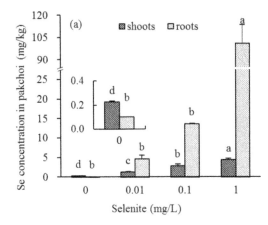

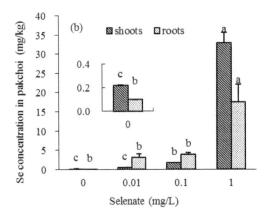

Figure 1. Concentrations of Se in shoots and roots of pakchoi treated with selenite (a) or selenate (b).

highest values for shoot and root biomass were measured concentration at the rate of 0.1 mg/L (Table 1). In general, the results indicate that at low concentrations selenite showed beneficial effects on the plant's growth, and pakchoi tended to tolerate higher concentrations of selenate than selenite in solution.

As shown in Figure 1, the concentration of Se in the shoots and roots highly depended on the chemical form and concentration of Se applied to the solution. Overall, at the Se concentrations of 0.01, 0.1 and 1 mg/L, the total Se uptake by pakchoi (shoots and roots) was 1.6, 2.9 and 2.1 greater, respectively, in selenite treatment than those in selenate treatment. These results suggest that selenite was likely more available for plant uptake than selenate under hydroponics-growing conditions.

Moreover, Se mainly accumulated in roots of pak choi in selenite-treated solution (Fig. 1a). In contrast to, selenate, most of Se accumulated in roots at low concentrations of selenate. When selenate concentrations were as high as 1 mg/L, Se principally accumulated in shoots (Fig. 1b). In addition, Se concentrations attained their maximum concentrations in shoots and roots after exposed to 1 mg Se/L of either selenite or selenate.

4 CONCLUSIONS

Selenite is more toxic than selenate under hydroponic growing conditions. A level of selenite as low as 0.01 mg Se/L enhanced shoot biomass of pakchoi, which progressively decreased when exposed to higher Se concentrations. However, Se-enriched pakchoi tolerated up to 1 mg Se/L of selenate without obvious growth reduction after a short-time exposure. Furthermore, selenite was more readily absorbed by plants than selenate under hydroponic conditions.

REFERENCES

Seppänen, M.M., Kontturi, J., Heras, I.L., Madrid, Y., Cámara, C. & Hartikainen, H. 2010. Agronomic biofortification of brassica with selenium-enrichment of SeMet and its identification in brassica seeds and meal. *Plant and Soil* 337(1–2): 273–283.

Liu, X.W., Zhao, Z.Q., Duan, B.H., Hu, C.X., Zhao, X.H. & Guo, Z.H. 2015. Effect of applied sulphur on the uptake by wheat of selenium applied as selenite. *Plant and Soil* 386(1–2): 35–45.

Kamei-Ishikawa, N., Tagami, K. & Uchida, S. 2007. Sorption kinetics of selenium on humic acid. *Journal of Radioanalytical and Nuclear Chemistry.* 274(3): 555–561.

Fujita, M. Ike, M. Hashimoto, R. Nakagawa, T. Yamaguchi, K. & Soda, S.O. 2005. Characterizing kinetics of transport and transformation of selenium in water-sediment microcosm free from selenium contamination using a simple mathematical model. *Chemosphere* 58(6): 705–714.

Rios, J.J. Blasco, B. Leyva, R. E. Sanchez-Rodriguez, M.M. Rubio-Wilhelmi, L. Romero, et al. 2013. Nutritional balance changes in lettuce plant grown under different doses and forms of selenium. *Journal of Plant Nutrition* 36(9): 1344–1354.

de Souza, M.P., Pilon-Smits, E.A., Lytle, C.M., Hwang, S., Tai, J., Honma, T.S.U., et al. 1998. Rate-limiting steps in selenium assimilation and volatilization by Indian mustard. *Plant Physiology* 117(4): 1487–1494.

Does selenium hyperaccumulation affect the plant microbiome?

A.T. Cochran, J. Bauer, R.J.B. Reynolds & E.A.H. Pilon-Smits*
Biology Department, Colorado State University, Fort Collins, CO, USA

M. Sura-de Jong, K. Richterova & L. Musilova
Department of Biochemistry and Microbiology, University of Chemistry and Technology in Prague, Prague, Czech Republic

1 INTRODUCTION

Selenium (Se) is an essential micronutrient for many animals, including humans, but can be a toxin and pollutant at higher levels. Selenium is not necessary for the survival of plants, but it has been shown to promote plant growth and antioxidant activity (Hartikainen, 2005). Hyperaccumulators are plants that accumulate toxic elements to at least 100-fold higher than other plant species (Baker & Brooks, 1989). For example, hyperaccumulator *Stanleya pinnata* (Brassicaceae) and *Astragalus bisulcatus* (Fabaceae) can bioconcentrate and tolerate Se to 0.5% and 1.5% of their dry weight, respectively (Galeas et al., 2007). This ability has been proposed to serve the plant as a defense mechanism against pathogens and herbivory activities (Boyd & Martens, 1992).

While increasingly seen as an extension of the host genome, the majority of plant microbiomes are not well understood. The microbiome is made up of endosphere, rhizosphere, and phyllosphere microbes, each with their own intercommunity interactions. The composition of these plant microbiomes vary depending on host and are affected by biotic and abiotic conditions (Turner et al., 2013). Endophytes are microbes that live inside of plants in a mutualistic relationship (Alford et al., 2010). The endophytes protect the plant from herbivores and pathogens and may also promote plant growth via IAA production, nitrogen fixation, phosphate solubilization, and siderophore production (Weyens et al., 2009). For its part, the plant provides the endophytes nutrients and a safe habitat.

Plant-microbe interactions of hyperaccumulators are a relatively less explored area (Alford et al., 2010). Since plants live in a close relationship with microbes, the question to be asked is how hyperaccumulation is affected by these plant-associated microbes? Conversely, another interesting question is: how does hyperaccumulation affect microbial density and community composition? We have investigated these questions in this study. The results are among the first to characterize the endophytic microbiomes of hyperaccumulator plants.

2 MATERIALS AND METHODS

Fresh plant materials (roots, stems, and leaves) were collected from *A. bisulcatus* and *S. pinnata* grown at a high Se site in Fort Collins, CO, USA. Selenium concentrations were measured using ICP-OES according to Fassel (1978) following acid digestion. Other fresh plant leaves and roots were surface-sterilized and endophytes were isolated from the tissues by following the method previously described by Sura de Jong et al. (2015). Selenium tolerance was measured by streaking isolates on LB agar plates containing various concentrations of selenate and selenite. Selected isolates were sequenced via MS MALDI-TOF and identified by 16S rDNA based on Se tolerance, geography, and morphology. T-RFLP was performed on the surface-sterilized plant material, following DNA extraction, as described by Sura de Jong et al. (2015).

3 RESULTS

There were 66 bacterial endophytes isolated from hyperaccumulator *S. pinnata* and *A. bisulcatus* growing in their natural seleniferous habitats, including seven genera, with *Bacillus, Pantoea,* and *Pseudomonas* being the most dominant (Fig. 1).

Isolates tested on pure cultures showed the evidence of plant growth promoting properties, including IAA or siderophore production, phosphate solubilization, or nitrogen fixation. Indeed, when seven selected isolates were inoculated to plants, the bacterial inoculation significantly enhanced biomass production by up to 3 folds (Sura-de Jong et al., 2015). The isolates were also characterized for Se tolerance.

Overall, the isolates showed remarkable selenate and selenite tolerance, up to 200 mM or beyond. Root isolates tended to be more Se tolerant than shoot isolates. TRFLP results showed that hyperaccumulator and non-hyperaccumulator plants had no significance difference in microbial diversity. This indicates that hyperaccumulator plants do not suffer from a decrease in microbial community from high levels of Se.

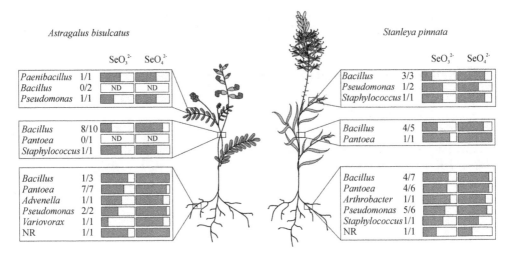

Figure 1. Tolerance of endophytic isolates to selenite and selenate taken from *A. bisulcatus* and *S. pinnata* (Sura de Jong et al., 2015).

The microbial composition of the hyperaccumulator and non-hyperaccumulator microbiomes is currently under investigation using Illumina sequencing.

ACKNOWLEDGEMENTS

This work was supported in part by grant KONTAKT LH12087 of Czech Ministry of Education, Youth and Sports.

REFERENCES

Alford, E.A., Pilon-Smits, E.A.H. & Paschke, M. 2010. Metallophytes – A view from the rhizosphere. *Plant and Soil.* 337: 33–50.

Baker, A.J.M. & Brooks, R.R. 1989. Terrestrial higher plants which hyperaccumulate metallic elements – A review of their distribution, ecology, and phytochemistry. *Biorecovery.* 1: 81–126.

Boyd, R.S., & Martens S.N. 1992. The rason d'être for metal hyperaccumulation by plants. pp. 279–289. In: A.J.M. Baker, J. Proctor, & R.D. Reeves (eds.), The Vegetation of Ultramafic (Serpentine) Soils. Andover, UK: Intercept.

Fassel, V.A. 1978. Quantitative elemental analyses by plasma emission spectroscopy. Science. 202, 183–191.

Galeas, M.L., Zhang, L.H., Freeman, J.L., Wegner, M. & Pilon-Smits, E.A.H. 2007. Seasonal fluctuations of selenium and sulfur accumulation in selenium HA and related non-accumulators. *New Phytologist.* 173: 517–525.

Hartikainen, H. 2005. Biogeochemistry of selenium and its impact on food chain quality and human health. *Journal of Trace Elements in Medicine and Biology.* 18: 309–318.

Sura-de Jong, M., Reynolds, R.J., Ricterova, K., Musilova, L., Hrochova, I., Frantik, T., *et al.* 2015. Selenium hyperaccumulators habor a diverse endophytic bacterial community characterized by extreme selenium tolerance and plant growth promoting properties. *Front. Plant Sci.* 6: 113.

Turner, T.R., James, E.K. & Poole, P.S. 2013. The Plant Microbiome. *Genome Biology.* 14: 209.

Weyens N., van der Lelie D., Taghavi S., Newman L. & Vangronsveld J. 2009. Exploiting plant–microbe partnerships to improve biomass production and remediation. *Trends in Biotechnology.* 27: 591–598.

Effects of sulfur and selenium interaction on pakchoi growth and selenium accumulation

S.Y. Qin, W.L. Zhao, Z. Li & D.L. Liang*

College of Resources and Environmental, Northwest A&F University, Key Laboratory of Plant Nutrition and the Agri-environment in Northwest China, Ministry of Agriculture, Yangling, Shaanxi, China

1 INTRODUCTION

Although selenium (Se) has not yet been proved to be an essential element for plants, it was observed to be conducive to some plant growth at certain concentrations (Wu et al., 2009). As an essential nutrient for humans and animals, dietary supplementation has become the most popular way of Se intake. Daily diet is the main source of Se intake (Rayman et al., 2005). Sulphur (S) and Se are group-six elements, and have similar physical and chemical properties. Some researchers found that there was competition in the plant's uptake of S and Se because they shared the same transport channels – via sulfate transporter (Stroud et al., 2010). Synergistic effects between sulfate and selenate absorption have also been observed in some studies (Ríos et al., 2008). In conclusion, the interaction between sulfate and selenate was not sure yet.

In this study, pakchoi was cultured in a sand culture experiment to determine effects of interaction between sulfate and selenate on pakchoi growth and Se absorption.

2 MATERIALS AND METHODS

Pakchoi (*Brassica chinensis*) were seeded in 20 cm × 40 cm plastic tubs and cultured with nutrient solution referring to Li et al. (2003) for three weeks. Three weeks later, same-sized pakchoi seedlings were selected and transplanted to plastic containers filled with quartz sand ($1/2$ 0.5–1 cm diameter and $1/2$ 1–2 cm diameter). Each container was limited to six seedlings. Nutrient solution containing S and Se treatments was added for the 9 days growth of pakchoi before harvest. Shoots and roots were harvested separately. The S and Se treatments were mixed S from $Na_2SO_4 \cdot 10H_2O$ and Se from $Na_2SeO_4 \cdot 10H_2O$ solutions at different concentrations (S: 0, 0.1, 1.0 mg/L; Se: 0, 0.1, 1.0, 10.0 mg/L). Root length, shoot height, root dry weight, shoot dry weight were measured. Samples were acid-digested in $HNO_3 - HClO_4$ (volume ratio of 4:1), and reduced by 6 M HCl. Concentrations of Se in sample solutions were determined by hydride atomic fluorescence spectrometry. Data were analyzed by SPSS and Microsoft Excel 2010.

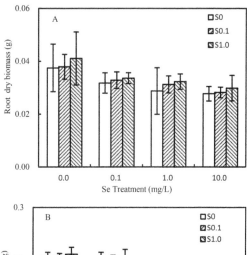

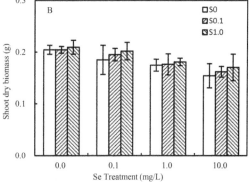

Figure 1. The influence of sulfur and selenium on shoot biomass (A) and root biomass (B).

3 RESULTS AND DISCUSSION

3.1 The influence of S and Se on growth of pakchoi

Single S treatment promoted the growth of pakchoi, reflected by the increase of the root length and shoot height with increasing S concentrations (not given). The root length of pakchoi in 1.0 mg/L Se treatment increased by 14.4% compared to that of the control. In contrast, the root length and shoot height of pakchoi decreased with raising of Se concentrations, indicating Se supplement at this experimental concentrations inhibited the growth of pakchoi. When exposed to

3.2 The influence of S and Se interaction on Se concentration of pakchoi

Concentrations of Se in both roots and shoots increased significantly with exogenous Se addition (Fig. 2). While S treatment decreased the shoot and root Se content, especially at high Se treatments. This observation indicated that S could also reduce the stress of additional Se by inhibiting the absorption of Se. Researchers have found that sulphate and selenate shared the same transporter for uptake of Se and S (Harris et al., 2014). At low S concentrations, sulfate transporters increased in the stimulation of exogenous selenium, thus showed a synergistic effect between the absorption of them; at high S concentrations, root epidermal cells preferentially absorbed S since it had a higher affinity. Consequently, antagonism between Se and S occurred, the absorption of Se decreased.

4 CONCLUSIONS

Selenate supplementations of 0.1, 1.0, 10.0 mg/L inhibited the growth of *pakchoi*, as shown by decreases in root length, shoot height and dry biomass. In addition to its nutritional role, sulfur can also alleviate the stress of exogenous selenate by inhibiting the absorption of Se, which results in an increase of root length, shoot height and biomass, and a decrease in Se content.

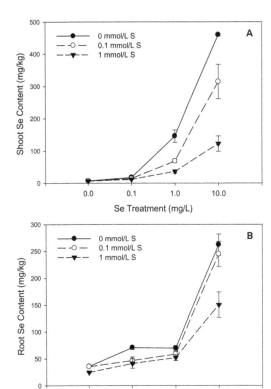

Figure 2. The influence of sulfur and selenium on shoot Se content (A) and root Se content (B).

Se-S mixture, the root length and shoot height of pakchoi also increased with the increase in S concentration, indicating that S could alleviate the stress of Se. The effects of S and Se on biomass of both shoots and roots of pakchoi were similar to that of root length and shoot height. Compared to the same Se concentration without S treatment, S supplementation slightly promoted biomass accumulation of pakchoi. On the contrast, amended Se showed negative effects on pakchoi biomass compared to the same S concentration without Se treatment. When exposed to 10 mg Se/L, inhibition rate of shoot and root biomass accumulation reached up to 35.3% and 32.5% compared to the non-Se treatment, respectively.

For the S and Se combined treatment, both shoot and root biomass increased with an increase of S concentration under the same Se treatment compared to the non-S treatment, indicating that S could alleviate the stress caused by Se (Fig. 1). In contrast, adding Se inhibited root and shoot dry biomass production.

REFERENCES

Harris, J., Schneberg, K.A., & Pilon-Smits, E.A. 2014. Sulphur–selenium–molybdenum interactions distinguish selenium hyperaccumulator *Stanleya pinnata* from non-hyperaccumulator Brassica juncea (*Brassicaceae*). *Planta* 239(2): 479–491.

Li, D.C., Zhu, J., & Tou, C.H. 2003. Effects of selenium on the growth and nutrient content at different sulphur level in pakchoi. *Journal of Zhejiang University (Agriculture and Life Sciences)*, 29(4): 402–406.

Rayman, M.P. 2005. Selenium in cancer prevention: a review of the evidence and mechanism of action. *Proceedings of the Nutrition Society*, 64 (04): 527–542.

Ríos, J.J., Blasco, B., Cervilla, L.M., Rubio-Wilhelmi, M.M., Ruiz, J.M., & Romero, L. 2008. Regulation of sulphur assimilation in lettuce plants in the presence of selenium. *Plant Growth Regulation*, 56(1): 43–51.

Stroud, J., Broadley, M., Foot, I., Fairweather-Tait, S, Hart, D., Hurst, R., et al. 2010. Soil factors affecting selenium concentration in wheat grain and the fate and speciation of Se fertilisers applied to soil. *Plant Soil*, 332(1–2): 19–30.

Wu, X.P., Liang, D.L., Bao, J.D., & Xue, R.L. 2009. Effects of different concentrations of selenate and selenite on growth and physiology of *pakchoi*. *Journal of Environmental Science*, 29(10): 2163–2171.

Are all Brazil nuts selenium-rich?

E.C. da Silva Júnior*, L.R.G. Guilherme, G. Lopes & G.A. de Souza
Department of Soil Science, Federal University of Lavras, Minas Gerais, Brazil

K.E. da Silva & R.M.B. de Lima
Researcher at Embrapa Amazônia Ocidental, Manaus, Amazonas, Brazil

M.C. Guedes
Researcher at Embrapa Amapá, Macapá, Amapá, Brazil

L.H.O. Wadt
Researcher at Embrapa Rondônia, Porto Velho, Rondônia, Brazil

A.R. dos Reis
Biosystems Engineering–São Paulo State University, Tupã, São Paulo, Brazil

1 INTRODUCTION

Brazil nuts (*Bertholletia excelsa, Lecythidaceae* family) from the Amazon region of Brazil are consumed worldwide, and are known as the richest food in selenium (Se). Depending on the soil Se content, concentrations of Se range from 5 to 512 mg/kg (Dumont et al., 2006; Chang et al., 1995).

Brazil nuts are a convenient dietary source of Se. The dietary consumption of one nut per day could avoid the need for other Se-fortified foods or expensive Se food supplements (Thomsom et al., 2008). It is known that the variation of Se concentration in Brazil nuts results from different growing conditions (such as soil Se contents), as well as among individual nuts (Secor & Lisk, 1989; Chang et al., 1995). However, there are few studies that characterize these local and regional variations of Se concentrations. Thus, this study aimed to evaluate the concentration of Se in Brazil nuts growing in different areas of the Amazon Region. This will show how the Se variability in soil, as well as genotypic variability impacts the Se concentration in Brazil nuts.

2 MATERIALS AND METHODS

Samples of Brazil nuts were obtained from native areas and crops in the states of Mato Grosso, Acre, Amazonas and Amapá. These nuts were collected from five plants in each state. Additionally, two different plants of 4 clones were collected from the Aruanã farm in Itacoatiara-Amazonas to compare the concentration of Se to almonds obtained locally. All samples were collected in January, 2015.

Acid digestion was performed in the laboratory of Fertility and Plant Nutrition in the Department of Soil Science at Federal University of Lavras. The samples were oven-dried at 60°C till constant weight, ground and 0.5 g of each material were placed in tubes for digestion in digesters blocks in triplicate, as well as standards reference material White Clover (BCR-402) and Plankton (BCR-414). Six mL of a mixture composed of nitric and perchloric acid in the ratio 2:1 (v/v), respectively, were added and allowed to stand at room temperature overnight and then digested at 200°C for 2 hours. After completing digestion, 10 mL of distilled water were added to the extracts. Selenium determinations were performed using graphite furnace-atomic absorption spectrophotometer. The recovery rate of the standard reference material was at an average (n = 3) 69.18% for White Clover (BCR-402) and 83.59% for Plankton (BCR-414).

3 RESULTS AND DISCUSSION

The results show that concentrations of Se in Brazil nuts varied significantly among different sampling places (Fig. 1). The average Se concentrations in nuts collected from Amazonas (AM) and Amapá (AP) were about 30-fold higher than those from Mato Grosso (MT) and Acre (AC). This regional variation of Se concentration in Brazil nut is in agreement with the previous research finding reported by Chang et al. (1995). The concentrations of Se in the almonds collected from the Acre-Rondônia region were 0.03–31.7 mg/kg (fresh weight) but 1.25–512.0 mg/kg (fresh weight) from the Manaus-Belém region. The almond Se concentrations from the four sampling regions showed a descending order of: Amazonas (97.785 mg/kg) > Amapá (82.920 mg/kg) > Acre (3.495 mg/kg) > Mato Grosso (2.084 mg/kg) (Fig. 1).

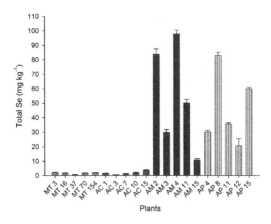

Figure 1. Selenium concentrations in Brazil nuts almonds from different parts of the Amazon region.

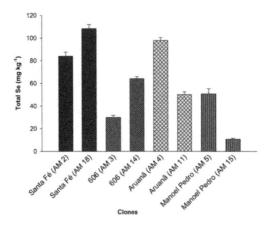

Figure 2. Selenium concentrations in Brazil nuts almonds of different clones grown in Itacoatiara, Amazonas.

The concentrations of Se accumulated in Brazil nuts of different farm clones in Aruanã were shown in Figure 2. Despite of being grown in relatively similar soils, different farm clones had significant variation of Se concentration, which suggests that there are other genetic factors that associated with the plant's ability of accumulating Se in the almonds.

With regard to intrinsic factors associated with the plant, is necessary to carry out more research. The correlations between levels of Se in soil and almonds needs further development. Additionally, factors contributing to a greater or lesser uptake of Se in Brazil nuts as well as the different clones of trees in the Amazon area need to be studied in order to better understand the plant's ability in absorb available Se in the soil and to translocate to the seeds.

4 CONCLUSIONS

Brazil nuts have regional and local variations in their Se concentration, which are influenced by both the soil Se concentration as well as the genotypic ability of each plant to accumulate Se.

REFERENCES

Chang, J.C., Gutenmann, W.H., Reid, C.M., & Lisk, D.J. 1995. Selenium content of brasil nuts from two geografic locations in Brasil. *Chemosphere* 30(4): 801–802.

Dumont, E., L. De Pauw, F. Vanhaecke, & R. Cornelis. 2006. Speciation of Se in Bertholletia excelsa (Brazil nut): A hard nut to crack? *Food Chemistry* 95(4): 684–692.

Ferreira, D.F. 2011. SISVAR: a computer statistical analysis system. *Ciência & Agrotecnologia* 35(6): 1039–1042.

Secor, C.L. & Lisk, D.J. 1989. Variation in the selenium content of individual brazil nuts. *Journal of food safety* 9: 279–281.

Thomson, C.D., Chisholm, A., McLachlan, S.K., & Campbell, J.M. 2008. Brazil nuts: an effective way to improve selenium status. *Am J Clin Nutr* 87: 379–384.

Accumulation of mercury and selenium by *Oryza sativa* from the vicinity of secondary copper smelters in Fuyang, Zhejiang, China

Xuebin Yin*
State Key Laboratory of Soil and Sustainable Agriculture, Institute of Soil Science, Chinese Academy of Sciences, Nanjing, China; School of Earth and Space Sciences, University of Science and Technology of China, Hefei, China

Jing Song, Zhibo Li & Wei Qian
State Key Laboratory of Soil and Sustainable Agriculture, Institute of Soil Science, Chinese Academy of Sciences, Nanjing, China

Chunxia Yao
Institute for Agro-Product Quality Standards and Testing Technologies, Shanghai Academy of Agricultural Sciences, Shanghai, China

Yongming Luo
Yantai Institute of Coastal Zone Research, Chinese Academy of Sciences, Yantai, China

Linxi Yuan
School of Earth and Space Sciences, University of Science and Technology of China, Hefei, China

1 INTRODUCTION

Mercury (Hg) is toxic to animals, particularly in the form of monomethyl Hg (O'Driscoll et al., 2005), however, the toxicity of Hg can be alleviated by selenium (Se) compounds (Chen et al., 2006). Usually, the Se content in food has been constituted to determine the severity of Hg toxicity, consequently, it was always included in the investigation of soil pollution and risk assessment of agricultural environment (Horvat et al. 2003). In this study, the city of Fuyang (one of the largest centers of secondary copper (Cu) smelting) in Zhejiang Province, China, was selected as the study area to investigate Hg and Se distributions and possible interactions in rice (*Oryza sativa*).

2 MATERIALS AND METHODS

The study area (29°55′1″ – 29°58′13″ N, 119°53′56″– 119°56′4″ E) with an area of approximately 10.9 km² is located in Fuyang. The sampling sites are situated in a long and narrow valley with elevations varying from 2 to 144 m, where 11 secondary Cu smelters are located. A total of 53 paddy soil samples (0–15 cm), 53 bran samples, and 53 crude rice samples were collected. Total Hg and total Se contents were determined via Atomic Fluorescence Spectrometer (AFS) after acid digestion (Yin et al., 2007).

3 RESULTS AND DISCUSSION

The geometric mean of the Hg concentrations in paddy soils was $251 \pm 164\,\mu g/kg$, over 27 times

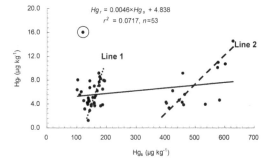

Figure 1. Comparison of crude rice Hg (Hg_r) and paddy soil Hg concentrations (Hg_s).

the background value ($9 \pm 1\,\mu g/kg$), and Se was $1049 \pm 816\,\mu g/kg$, 5 times the background value ($212 \pm 54\,\mu g/kg$) (Yin et al. 2007). Soil Hg concentrations in approximately 30% of the samples exceeds the limit of $300\,\mu g/kg$ Hg in soils with pH < 6.5 (environmental quality standard for soils, GB15618-1995, State Environmental Protection Administration of China). Most concentrations of Hg ($6.0 \pm 2.8\,\mu g/kg$) in crude rice are lower than the environmental quality standard for foodstuffs (Hg: $20\,\mu g/kg$), but nearly 10% of the samples have values exceeds the limit for rice Hg according to the Green Food Quality Standard of China (GB18406.1-2001). The crude rice have Se concentrations with $214 \pm 476\,\mu g/kg$ Generally, the bran Hg concentrations ($13.6 \pm 5.3\,\mu g/kg$) are twice those of crude rice, whereas they have almost equal concentrations of Se ($208 \pm 366\,\mu g/kg$) in bran.

The relationship between Hg_r in rice, and Hg_s in soil, is plotted in Figure 1. Although the samples do not

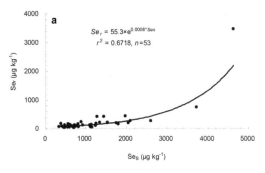

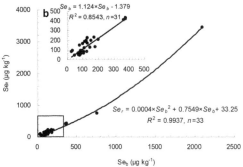

Figure 2. Relationship (a) between Se in soils (Se_s) and Se in cruderice (Se_r) and (b) between Se in crude rice (Se_r) and Se in bran (Se_b).

show a clear trend as a whole, the section of Hg_s over 300 μg/kg has an apparently linear correlation with Hg_r as shown by Line 2, while for Hg < 300 μg/kg, this relationship is less pronounced (Line 1). These findings suggest that the soil Hg level appears to play a more important role in determining the accumulation of Hg in rice in heavily polluted areas than slightly polluted areas.

In terms of Se, the r values of the Se concentrations in paddy soil (Se_s) and rice bran (Se_b), and the Se_s–Se concentrations in crude rice (Se_r) are 0.763 and 0.864, respectively, and an exponential increment in the Se content in rice follows with an increasing Se content in paddy soils, as shown in Fig 2 (a). Similarly, Fig 2 (b) shows a significant positive correlation between rice bran Se and crude rice Se. This observation suggests that the uptake of Se by rice plants in paddy soil is more efficient, which will result in more Se transferred to rice grains.

The Hg–Se correlation coefficient (r) in paddy soil, rice bran, and crude rice is 0.112 ($p = 0.429$, $n = 53$), −0.101 ($p = 0.584$, $n = 33$), and −0.175 ($p = 0.214$, $n = 53$), respectively. A p-value of >0.1 indicates an insignificant correlation between them. Similarly, in Guizhou Province, Horvat et al. (2003) did not find a significant Hg–Se correlation in the sampled rice plants. Nevertheless, the slightly negative correlation between Hg_r and Se_r (or Hg_b and Se_b) indicates that Se could somewhat reduce the uptake of Hg in rice. This possibility is consistent with the previous reports on tomato (*Lycopersicum esculentum*), radish (*Raphanus sativus*) (Shanker et al. 1996a, b), and medical plant (*Portulaca oleracea* Linn) (Thangavel et al. 1999), as well as with hydroponic experiments (Zhang et al. 2006). The mechanism of the interaction between two elements is assumed as the formation of insoluble compounds of mercuric sulfide (HgSe) with a lower log k_{sp} value of 52 (25°C) around the rhizosphere. Accordingly, in contrast to the positive correlation between Hg and Se in animals, a negative Hg–Se correlation occurs frequently in the aboveground part of the plant.

4 CONCLUSIONS

The soil Hg and soil Se are up to 251 ± 164 and 1049 ± 816 μg/kg ($n = 53$), respectively, indicating that paddy soil has been substantially contaminated by smelting-related Hg and Se. However, the Hg_T and Se_T contents in crude rice ranged from 1.2 to 16 μg/kg and 42 to 3460 μg/kg, respectively, though Hg–Se interactions in rice are not evident.

ACKNOWLEDGEMENTS

This work was supported by National Natural Science Foundation of China (NNSFC31400091), Natural Science Foundation of Jiangsu Province (BK2012195, BK2012202), and Applied Basic Research Project of Suzhou (SYN201306).

REFERENCES

Chen, C.Y., Qu, L.Y., Zhao, J.J., Liu, S.P., Deng, G.L., Li, B., et al. 2006. Accumulation of mercury, selenium and their binding proteins in porcine kidney and liver from mercury-exposed areas with the investigation of their redox responses. *Sci Total Environ* 366 (2–3): 627–637.

Horvat, M., Nolde, N., Fajon, V., Jereb, V., Logar, M., Lojen, S., et al. 2003. Total mercury, methylmercury and selenium in mercury polluted areas in the province Guizhou, China. *Sci Total Environ* 304: 231–256.

O'Driscoll, N.J., Rencz, A.N., & Lean, D.R.S. 2005. The biogeochemistry and fate of mercury in natural environments (Chapter 14). In: Sigel, A., Sigel, H., and Sigel, R.K.O. (editors). Metal Ions in Biological Systems (Volume 43). Marcel Dekker, Inc., New York. pp. 221–238.

Shanker, K., Mishra, S., Srivastava, S., Srivastava, R., Dass, S., Prakash, S., et al. 1996a. Study of mercury-selenium (Hg-Se) interactions and their impact on Hg uptake by the radish (*Raphanus sativus*) plant. *Food Chem Toxicol.* 34(9): 883–886.

Shanker, K., Mishra, S., Srivastava, S., Srivastava, R., Daas, S., Prakash, S., et al. 1996b. Effect of selenite and selenate on plant uptake and translocation of mercury by tomato (*Lycopersicum esculentum*). *Plant Soil* 183(2): 233–238.

Thangavel, P., Shahira Sulthana, A., & Subburam, V. 1999. Interactive effects of selenium and mercury on the restoration potential of leaves of the medicinal plant, *Portulaca oleracea* Linn. *Sci Total Environ.* 243(244): 1–8.

Yin, X.B., Sun, L.G., Zhu, R.B., Liu, X.D., Ruan, D.Y., & Wang, Y.H. 2007. Mercury-selenium association in Antarctic seal hairs and animal excrements over the past 1,500 years. *Environ Toxicol Chem* 26(3): 381–386.

Zhang, L.H., Shi, W.M., & Wang, X.C. 2006. Difference in selenite absorption between high- and low-selenium rice cultivars and its mechanism. *Plant Soil* 282: 183–193.

Selenium and mycorrhiza on grass yield and selenium content

S.M. Bamberg* & M.A.C. Carneiro
Federal University of Lavras, Lavras, Minas Gerais, Brazil

S.J. Ramos
Vale Institute of Technology, Ouro Preto, Minas Gerais, Brazil

J.O. Siqueira
Vale Institute of Technology, Belém, Pará, Brazil

1 INTRODUCTION

Selenium (Se) has an important function in physiological processes in humans and animals. Brazil is the world's largest exporter of beef, and has around 172 million hectares of grasslands that support cattle herds of approximately 205 million heads. About 60% of the total area of pastures is composed by grasses of the Brachiaria genus. Animals fed with this grass characterize a large part of the Brazilian beef cattle production, but the Brazilian forage pastures do not provide adequate dietary Se for livestock (Reis et al., 2009).

The arbuscular mycorrhiza fungi (AMF) are microorganism that can enhance Se content in plants (Durán et al., 2013). The aim of this work was to evaluate the addition of Se and mycorrhiza effect on pasture grass yield and Se content.

2 MATERIALS AND METHODS

The experiment was carried out in a greenhouse located at Federal University of Lavras, for 60 days in a randomized design (5x2 scheme), with three replications. The Oxisol soil was collected from 20 cm depth. After air drying, the soil was sieved through a 2 mm mesh, physical and chemical analyses were performed (Table 1).

Based on chemical soil composition analysis, liming was performed to raise the base saturation to 50% using calcined lime. Grass plants were grown in pots containing 5 kg of soil and each pot received macronutrient and micronutrient fertilizer at the following rates (g/dm^3): 0.66 $NH_4H_2PO_4$, 0.31 KH_2PO_4, 0.14 KNO_3, 0.26 NH_4NO_3, 0.38 $MgSO_4 \cdot 7H_2O$, 2.86 H_3BO_3, 5.9 $CuSO_4 \cdot 5H_2O$, 22 $ZnSO_4 \cdot 7H_2O$, and 1.3 $(NH_4)_6Mo_7O_{24} \cdot 4H_2O$.

Two AMF species, *Acaulospora morrowiae* and *Claroideoglomus etunicatum*, were applied in the soil before sowing (thousand spores inoculum). On each pot, twenty seeds of *Brachiaria decumbens* were planted, and five sodium selenate doses were applied: 0.0, 0.5, 1.0, 3.0, and 6.0 mg/kg (Na_2SeO_4, Sigma-Aldrich, St. Louis, MO). Fifty days after germination, plants were harvested and dried in a forced-air drying oven at 55–60°C for 3–5 days. Shoot biomass were weighed to determinate the production per pot and then crushed for chemical digestion.

Total Se content was determined in a PerkinElmer Analyst 800 atomic absorption spectrophotometer. Once dried, tissues (approximately 500 mg) were acid digested in 10 mL HNO_3 in Teflon PTFE flasks, and submitted to 0.76 MPa for 10 min in a microwave oven. After cooling to room temperature, the extract was filtered and diluted by adding 5 mL of bi-distilled water.

Statistical results were obtained by variance and regression analysis using SISVAR (Ferreira, 2006). Averages of components of the factorial treatments were compared by Tukey test at 5% probability.

Table 1. Physical and chemical analyses.

	Unit	Value
pH$_{(H2O)}$	–	4.9
H + Al	cmol$_c$/dm^3	3.2
Al	cmol$_c$/dm^3	0.6
Ca	cmol$_c$/dm^3	0.1
Mg	cmol$_c$/dm^{-3}	0.1
K	mg/dm^3	48.0
Na	mg/dm^3	0.0
P	mg/dm^3	0.8
Se	mg/dm^3	0.06
OM	%	1.4
Sand	%	73
Loam	%	2
Clay	%	25

3 RESULTS AND DISCUSSION

Depending on the Se doses applied to the soil and the presence of mycorrhiza, different responses in shoot

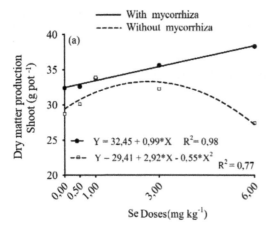

Figure 1. Shoot dry matter of *Brchiaria decumbens* treated to different doses of sodium selenate.

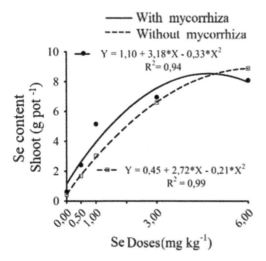

Figure 2. Shoot Se content of *Brachiaria decumbens* treated to different doses of sodium selenate.

growth were observed ($p < 0.05$, Fig. 1). The shoot dry matter showed a linear increase, to higher Se doses. For non-inoculated AMF plants, and at the highest Se dose supplied, there was a reduction in shoot yield. AMF has ability to extend the root's area absorption by the fungal hyphae, thus allowing the plants to absorb more nutrients and water (Smith & Read, 2008). This benefit probably contributed to the better grass growth. Ramos et al. (2012), also found increase in Brachiaria plant yield, when they applied selenate doses higher than 3 mg/kg. The Se application increased the Se content in brachiaria leaves. At lower Se treatments, AMF was more effective for increasing Se levels in grass (Fig. 2).

4 CONCLUSIONS

The beneficial effects of mycorrhizae on Brachiaria growth was observed for all Se doses applied. Also, we found that mycorrhiza inoculation can contribute to increased Se levels in Brachiaria plants, especially at lower Se doses.

REFERENCES

Durán, P., J.J. Acuña, M.A. Jorquera, R. Azcón, F. Borie, & P. Cornejoa, et al. 2013. Enhanced selenium content in wheat grain by co-inoculation of selenobacteria and arbuscular mycorrhizal fungi: A preliminary study as a potential Se biofortification strategy. *Journal of Cereal Science*, 57(3): 275–280.

Ferreira, Daniel Furtado. 2011. Sisvar: a computer statistical analysis system. Ciência e Agrotecnologia (UFLA), Lavras,35(6): 1039–1042.

Ramos, S.J. Ávila, F.W., Boldrin, P.F., Pereira, F.J., Castro, E.M., Faquin, V., et al. 2012. Response of brachiaria grass to selenium forms applied in a tropical soil. *Plant, Soil and Environment*, Czech Academy of Agricultural Sciences, Czech Republic, 58(11): 521–527.

Reis, L.S.L.S., Chiacchio, S.B., Pardo, R.T., Couto, R., Oba, E., & Kronka, S.N. 2009. Effect of the supplementation with selenium on serum concentration of creatine kinase in cattle. Archivos de Zootecnia, 58(11): 753–756. (In Portuguese).

Smith, S.E. & Read, D.J. 2008. *Mycorrhizal Symbiosis* 3.ed ed., London: Academic Press.

High selenium content reduces cadmium uptake in *Cardamine hupingshanesis* (Brassicaceae)

Z.Y. Bao* & H. Tian
School of Earth Sciences, China University of Geosciences, Wuhan, China

Y.H. You & C.H. Wei
Faculty of Materials Science and Chemistry, China University of Geosciences, Wuhan, China

1 INTRODUCTION

Selenium (Se) is an essential micronutrient for animal and human health. Selenium deficiency appear to be a widespread problem in China, with nearly 72% of its total land area is Se deficient (Lenz & Lens, 2009; Tan et al., 2002). Numerous health-related diseases are associated with a low intake of Se (Thomson, 2004). Consequently, Se-biofortification strategies have attracted much research attention around the world. However, the distribution and contents of other trace elements in Se-enriched plants or food products also deserve their attention. For example, *Cardamine hupingshanesis* (Brassicaceae) has been identified as a Se-hyperaccumulator plant in Enshi, China (Yuan et al., 2013). However, this species can also hyper-accumulate other toxic metals. In this study, the accumulation of arsenic (As), cadmium (Cd), zinc (Zn), and manganese (Mn) in *Cardamine hupingshanesis* were investigated under the greenhouse conditions, along with the enrichment and distribution of Se in this plant.

2 MATERIALS AND METHODS

2.1 *Experimental design and sampling*

Cardamine hupingshanesis seeds were collected from Enshi in Huber province, and the plants were cultivated in the soil fertilized with Na_2SeO_3 under greenhouse conditions in October 2013. Six treatment levels of Se fertilizer: 0, 70, 150, 300, 750, 1500 mg Na_2SeO_3 per unit (3.5 m^2) and 3 replicates per treatment were performed. The homogenized topsoil from each treatment of each unit was collected before seeding. After 5 months (March 2014, Fig. 1), plants were collected from each treatment, then washed, oven-dried, and crushed for Se analysis.

2.2 *Sample analysis*

A 0.05 g dry sample (roots, stems and leaves, respectively) was acid-digested in 2 mL HNO_3 and 1 mL

Figure 1. *Cardamine hupingshanesis* in the greenhouse.

H_2O_2 for 10 hours at room temperature, then placed in closed Teflon vessels at 140°C for 8 hours. In addition, soil samples were also acid-digested using the same methods, and prepared for Se analysis.

Selenium concentrations were determined by hydride generation atomic fluorescence spectrometry (AFS-933, Titan Instruments Co., Ltd, Beijing). Other trace elements (As, Cd, Zn, and Mn) were determined by inductively-coupled plasma mass spectrometry (ICP-MS, ELAN DRC-e, PerkinElmer, USA). The quality control samples included method blanks, certified materials (GSB-5 and GSS-5), and triplicates.

3 RESULTS AND DISSCUSSION

3.1 *Selenium contents in topsoil and plant tissues*

Selenium concentrations in topsoil and different plant tissues (roots, stems, and leaves) of *Cardamine hupingshanesis* are shown in Figure 2. The Se concentrations in soil were 0.15 ± 0.04 mg/kg in the control and 7.41 ± 0.22 mg/kg when 1500 mg Na_2SeO_3 per m^2 land was added into the topsoil; the correlation coefficient (r^2) between these two variables was 0.9649. As expected, Se concentrations in plant tissue increased with increasing the addition of Na_2SeO_3 to soils. When the treatment level was 1500 mg/m^2 of Na_2SeO_3, Se concentrations were highest in roots (32.87 ± 1.17 mg/kg), followed by stems (22.88 ± 0.33 mg/kg) and leaves (16.14 ± 1.23 mg/kg). Also, high correlation coefficient values were obtained between Se concentrations in soils and roots (0.9939), stems (0.9957),

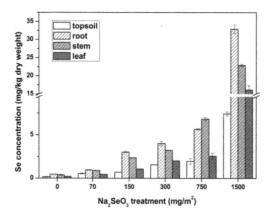

Figure 2. Se concentrations and distribution in different tissues of *Cardamine hupingshanesis*. Data shown are means ± standard deviation (n = 3).

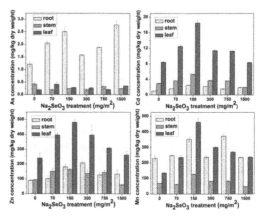

Figure 3. Distributions of As, Cd, Zn and Mn in plant tissues of *Cardamine hupingshanesis*. Data shown are means ± one standard deviation (n = 3).

and leaves (0.9944). In addition, the distribution of Se in plants was in the following descending order: roots > stems > leaves.

3.2 Trace elements (As, Cd, Zn, Mn) in plant tissues

The concentrations of As, Cd, Zn, and Mn were also analyzed in plant tissues (Fig. 3). The distribution of As was similar to that of Se; roots > leaves > stems. Interestingly, the distribution and concentration of Cd were completely opposite to that of Se. A similar phenomenon was also observed in a hydroponic experiment conducted involving rice with Se and Cd in solution (Lin et al., 2012). The concentration of Cd declined in three plant tissues (root, stem, and leaves) after the addition of Se fertilizer at a rate of 150 mg/m^2. This observation indicates a synergistic relationship between Cd and Se up to 150 mg Na$_2$SeO$_3$/m^2, and then Cd concentrations decreased with subsequently higher Se treatments. Similar results of Zn and Mn concentrations were observed in each tissue. More studies are needed to interpret the mechanisms involved with these accumulation these processes in *Cardamine hupingshanesis*.

4 CONCLUSION

The Se concentrations increased with the addition of Se fertilizer, showing a descending order: roots > stems > leaves. In addition, the Se treatment showed inhibitory effects on the absorption of Cd, Zn, and Mn in *Cardamine hupingshanesis*.

REFERENCES

Lenz, M. & Lens, P.N. 2009. The essential toxin: the changing perception of selenium in environmental sciences. Sci Total Environ, 407: 3620–3633.

Lin, L., Zhou, W., Dai, H., Cao, F., Zhang, G. & Wu, F. 2012. Selenium reduces cadmium uptake and mitigates cadmium toxicity in rice. J Hazard Mater, 235–236: 343–351.

Tan, J.A., Zhu, W., Wang, W., Li, R., Hou, S., Wang, D. et al. 2002. Selenium in soil and endemic diseases in China. Science of the Total Environment, 284: 227–235.

Thomson, C.D. 2004. Assessment of requirements for selenium and adequacy of selenium status: a review. Eur J Clin Nutr, 58: 391–402.

Yuan, L., Zhu, Y., Lin, Z.-Q., Banuelos, G., Li, W. & Yin, X. 2013. A novel selenocystine-accumulating plant in selenium-mine drainage area in Enshi, China. PLOS ONE, 8: e65615. doi:10.1371/journal.pone.0065615.

Selenium accumulation and its effects on heavy metal elements in garlic

A.Q. Gu, Y.Y. Luo, H. Tian & Z.Y. Bao*
School of Earth Sciences, China University of Geosciences, Wuhan, China

C.H. Wei
Faculty of Materials Science and Chemistry, China University of Geosciences, Wuhan, China

X.L. Chen
Zhejiang Institute of Geological Survey, Zhejiang, China

1 INTRODUCTION

Selenium (Se) is an essential trace element for animals and humans (Zhao et al., 2013). For humans and animals their main source of Se is from diet. As a widely accepted condiment, garlic (*Allium sativum* L.) has also acquired a reputation in Asiatic and Western cultures as a prophylactic and therapeutic medical agent (Song & Milner, 2001). Previous work has shown that garlic accumulated Se in Se contained soils (Larsen et al., 2006). In the present study, we determined the uptake of Se and other heavy metals in garlic grown in Se treated soil.

2 MATERIALS AND METHODS

Garlic was planted in a standardized greenhouse in September 2013, and harvested in May 2014. A total of 15 separate plots (3.5 m^2 per plot) were used for planting the garlic. Every three plots were designed as one group (five groups in total) and labeled as either G1, G2, G3, G4, or G5 which corresponded to 0, 0.25, 0.50, 1.0, 2.0 mg Se/kg added as sodium selenite to soil, respectively, prior to planting garlic.

Nine months later, garlic was harvested, chopped, and washed carefully for 3–4 times with ultra-pure water. The garlic samples were dried in a drying oven at 50°C; crushed to fine powder, and stored at 4°C. Then, 0.2 g powder was put in a PTFE digestion vessel, and 2 ml HNO_3 (BV-III grade) and 2 ml H_2O_2 (GR grade) (Zhao et al., 2013) were added at room temperature for 12 h. After that, samples in the PTFE digestion vessels were placed in sealed steel tanks at 140°C for 8 hours to ensure complete digestion. The digested solutions were stored at 4°C until analyzed for Se and other elements using HG-AFS and ICP-MS. The cabbage standard reference GBW10014 (GSB-5) was used as quality control and the concentrations of Se, Cd, Zn, and Mn were all within the standard range of recommendation.

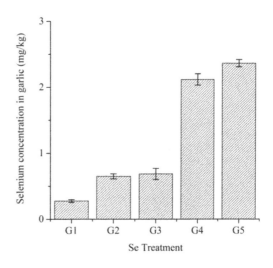

Figure 1. Selenium content in garlic bulb in each group.

3 RESULTS AND DISCUSSION

3.1 *The Se content in garlic*

The results showed that the Se concentrations increased with increasing Se additions to the soil (Fig. 1). The highest Se content (2.36 ± 0.05 mg/kg) was measured in G5 compared to group G1 (0.28 ± 0.02 mg/kg).

3.2 *The effect of Se on Cd accumulation*

We did not observe an increasing Cd concentration with increased Se added to the soil (Fig. 2). The Cd concentration showed positive relationships from G1 to G3 and negative relationships from G3 to G5, and reached the peak at the G3.

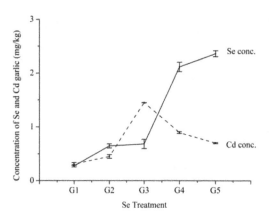

Figure 2. Concentrations of Se and Cd in garlic tissues.

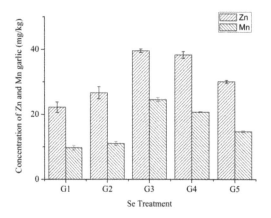

Figure 3. Concentrations of Zn and Mn in garlic tissues.

3.3 *The effect of Se on Mn and Zn concentration in garlic*

Generally, Zn and Mn in garlic also showed similar pattern of accumulation as Cd (positive relationship from G1 to G3 and negative relationship from G3 to G5) (Fig. 3). The variation tendency of Zn concentration was similar to that of Mn and Cd, but the concentration of Zn was higher.

4 CONCLUSIONS

The Se concentration increased in garlic with increasing Se added to the soil. Initially, the concentrations of Cd, Mn, and Zn in garlic showed a positive relationship with the Se concentration in soil (G1 to G3) and then negative relationship (in G3 to G5) (Guo, 1994). This phenomenon showed that improving the Se concentration in soil can increase Se concentration and eventually decrease heavy metal accumulation in garlic. Further study is, however, needed to more clearly interpret this mechanism.

REFERENCES

Guo, Y. & Wu, J. 1994. Influnce of Zn, Mn, Cu to Se for garlic uptake. Acta Agriculturae Universitatis Pekinensis, 20: 83–87.

Larsen, E.H., Lobinski, R., Burger-Meyer, K., Hansen, M., Ruzik, R., & Mazurowska, L. 2006. Uptake and speciation of selenium in garlic cultivated in soil amended with symbiotic fungi (mycorrhiza) and selenate. Anal Bioanal Chem, 385: 1098–108.

Song, K. & Milner, J. A. 2001. The influence of heating on the anticancer properties of garlic. Journal of Nutrition, 131: 1054S–1057S.

Zhao, J., Gao, Y., Li, Y., Hu, Y., Peng, X., Dong, Y., et al.. 2013. Selenium inhibits the phytotoxicity of mercury in garlic (*Allium sativum*). Environmental Research, 125: 75–81.

Effect of selenium on cadmium uptake and translocation by rice seedlings

Y.N. Wan, S.L. Yuan, Z. Luo & H.F. Li*
College of Resources and Environmental Sciences, China Agricultural University, Beijing, China

1 INTRODUCTION

Cadmium (Cd) is one of the most harmful and widespread heavy metals in agricultural soils. Soil pollution has led to Cd accumulation in plants, especially in rice, which may result in a serious risk on human health. Therefore, it's important to develop an efficient way to prevent Cd accumulation in rice. The antagonism of Se on Cd stress has been observed in many different plants (Lin et al., 2012; Issam et al., 2014), as well as selenium's impact on reducing abiotic stresses of plants (Feng et al., 2013). Thus, the present study was conducted to investigate the effect of Se on Cd uptake and transportation. Specifically, we aim to address influence of Se both as a pretreatment and when simultaneously applied with Cd.

2 MATERIALS AND METHODS

Rice seedlings (*Oryza sativa* L., Zhunliangyou 608) were planted in plastic pots filled with 1/2 Kimuar nutrient solution (pH 5.5) for 35 days. The seedlings were transferred to 2.5 L pots containing nutrient solution, in which Se (5 µM, Na_2SeO_3 or Na_2SeO_4) or Cd (5 µM, $3CdSO_4·8H_2O$) was added to form 10 treatments: (1) Control; (2) SeIV; (3) SeVI; (4) Cd; (5) Cd + SeIV; (6) Cd + SeVI; (7) *pre*-SeIV + Cd; (8) *pre*-SeVI + Cd (2 d SeIV or SeVI pretreatment before Cd exposure); (9) *pre*-Cd + SeIV; (10) *pre*-Cd + SeVI (2 d Cd pretreatment before SeIV or SeVI exposure). Two days later, shoots and roots were harvested, seperated, washed, dried, and powdered. The powdered samples were acid-digested (HNO_3) using microwave, and the total Se and Cd concentrations in the digestes were determined by ICP-AFS (AFS-920, Beijing Jitian Instruments Co., Ltd., Beijing, China) and ICP-MS (Agilent ICP-MS 7500ce, Santa Clara, CA, USA). All results were expressed as an average of three replications. Treatment effects were determined by analyzes of variance using SAS software.

3 RESULTS AND DISCUSSION

3.1 Effect of pretreatment and simultaneous application of Cd and Se on their uptake by rice seedlings

Compared with the Cd treatment alone, the addition of selenite or selenate significantly decreased Cd uptake by 25% and 24% 2 days after the treatment,

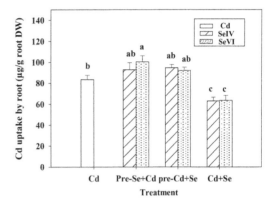

Figure 1. Effect of simultaneous application and pretreatment of Se on Cd uptake by rice seedlings exposed to Cd. The different letters indicate statistically significant differences among treatments at the $P < 0.05$.

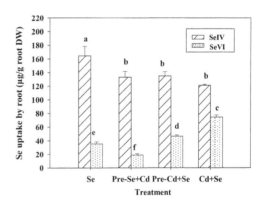

Figure 2. Effect of simultaneous application and pretreatment of Cd on Se uptake by rice seedlings exposed to Se. The different letters indicate statistically significant differences among treatments at $p < 0.05$.

respectively. In contrast to simultaneous application, pre-selenite or pre-selenate treatment increased Cd uptake by 11% and by 20%, respectively (Fig. 1).

In comparing with the Se alone treatment, the addition of Cd significantly decreased Se uptake by 26% in the selenite treatment, but increased by 10% in selenate treatment. Similar results were obtained in the pre-Cd treatment, where Se uptake decreased by 18% in selenite treatment, and increased by 30.9% in selenate treatment. Furthermore, in the pre-selenite

Table 1. Effect of Se on distribution of Cd in rice seedlings. The different letters in the same column indicate statistically significant differences among treatments ($p < 0.05$, $n = 3$).

Treatment	Distribution Ratio (%)		Transfer Factor
	Roots	Shoots	
Cd	67.5 ± 1.3 c	32.5 ± 1.3 c	0.12 ± 0 a
Cd + SeIV	85.6 ± 1.7 a	14.4 ± 1.7 a	0.05 ± 0.01 b
Cd + SeVI	68.0 ± 1.9 bc	32.0 ± 1.9 bc	0.12 ± 0.01 a
pre-SeIV + Cd	74.3 ± 3.8 b	25.7 ± 3.8 b	0.10 ± 0.02 a
pre-SeVI + Cd	69.8 ± 0.7 bc	31.2 ± 0.7 bc	0.10 ± 0.01 a

Table 2. Effect of Cd on distribution of Se in rice seedlings. The different letters in the same column indicate statistically significant differences among treatments ($p < 0.05$, $n = 3$).

Treatment	Distribution Ratio (%)		Transfer Factor
	Roots	Shoots	
SeIV	81.9 ± 1.5 a	18.1 ± 1.5 c	0.07 ± 0.01 a
SeVI	18.9 ± 1.3 c	81.1 ± 1.3 a	1.14 ± 0.09 b
Cd + SeIV	80.7 ± 2.2 a	19.3 ± 2.2 c	0.07 ± 0.01 a
Cd + SeVI	19.4 ± 1.8 bc	80.6 ± 1.8 ab	1.11 ± 0.10 b
pre-Cd + SeIV	80.5 ± 2.0 a	19.5 ± 2.0 c	0.07 ± 0.01 a
pre-Cd + SeVI	25.2 ± 3.9 b	74.8 ± 3.9 b	0.77 ± 0.13 b

or pre-selenate treatment, Cd addition decreased Se uptake by 19% and 47%, respectively (Fig. 2).

3.2 *Effect of pretreatment and simultaneous application of Cd and Se on their distribution in rice seedlings*

Compared with Cd treatment alone, selenite addition decreased the Cd distribution in shoots by 56% and 21% in simultaneous application and pre-selenite treatment, respectively. The Cd transfer factor (TP) of rice seedlings decreased markedly in the simultaneous application of selenite and Cd, but there are no significant change in the pre-selenite treatment. Furthermore, the addition of selenate did not significantly affect the distribution and TP of Cd in shoots (Table 1). As for the effect of Cd on Se distribution, the addition of Cd did not affect the distribution of Se in shoots with the exception of pre-Cd+SeVI treatment (Table 2).

By having both Cd and Se in the solution, it can reduce the Cd uptake by roots, which was similar to the previous studies (Maria *et al.*, 2008; Lin *et al.*, 2012). However, pre-Se treatment increased the Cd uptake, which is contrary to previous results observed by Lin *et al.* (2012) and Issam *et al.* (2014). The role of Se in alleviating the toxicity of heavy metals depended on ionic species, Se dose, Se valence and plant species (Chen, 2014). There are reports that found that selenite was more effective in decreasing Cd uptake by plants than selenate (Shanker, 1995; Muñoz, 2007).

4 CONCLUSIONS

Simultaneous application of Cd and Se markedly reduced Cd uptake by roots, however pre-Se and pre-Cd treatments slightly increased the uptake of Cd. Furthermore, selenite addition decreased the Cd distribution in shoots, but selenate did not significantly affect the distribution of Cd. Selenite may be more effective in decreasing Cd transportation from root to shoot than selenate. Furthermore, Cd addition also affected the uptake of Se by rice seedlings.

REFERENCES

Chen, M.X, Cao, L., Song, X.Z., Wang, X.Y., Qian, Q.P., & Liu, W. 2014. Effect of iron plaque and selenium on cadmium uptake and translocation in rice seedlings (*Oryza sativa*) grown in solution culture. *International Journal of Agriculture and Biology*, 16(6): 1159–1164.

Feng, R.W., Wei, C.Y., & Tu, S.X. 2013. The roles of selenium in protecting plants against abiotic stresses. *Environmental and Experimental Botany*, 87: 58–68.

Filek, M., Keskinen, R., Hartikainen, H., Szarejko, I., Janiak, A., & Miszalski, Z. 2008. The protective role of selenium in rape seedlings subjected to cadmium stress [J]. *Journal of Plant Physiology*, 165: 833–844.

Issam, S., Yacine, C., & Wahbi, D. 2014. Selenium alleviates cadmium toxicity by preventing oxidative stress in sunflower (*Helianthus annuus*) seedlings. *Journal of Plant Physiology*, 171: 85–91.

Lin, L., Zhou, W.H., Dai, H.X., Cao, F.B., Zhang, G.P., & Wu, F.B. 2012. Selenium reduces cadmium uptake and mitigates cadmium toxicity in rice [J]. *Journal of Hazardous Materials*, 235-236: 343–351.

Muñoz, A.H.S., Wrobel, K., Corona, J.F.G., & Wrobel, K. 2007. The protective effect of selenium inorganic forms against cadmium and silver toxicity in mycelia of Pleurotus ostreatus. *Mycological research*, 111: 626–632.

Shanker, K., Mishra, S., Srivastava, S., Rohit, S., Dass, S., & Prakash, S. 1995. Effect of selenite and selenate on plant uptake of cadmium by kidney bean (*Phaseolus mungo*) with reference to Cd-Se interaction. *Chemical Speciation and Bioavailability* 7(3): 97–100.

Selenium biofortification

Environmental pathways and dietary intake of selenium in a selenium rich rural community in China: A natural biofortification case study

Gary S. Bañuelos*
USDA-Agricultural Research Service, San Joaquin Valley Agricultural Sciences Center, Parlier, CA, USA

Jiefu Tang, Yuzhu Hou, Xuebin Yin & Linxi Yuan
School of Nanoscience & Advanced Lab for Selenium and Human Health, USTC, China;
Jiangsu Bio-Engineering Research Centre on Selenium, China

1 INTRODUCTION

Diet plays a significant role in the Chinese concept of health and disease. In overall terms, plant-derived foods provide about 93% of energy, 87% of protein and 55% of fat to the Chinese (Sun et al., 2014). The obvious result of "good health" can be observed in long-lived individuals inhabiting specific areas in China, e.g., Bama, Guangxi Province; Enshi, Hubei Province and Shitai, Anhui province (Liu & Li, 1986). Not only are the inhabitants consuming locally grown food commodities but their normal active life activities, e.g., food gathering, strenuous walking, climbing, and carrying activities, and experiencing no city stress, all likely contribute to their longevity. Another factor that appears to naturally exist in these areas, as well as in plants and other food products, is sufficient levels of natural-occurring selenium (Se). This is an important environmental finding since Se deficiencies are present in over 72% of the population of China (Wu et al., 2015). In a rural region within Anhui Province, a recent survey conducted by Anhui Bureau of Geological Survey (unpublished data) showed that there are about 407 km² within Shitai county that are considered Se rich. In this county, there are two unique villages called "Dashan" and "Xianyu" that are both known for their very low or non-existent rates of cancer. In addition, there is a high percentage (about 12%; unpublished data) of both their populations who are over 80 years old. Is there also a role for Se and/or a relationship between Se intake and longevity? Our objective was to eventually determine the sources of Se for the typical inhabitants in this region, and calculate the Se daily intake based on diet investigation.

2 MATERIALS AND METHODS

Shitai County (N 29°59′–30°24′, E117°12′–117°59′) is located in the southern part of Anhui Province. To trace the Se intake differences, four villages and the county town were selected as study sites (Figure 1). Soil samples were randomly collected in triplicate from

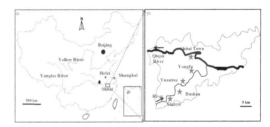

Figure 1. The study site location and sampling sites.

0–30 cm at four different village sites: Dashan, Xianyu, Yuantou, and Yongfu. In addition, samples were collected from different crops (corn, rice, soy bean), vegetables (pumpkin, eggplant, beninscasa, luffa, potato, radish, yam bean, spinach, potherb mustard, garlic), animal samples (pork, fish, chicken, egg), and some drinking water sources. Furthermore, hair samples were also collected from hair cutting saloons located in the villages from males and females representing different age groups. The soil and plant samples were dried at 65°C and ground to pass through a 200 mesh sieve, while hair samples were cut into less 5 mm pieces. All treated samples were acid-digested and analyzed for total Se by atomic fluorescence spectrometry (AFS) (Kulp & Pratt, 2004).

3 RESULTS

All data are preliminary and more analyses are currently in process. Selenium concentrations in water samples ranged from 2 to 4 μg/L in Dashan, Xianyu, Yongfu, Yuantou and Shitai, respectively. Total Se concentrations in soils were as follows in: Dashan (1779 ± 639 μg/kg DW, n = 12), Xianyu (1149 ± 918 μg/kg DW, n = 5), Yongfu (560 ± 109 μg/kg, n = 2), and Yuantou (363 ± 140 μg/kg DW, n = 6); foods in Shitai town came from out of this county, so no soil sample was collected. Selenium concentrations in food products (primarily from Dashan village) and limited samples from Shitai town are shown in Table 1. There were

Table 1. Mean selenium concentration in foods (μg/kg) and calculated daily Se intake (μg) for two villages.

Food (μg/kg)	Dashan	Shitai
Crops (DW)		
Corn	236	26
Rice	699	38
Soy bean	706	120
Vegetables (FW)		
Pumpkin	22	16
Eggplant	62	–†
Benincasa	180	17
Luffa	139	64
Potato	207	21
Radish	322	97
Yam bean	305	105
Spinach	120	58
Potherb mustard	147	–†
Garlic	755	51
Pork	54	61
Fish	1215	235
Chicken meat	203	–†
Chicken kidney	635	–†
Chicken egg	441	191
Daily intake (μg)	115	52

†Data are not currently available.

Table 2. Mean selenium concentrations (μg/kg) in hair samples collected from different age populations of females (F) and males (M) in two villages.

	Population ages in years			
	<18	15–30	30–60	>60
Dashan				
F	522	552	546	576
M	481	542	713	612
Shitai				
F	–†	132	333	–†
M	–†	257	220	263

†Data are not currently available.

significant differences in Se concentrations of hair collected between the two villages (Table 2).

4 DISCUSSION

Among the many health and genetic factors that likely contribute to longevity in this region, Se nutrition may also play a role. Selenium uptake by vegetation will naturally-occur and Se absorbed by the inhabitants will also be strongly influenced by food type, food processing, cooking technique, or even the use of Se-enriched coal products for cooking. The natural Se biofortification process is dependent upon the heterogeneously distribution of soil Se (Blazina et al., 2014). In this regard, underlying bedrock geology has generally been viewed as the primary source of Se in soils (Nriagu & Pacyna, 1988), and in such areas, one should also consider atmospheric deposition of volatile Se. More intensive multi-disciplined research is currently in progress to more clearly elucidate the natural sources of Se and how the environmental factors, e.g., soil, water and atmosphere, may contribute to a natural biofortification of Se in the diets of rural inhabitants within the Anhui province.

Disclaimer: mention of trade names or commercial products in this publication is solely for the purpose of providing specific information and does not imply recommendation or endorsement by the US Department of Agriculture. USDA is an equal opportunity provider and employer.

REFERENCES

Blazina, T., Sun, Y., Voegelin, A., Lenz, M., Berg, M. & Winkel, L.H.E. 2014. Terrestrial selenium distribution in China is potentially linked to monsoonal climate. *Nature Communications*, 5. 4717.
Kulp, T.R. & Pratt, L.M. 2004. Speciation and weathering of selenium in Upper Cretaceous chalk and shale from South Dakota and Wyoming, USA. *Geochimica et Cosmochimica Acta*, 68, 3687–3701.
Liu, B.-S. & Li, H. 1986. General situation and features of geriatric epidemiological study in China. *Chinese Medical Journal*, 99, 619–627.
Nriagu, J.O. & Pacyna, J.M. 1988. Quantitative assessment of worldwide contamination of air, water and soils by trace metals. *Nature*, 333, 134–139.
Sun, J., Buys, N.J. & Hills, A.P. 2014. Dietary pattern and its association with the prevalence of obesity, hypertension and other cardiovascular risk factors among Chinese older adults. *International Journal of Environmental Research and Public Health*, 11, 3956–3971.
Wu, Z., Bañuelos, G.S., Lin, Z.Q., Yin, X., Yuan, L., Liu, Y., et al. 2015. Biofortification and phytoremediation of selenium in China. *Frontiers in Plant Science*, 6: 136.

Effects of agronomic biofortification of maize and legumes with selenium on selenium concentration and selenium recovery in two cropping systems in Malawi

A.D.C. Chilimba*
Ministry of Agriculture and Food Security, Department of Agricultural Research Services, Lunyangwa Research Station, Mzuzu, Malawi

S.D. Young & E.M. Joy
University of Nottingham, School of Biosciences, Sutton Bonington Campus, Loughborough, UK

1 INTRODUCTION

Selenium (Se) is an essential element for humans and is derived primarily from dietary sources (Fair-weather-Tait et al., 2011). Habitual sub-optimal dietary Se intake leads to reduced Se status, which is associated with a range of adverse health outcomes including cardiovascular disorders, impaired immune function, and some forms of cancer (Fair-weather-Tait et al., 2011). Sub-optimal dietary Se intake is widespread in Malawi due to low levels of plant-available Se in the predominantly low-pH soils and narrow food choices, including little animal-source products (Chilimba et al., 2011). Chilimba et al. (2012) reported a linear response to Se application in maize grain Se concentration. An annual application of 5 g/ha of Se to maize crops grown on low-pH soils in Malawi can raise average dietary Se intake by 26–37 µg per person per day, greatly reducing risks of dietary Se deficiency. Agronomic biofortification via Se-enriched fertilisers might therefore be a cost-effective way to address widespread Se deficiency in Malawi and could be adopted as national policy – as in Finland, where dietary Se intake has been increased (Eurola et al., 2004). Current knowledge suggests that Se recovery by crops is inefficient. However, most recovery studies have been conducted using sole crops, whereas many smallholder farmers intercrop maize with legumes. It is likely that Se recovery by different crops varies and that intercropping systems may also influence Se recovery by individual species.

2 MATERIALS AND METHODS

The experimental treatments consisted of five cropping systems: monocrop maize, monocrop groundnuts, monocrop soybean, intercrop maize/groundnuts and intercrop maize/soybean. There were two application rates of Se: 0 and 10 g/ha. Experiments were conducted at three different sites: Chitedze (Chromic luvisol, Central region), Zombwe (Lixisol, Northern region) and Lunyangwa (Ferralsol, Northern region). The experimental plots were laid out in a randomized complete block design with three replicates at each growing site.

The maize, soybean and groundnut varieties used were SC627, Ocepara-4 and CG7 respectively. Maize seeds were planted at 75 cm apart and intercropped legumes of soybean and groundnuts were planted between the maize planting stations. The maize crop was applied with 200 kg/ha of 23.10.5.+6S +1.0Zn (N.P$_2$O$_5$.K$_2$O) applied to the crop as a basal dressing soon after seed emergence and a top dressing of urea at 100 kg/ha of N applied two weeks after basal dressing. Sodium selenate (Na$_2$SeO$_4$) was applied as a liquid drench with two levels of application, 0 and 10 g/ha. Sub-samples of biomass and grain were collected and dried in drying ovens at Lunyangwa and Chitedze research stations at 65°C for 24 h and analyzed for Se by ICP-MS (X-Series[II], Thermo Fisher Scientific Inc., Waltham, MA, USA) using a hydrogen reaction cell.

3 RESULTS AND DISCUSSION

Application of Se at 10 g/ha significantly increased Se concentration of all crops ($p < 0.001$). There were also significant differences between crop types ($p < 0.001$). Selenium concentrations were highest in soybean grain followed by groundnut and maize grain in applied (10 g/ha) and control treatments (0 g/ha) (Fig. 1), and the trend for total Se recovered by the respective crop per ha was similar to Se concentration for all growing sites (Fig. 2).

Thus, it appears that farmers who cultivate groundnut and soybean could substantially elevate their Se intake through biofortification. Average dietary Se intake would range between 61 and 71 µg/cap/d, which is likely to be optimum for most people (Fairweather-Tait et al., 2011). The Se concentration in

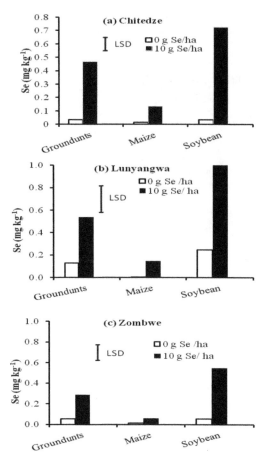

Figure 1. Effects of Se application and crop type on Se concentration at three growing sites (a) Chitedze, (b) Lunyangwa, and (c) Zombwe.

Se-fortified maize grain was similar to that in control crops of soybean and groundnuts at Lunyangwa and Zombwe (Fig. 1). This observation clearly indicates that legumes are more efficient in extracting native soil Se than maize.

4 CONCLUSIONS

Selenium application of 10 g/ha in the form of a sodium selenate liquid drench was effective in increasing Se concentration of maize grain, groundnut kernel and soybean grain. Considering only the edible portion of crops, soybean provided the greatest supply of Se followed by groundnuts and then maize. Agronomic biofortification of legumes and maize crops with Se-containing fertilizer can supply adequate intake of Se to the population of Malawi within diversified cropping systems.

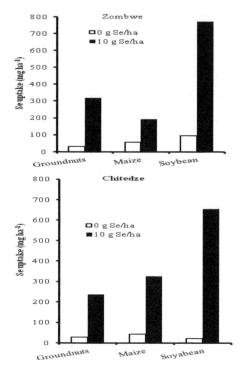

Figure 2. Effects of Se application and crop type on Se accumulation edible portion of maize, groundnuts and soybean at two sites, (a) Zombwe and (b) Chitedze.

REFERENCES

Chilimba, A.D.C., Young, S.D., Black, C.R., Rogerson, K.B., Ander, E.L., Watts, M., et al. 2011. Maize grain and soil surveys reveal suboptimal dietary selenium intake is widespread in Malawi. *Scientific Reports:* 1: 72.

Chilimba, A.D.C., Young, S.D., Black, C.R., Meacham, M.C., Lammel, J., & Broadley, M.R. 2012. Agronomic biofortification of maize with selenium (Se) in Malawi. *Field Crops Research:* 125: 118–128.

Eurola, M., Hietaniemi, V., Kontturi, M., Tuuri, H., Kangas, A., Niskanen, M., et al. 2004. Selenium content of Finnish oats in 1997–1999: effect of cultivars and cultivation techniques. Agricultural and Food Science 13: 46–53.

Fairweather-Tait, S.J., Bao, Y., Broadley, M.R., Collings, R., Ford, D., Hesketh, J., et al. 2011. Selenium in human health and disease. *Antioxidants and Redox Signaling:* 14: 1337–1383.

Potential roles of underutilized crops/trees in selenium nutrition in Malawi

Diriba B. Kumssa, Edward J.M. Joy, Scott D. Young & Martin R. Broadley*
School of Biosciences, University of Nottingham, Sutton Bonington, Loughborough, UK

E. Louise Ander & Michael J. Watts
Centre for Environmental Geochemistry, British Geological Survey, Keyworth, Nottingham, UK

Sue Walker
Crops For the Future, The University of Nottingham Malaysia Campus, Jalan Broga, Semenyih, Selangor Darul Ehsan, Malaysia

1 INTRODUCTION

Selenium (Se) is an essential mineral micronutrient in humans. Human dietary intakes and deficiencies of Se vary to a large extent due to the variation in soil geochemistry (Ge & Yang, 1993; Combs, 2001; Joy et al., 2015a). In Malawi, Se deficiency is likely to be very widespread (Joy et al., 2015a) with a supply of 33.6 μg/*capita*/d, based on United Nations Food and Agriculture Organization (FAO) Food Balance Sheets (FBS) (Joy et al., 2014). The Estimated Average Requirement (EAR) for Se in humans ranges from 17 μg/d for children aged 1–3 y to 59 μg/d for lactating women aged 31–50 y (IOM, 2000). The EAR for Se is an intake level that fulfils the health requirements of half of the individuals in a given population (IOM, 2000). The aim of this study was to estimate the percentage contribution of non-cereal dietary sources to Se nutrition in Malawi population with emphasis on some under-utilized crop/tree (UUCT) species.

2 MATERIALS AND METHODS

The National Statistics Office of Malawi conducted the Third Integrated Household Survey (IHS3) in 2010/11 (NSO, 2012). Household food consumption data from >12,500 households was collected during this survey. In the food consumption module, interviewees were asked to recall the food consumed by their household during the past 7 d from a list of 112 items (e.g. 'Maize *ufa* refined (fine flour)', 'Dried fish', *etc.*). These consumption data were cleaned of errors, and combined with local food composition data (Joy et al., 2015a) to estimate the intake of Se from various foods in the Malawian population.

3 RESULTS AND DISCUSSION

In Malawi, the median Se intake was 23.5 μg/*capita*/d in 2010/11 (Joy et al., 2015b); the EAR was

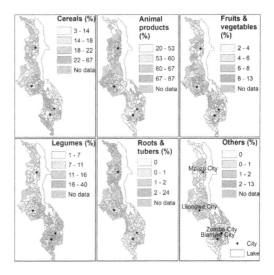

Figure 1. Percentage contribution of food groups to mean dietary Se in Malawian population by EPA.

36 μg/*capita*/d. Animal products (AP) and cereals (C) were the major sources of dietary Se contributing 20–87% and 3–67%, respectively, at Extension Planning Area (EPA) (Fig. 1). Fruits and vegetables (FV), legumes (L), roots and tubers (RT) contributed 2–13%, 1–40%, 0–24% of dietary Se, respectively. Overall, AP, C, FV, and L contributed ~60, 18, 6, 14, and 2%, respectively, to Se nutrition (Figure 2).

Among the foods reported to be consumed by the interviewees during the IHS3, *C. sesamoides*, *M. oleifera* and fresh water fish were the foods with the highest concentration of Se, providing ~900, 800, and 700 μg/kg dry weight, respectively. However, according to the IHS3 dietary consumption, 74% of the population were at risk of Se deficiency during 2010–11 (Joy et al., 2015b) due to low supply of fresh water fish, *C. sesamoides* and *M. oleifera*, and other good sources of Se (e.g. eggs, beans, etc.). A family with 6 equivalent adult members would be able to

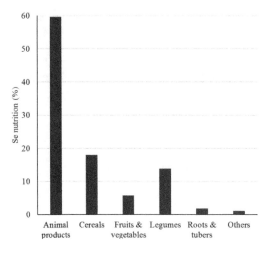

Figure 2. Contribution of various food groups to mean Se nutrition in Malawian population.

meet their daily Se nutritional requirement if they are able to obtain 500 g/d of *M. oleifera*. Despite being an excellent source of Se and other macro and micronutrients, IHS3, only nine out of >12,500 interviewed households reported using *M. oleifera* in their diet during 2010/11. On the other hand, ~8000 households reported eating pumpkin leaves (locally known as *Nkhwani/Mfutso*). The average consumption per household was 17 g/d, providing an average ~2 μg Se per household. If these households were consuming similar quantities of *C. sesamoides* and *M. oleifera*, the household level Se supply would have been ~15 and 13 μg, respectively.

4 CONCLUSIONS

The IHS3 dietary recall survey data indicate there is considerable risk dietary Se deficiency risk in the Malawian population (Joy *et al.*, 2015b). This observation is consistent with grain and food surveys (Joy *et al.*, 2014, 2015a), and direct assessment of biomarkers of Se status (Hurst *et al.*, 2013). Dietary diversification using UUCT species, for example, *M. oleifera* could play an important role in reducing Se deficiency. In addition to its high Se content, *M. oleifera* contains other micro- and macro nutrients, and is a fast growing tree that can survive on marginal lands. Almost all the tissues (leaves, pods, seeds, roots, flowers) of this plant are edible, and it produces leaves at the end of the dry season when food is scarce (Jahn, 1991; Odee, 1998; NRC, 2006). Unlike other vegetables, the leaves can be sun-dried and stored for long time as powder without a deterioration in most of the dietary nutritional content. The relevant ministries (i.e., agriculture, health, and education) could promote the production and utilization of UUCT species as valuable sources of Se and other nutrients in Malawi. Further research is needed to assess the interaction between soil properties and the nutritional composition of *Moringa* species.

REFERENCES

Combs, G.F. 2001. Selenium in global food systems. *British Journal of Nutrition* 85: 517–547.

Ge, K. & Yang, G. 1993. The epidemiology of selenium deficiency in the etiological study of endemic diseases in China. *American Journal of Clinical Nutrition* 57: 259S–263S.

Hurst, R., Siyame, E.W.P., Young, S.D., Chilimba, A.D.C., Joy, E.J.M., Black, C.R., *et al.* 2013. Soil-type influences human selenium status and underlies widespread selenium deficiency risks in Malawi. *Scientific Reports* 3: 1425.

IOM. 2000. Dietary Reference Intakes: Applications in Dietary Assessment. The National Academies Press: Washington, DC, USA.

Jahn, S.A.A. 1991. The traditional domestication of a multipurpose tree *Moringa stenopetala* (Bak. f.) Cuf. in the Ethiopian Rift Valley. *Ambio* 20, 244–247.

Joy, E.J.M., Ander, E.L., Young, S.D., Black, C.R., Watts, M.J., Chilimba, A.D.C. *et al.* 2014. Dietary mineral supplies in Africa. *Physiologia Plantarum* 151: 208–229.

Joy, E.J.M., Broadley, M.R., Young, S.D., Black, C.R., Chilimba, A.D.C., Ander, E.L., *et al.* 2015a. Soil type influences crop mineral composition in Malawi. *Science of the Total Environment* 505, 587–595.

Joy, E.J.M., Kumssa, D.B., Broadley, M.R., Watts, M.J., Young, S.D., Chilimba, A.D.C., *et al.* 2015b. Dietary mineral deficiencies in Malawi: spatial and socioeconomic assessment. *BMC Nutrition* (submitted in May 2015).

NRC. 2006. Moringa. In: Lost Crops of Africa: Vegetables. Vol. 2, pp. 377. National Academies Press: Washington, DC, USA.

NSO. 2012. Malawi Third Integrated Househod Survey (IHS3). National Statistcs Office (NSO), Zomba.

Odee, D. 1998. Forest biotechnology research in drylands of Kenya: the development of Moringa species. *Dryland Biodiversity* 2, 7–8.

Yu, S.Y., Zhu, Y.J. & Li, W.G. 1997. Protective role of selenium against hepatitis B virus and primary liver cancer in Qidong. *Biological Trace Element Research* 56, 117–124.

Necessity of biofortification with selenium of plants used as fodder and food in Romania

Radu Lăcătuşu*, Anca-Rovena Lăcătuşu, Mihaela Monica Stanciu-Burileanu & Mihaela Lungu
National Research & Development Institute for Soil Science, Agrochemistry and Environment Protection Bucharest, Romania

1 INTRODUCTION

Selenium (Se) deficiencies in animal nutrition have major implications for humans regarding the Se enrichment of some fodder and food sources. In the southeastern part of Romania, there have been reported Se deficiencies in the soil-wheat plant system, with potentially negative effects on animal and human nutrition (Lăcătuşu et al., 2002, 2012). As a result, Se is frequently administered in animal feed, and when necessary, Se is supplemented in human nutrition. To eliminate these issues related to Se-deficiencies, research has been performed on growing Se-rich plants in greenhouse and field experiments, and also, at an early phase, research on the possibility to get an organic fertilizer Se-rich, made of some organic waste composting. This study shows the necessity and possibility on biofortifying plants with Se that are used as fodder and vegetable foods.

2 MATERIALS AND METHODS

Effect of Se application on soil, seed and plant tissue was studied in alfalfa. The plants were grown in 8 liter pots using a soil material from the upper horizon (A) of a Cambic Chernozems under greenhouse conditions. Selenium treatments are presented in Tables 1, 2 and 3. Levels of N and P were equivalent to $N_{120}P_{80}$. Selenium was applied as K_2SeO_2. In the field study with Chernozems soil (located in the southeastern part of the country), wheat plants were sprayed at 9–10 stage (on the Feeks scale), with Na_2SeO_3 solutions (sodium selenite) containing rates of either 1, 10, or 20g Na_2SeO_3/ha.

Selenium content in alfalfa plants and wheat grain samples were analyzed using AAS, hydride version, and the gained results were statistically evaluated by the analysis of variance with LSD test.

3 RESULTS AND DISCUSSION

Extensive research surveys have been used for determining the Se content in soil-wheat plant systems in the south-eastern part of Romania; two areas: South-eastern Romanian Plain (first area) and Central and Southern Dobrogea (second area). Analysis data showed contrasting values; close to normal Se concentrations were measured in the first area and deficient Se concentrations were measured in the second area. As a consequence of soil Se concentrations, especially soluble soil Se, the Se content of wheat plants was also different between both areas. Although green wheat plants grown on soils accumulated some Se in the Dobrogea area and the Se content was on average almost 50% less than that of green wheat grown on soils of South-Eastern Romanian Plain. This marked difference was especially observed in the Se grain concentrations. Wheat grain grown on soils of south-eastern Romanian Plain contained a slightly below average Se concentrations (146 ± 189 μg/kg), characteristic for wheat grains harvested from 13 wheat-growing countries of the world (Kabata-Pendias & Pendias, 2001).

Analytical data obtained in the greenhouse study (Tables 1, 2, & 3) suggests a positive effect of Se, both on the development of alfalfa plants and the accumulation of the micronutrient in the plants. Thus, the accumulation of Se onto the soil using rates up to 10 mg/kg, has contributed to an increasing plant dry weight (up to 3.7 times), and increased Se concentration by 5.7 times (Table 1). Spraying seed with Se before planting with Se solutions containing from 1 to 5 mg/L also had a positive effect by increasing both plant mass three times and Se content by 3.8 times. Finally, spraying plants with a Se solution containing 40 mg/L, increased plant mass by 2.2 times and the Se concentration by 7.5 times.

The foliar application of Se on field-grown wheat plants during the stage of 9–10 (on the Feeks scale), showed a trend of increasing Se concentration in the wheat grain, but the treatments were not significant (Table 4). This, relatively lower Se accumulation rate observed on wheat grown under field conditions may be caused by the chemical form (selenite) in which Se was applied. According to Eurolla et al. (2003), Se applied as selenite was reportedly less effective than selenate for the plant accumulation of Se.

As an alternative to Se accumulation by sodium selenite application on plants or soil, a soil organic fertilizer made up by composting of three organic waste

Table 1. The effect of soil Se application on alfalfa dry mass and the Se concentration.

Dose of Se added onto soil (mg/kg)	Plant dry mass (g/pot)	Plant Se concentration (µg/kg)
0	14	130
1	20	175
2.5	22a	280a
5.0	35b	463b
7.5	47b	520b
10.0	52b	739c

Table 2. The effect of spraying seed with Se on alfalfa dry mass and the Se concentration.

Dose of Se added to seed (mg/kg)	Plant dry mass (g/pot)	Plant Se concentration (µg/kg)
0	14	130
1	25	210a
2.5	37a	314b
5.0	42b	497c

Table 3. The effect of Se application to the plant on the alfalfa dry mass and the Se concentration.

Dose of Se added to plant (mg/L)	Plant dry mass (g/pot)	Plant Se concentration (µg/kg)
0	14	130
40	32	976

Table 4. The effect of foliar fertilization of Se Se accumulation in wheat grains

Treatments on plants	Se concentration in wheat grains (µg/kg)
V1 – Control (0)	117.5
V2 – 1g Na_2SeO_3/ha	130.0
V3 – 10g Na_2SeO_3/ha	120.0
V4 – 20g Na_2SeO_3/ha	112.5
LSD 5%	*16*
LSD 1%	*23*
LSD 0.1%	*32*

(manure, sludge from wastewater treatment plant and marine algae) was obtained. The amount of the total Se was 786 µg/kg, and those of the soluble Se was 62 µg/kg of composted material. These values show that the use of this compost on soils deficient in Se may help to diminish the phenomenon of this microelement deficiency that prevails in the Black See coastal area where Dobrogea County is located.

4 CONCLUSIONS

Selenium deficiencies have been recorded in wheat (green plants and grains) cultivated in Central and Southern Dobrogea. In our greenhouse study, the application of Se from potassium selenite to the soil, on the seed and leaves of alfalfa contributed to an increase in dry mass of plants and their Se accumulation. The foliar application of Se as sodium selenite on the wheat plants, did not significantly increase Se content of wheat grain. Further research is necessary for determining the accurate dosage of Se application delivery to effectively biofortify food and feed products with Se for Se deficient areas like Dobrogea.

ACKNOWLEDGEMENTS

This work has been carried out through the Partnerships in priority areas – PN II program, developed with the support of Ministry of National Education-UEFISCDI, project no. PNII-PT-PCCA-2013-4-0675 – FEROW.

REFERENCES

Eurola, M., Alftham, G., Aro, A., Ekholm, P., Hietaniemi, V., Raino H., et al. 2003, Results of the Finnish selenium monitoring program 2000–2001, Agrifood Res. Reports, 36.

Kabata-Pendias, A. & Pendias, H. 2001. Trace Elements in Soils and Plants, CRC Press, Boca Raton, FL.

Lăcătuşu, R., Trepăduş, I., Mihaela, Lungu, Cârstea, St., Kovaksovics, B., & Craciun, L. 2002, Selenium abundance in some soils of Dobrogea (Romania) and ovine miodistrophy incidence, Trans. 21 Workshop "Macro- and Trace Elements", Jena, 114–119.

Lăcătuşu, R., Mihaela, L., Mihaela, M., Stanciu-Burileanu, A.R., Lăcătuşu, I., Rîşnoveanu, A., et al. 2012. Selenium in the soil-plant system of the Făgăraş Depression, Carpath. J. of Earth and Environmental Sciences, 7(4): 37–46.

The selenium content of organically produced foods in Finland

P. Ekholm*
University of Helsinki, Department of Food and Environmental Sciences, Helsinki, Finland

G. Alfthan
National Institute of Health and Welfare, Helsinki, Finland

M. Eurola
Natural Resources Institute, Jokioinen, Finland

1 INTRODUCTION

Fertilizers in Finland have been supplemented with sodium selenate since 1985. The selenium (Se) content of agricultural plants was found to be extremely low in the 1970's and the average daily Se intake in Finland was reported to be one of the lowest in the world (Varo & Koivistoinen, 1980). The Se supplementation of fertilizers has clearly increased the Se contents of all foods produced in Finland. The Se concentration in wheat flour is presently circa 0.15 mg/kg/dm, which is about 15 times higher than without Se fertilization. The Se concentrations of animal products have also increased. The Se concentrations of beef and pork are now 3 and 1.5 times higher than before the Se fertilization practice, respectively (Ekholm et al., 1990; Eurola et al., 1990; Alfthan et al., 2015). The present average daily Se intake is about 0.08 mg/day at energy level 10 MJ. Milk, cheese, meat, and meat products are the most important Se sources in the Finnish diet.

In organic food production, the use of Se supplemented fertilizers is not allowed in Finland. Because of this practice we can assume that the Se contents of the products of organic farming are lower than in conventional farming. In general, the Se content of soil and thus plant available Se are very low in Finland. The soils are acidic and bedrock is very old. Soil sorption characteristics and reducing conditions also decrease the amount of bioavailable Se in Finnish soils during autumn and winter. In this study, the Se contents in organic food samples were analysed and the results were compared with the conventionally produced samples. Also an estimation of the Se intake in different special diets has been projected.

2 MATERIALS AND METHODS

Organic food samples (wheat flour, rye flour, peas, beef, pork fillet, milk, cheese, eggs) were purchased from six retail food stores in the Forssa area in Central Finland. These stores represent the main distribution channels in Finland. Three samples of each food items were prepared and pooled in pairs to make three samples per food item. The samples were freeze dried, homogenized, and prepared as described below.

The dried samples were digested in a mixture of concentrated HNO_3, $HClO_4$, and H_2SO_4. Selenium was reduced to Se IV by 6 M HCl. After digestion, the samples were transferred into 25 mL volumetric flasks using fresh Q POD element water (Millipore). Selenium was analysed as hydride by ICP-OES equipment (iCAP 6000 Series Thermo Scientific) with two different lines 196.0 and 203.9 nm. The accuracy and the precision of the method was checked by using one in-house reference sample (wheat flour) and three official reference materials (Table 1).

Table 1. Accuracy and precision of the analytical method.

	Se Concentration (mg/kg)	
	Measured value	Certified value
SRM1567 Wheat flour	1.2 ± 0.3	1.1 ± 0.2
SRM1549 Milk powder	0.11 ± 0.02	0.11 ± 0.01
NBS1577a Bovine liver	0.76 ± 0.09	0.71 ± 0.07
Internal QC sample (Wheat flour)	0.043 ± 0.002	0.040 ± 0.003

3 RESULTS

The Se concentrations of organic foods were clearly lower in comparison with the conventional products (Table 2). The results are very similar compared to results obtained in the 1970's before the use of the Se supplemented fertilizers (Alfthan et al., 2011). The major difference is that today there is over 15 times more Se in wheat flour and 5 times more Se in rye

Table 2. Selenium concentrations of organically produced and conventional food samples.

Sample	Se Concentration (mg/kg, DW)	
	Organic farming	Conventional farming[†]
Wheat flour	<0.01	0.15
Rye flour	0.01	0.05
Pea	<0.01	–
Milk	0.13	0.22
Cheese	0.19	0.39
Eggs	0.64	0.88
Beef	0.34	0.55
Pork fillet	0.58	0.59

[†]Eurola et al. (2011)

flour than at earlier times. The effect of the Se supplementation is seen only within the same growing season. During the winter, sodium selenite is reduced and is no longer available to plants in the next year (Eurola et al., 1990). This observation is the main reason for the lower Se content of rye which is a winter cereal and is sowed in autumn. The difference in Se concentrations in animal products is not so considerable. Selenium deficiency diseases are a problem in organic husbandry and sufficient Se intake has to be cared for. This extra Se increases the Se concentrations of organic animal products makes the difference between the organic and conventional products lower.

4 CONCLUSIONS

The Se concentrations of all organically produced foods were clearly lower than in conventional farming. These differences certainly affect the Se intake. The average daily Se intake has been about 0.08 mg/day in recent years. Using mostly organically produced foods decreases the intake by circa 30% because of the relatively high Se content of animal products. Seventy percent of Se intake originates from the animal product in mixed diets. However, the vegan diets which are compiled only from organic foods may be deficient in Se.

REFERENCES

Alfthan, G., Aspila, P., Ekhom, P., Eurola, M., Hartikainen, H., Hero, H., et al. 2011. Nationwide supplementation of sodium selenate to commercial fertilizers. History and 20-years results from Finnish selenium monitoring program. 317–337. In: T. Thompson & L. Amoroso (eds), Food Based Approaches (FBAs) for Combating Micronutrient Deficiencies. Food and Agriculture Organization of the United Nations FAO and CAB International, FAO. (http://www.fao.org/docrep/013/am027e/am027e.pdf)

Alfthan, G., Eurola, M., Ekholm, P., Venäläinen, E-R., Rooth, T., Korkalainen, K., et al. 2015. Effects of nationwide addition of selenium to fertilizers on foods, and animal and human health in Finland: from deficiency to optimal selenium status of the population. *J Trace Elem Med Biol.* 31:142–147.

Ekholm, P., Ylinen, M., Koivistoinen, P. & Varo, P. 1990. Effects of general soil fertilization with sodium selenate in Finland on the selenium content of meat and fish. *J Agric Food Chem.* 38:695–698.

Eurola, M., Ekholm, P., Ylinen, M., Koivistoinen, P., & Varo, P. 1990. Effects on selenium fertilization on the selenium content of cereal grains, flour, and bread produced in Finland. *Cereal Chem.* 67:334–337.

Eurola, M., Alfthan, G., Ekholm, P., Root, T., Suoniitty, T., Venäläinen, E-R., et al. 2011. Report of the Selenium Working Group. MTT Raportti 35. Jokioinen, Finland. (in Finnish). (http://www.mtt.fi/mttraportti/pdf/mttraportti 35.pdf)

Varo, P. & Koivistoinen, P. 1980. Mineral element composition of Finnish foods. XII. General discussion and nutritional evaluation. *Acta Agric Scand*; Suppl 22: 165–171.

Effects of soil selenium ore powder application on rice growth and selenium accumulation

Wenjing He, Bin Du, Yaomei Luo, Huoyun Chen, Huan Liu, Sen Xiao & Danying Xing*
Institute of Application Technology for Crop Selenium Enrichment, Yangtze University, Jingzhou, Hubei, China; Hubei Collaborative Innovation Center for Grain Industry, Jingzhou, Hubei, China

Jianlong Xu*
Institute of Crop Sciences, Chinese Academy of Agricultural Sciences, Beijing, China; College of Agriculture, Yangtze University, Jingzhou, Hubei, China

1 INTRODUCTION

Selenium (Se) is essential to human body. Selenium deficiency may cause some diseases (Su, 1998). Organic selenium is the main form which benefits human body (Zhong & Wang, 2007). According to the principles of food chain, it is safe and cost-effective for human to absorb the organic Se. In southern part of China, people rely heavily on rice. Enshi, a city in Hubei, has abundant Se resource (Liu, 1993). Study the effects of soil Se ore powder application on rice growth and Se accumulation displays the significance of developing Se-enriched rice production (Xing, 2003). During 2013 and 2014, we conducted an experiment on two rice varieties in Fengcheng, China, where the soil selenium background concentration was 0.37 mg kg^{-1} to research the effect of Se ore powder (SOP) on rice growth and the its post-season effect.

2 MATERIALS AND METHODS

The experiment used Se ore powders (SOPs) produced in Enshi, Hubei, China, with an approximate Se concentration of 83 mg kg^{-1}. SOPs were applied on the early season rice cultivar Lingliangyou 211 and the second season rice cultivar Ziyou 218 for consecutive two years. The rice growing period of early season rice variety Lingliangyou 218 was 110 days and that for the second season rice variety Ziyou 218 was 107 days.

Different SOP application rates of 0, 10, 20, 30, 40 and 50 kg on the 667 m^2 (or 1 mu) plot were selected for a field experiment located in Fengcheng (28.1°N, 115.47°E), Jiangxi province China. The local soil Se background concentration was 0.37 mg/kg. Its annual temperature was 17.6–18.2°C and the sunshine hours were 1936 hours. The annual rainfall in Fengcheng ranged from 1670 mm to 1912 mm, and the frost-free season was 280 days. Randomized block design with three replications was adopted. The plot area for seeding is around 13.2 m^2. The Yangtze University Field Interval Method using bamboo frame and plastic film to isolate the plots was adopted in this field experiment. Guard row of one to two meters wide around the isolation plot was set up in the beginning of the experiment. Selenium ore powders were put on the first season rice. After the harvest of first season rice, tillage was conducted, plastic film was repaired, and the field was irrigated. The nitrogen phosphorus and potassium fertilizers were applied to plant the second season rice. The growth of the second season rice was observed in order to find the Se accumulation effect.

3 RESULTS AND DISSCUSSION

The economical characters and Se concentration were identified in grains with different amount of Se ore powder application (Table 1). The results of this study indicated that (1) Se ore powder application significantly affected the production and Se concentration of the first and second season rice. The Se ore powder could be used for Se-rich grain production; (2) SOPs reduced panicles per unit area (PUA) by 6.61% to 12.12%, while it increased filled grains per panicle (FGP) by 15.10% to 22.76%, seed setting rate (SSR) by 10.71% to 15.22%, thousand grain weight (TGW) by 1.5% to 3.37% and grain yield per unit area (GYA) by 5.67% to 13.22% for the early season rice Lingliangyou 211; (3) Post-effects of SOP on the second season rice Ziyou 218 included PUA reduction by 10.87% to 14.21%, while it increased FGP by 13.18% to 20.43%, SSR by 6.44% to 11.25%, TGW by 5.39% to 14.11% and GYA by 6.79% to 12.65%; (4) SOPs increased the Se concentrations in the grains and stems of the first season rice, and the average Se concentration varied from 0.0557 to 0.2268 mg/kg and from 0.1451 to 0.2797 mg/kg, respectively, from this two-year field experiment. Selenium concentrations in grains and stems increased accordingly with increasing the amount of SOPs applied, reaching highly significant correlations (r = 0.9969, p = 0.05 and 0.9915, p = 0.05, respectively). Selenium concentration was

Table 1. The economical characters and selenium concentration of rice with different rates of selenium ore powder application. Means with different letters are significantly different at p = 0.05.

Rice cultivar	SOPs rate (kg/667 m^2)	Economical Characters					Se Concentration	
		Ear number (10^4/667m^2)	Grain number per ear	Seed setting rate (%)	Thousand grain weight (g)	Grain yield (kg/667 m^2)	Stem (mg/kg)	Grain (mg/kg)
Ling Liang You 211		23.44	78	73.19	26.7	467.19 d	0.1451e	0.0557f
	10	20.82	91	81.93	27.5	518.13 ab	0.1541e	0.0945e
	20	21.30	95	84.32	27.2	508.95 b	0.1880d	0.1232d
	30	21.89	90	81.03	27.1	494.23 c	0.2221c	0.1607c
	40	21.67	88	84.33	27.5	493.67 c	0.2436b	0.2051b
	50	20.60	96	81.47	27.6	528.97 a	0.2797a	0.2268a
Zi You 218		18.30	84	65.88	24.1	352.52 c	0.1090f	0.0227f
	10	16.19	96	73.29	25.4	376.44 b	0.1210e	0.0682e
	20	16.31	97	71.78	26.1	389.82 ab	0.1462d	0.0947d
	30	15.56	99	72.57	26.2	386.43 ab	0.1913c	0.1165c
	40	15.70	101	70.12	26.2	391.25 a	0.2181b	0.1505b
	50	16.13	95	71.98	27.5	397.11 a	0.2474a	0.1903a

higher in stems than in grains for early season rice, and the difference of Se concentration between stems and grains became decreased according with the augment of the amount applied, reaching negatively significant correlation (r = −0.7975); (5) there were significant correlations between the amount of SOP and each of PUA, FGP, TGW and GYA in the second season rice, indicating that SOPs had stronger effects on above traits for the second season rice than the early season rice and showed obvious post-seasonal effect; (6) Post-effect of SOPs increased the Se concentrations in the grains and stems of the second season rice, and the average Se concentrations in the two years varied from 0.0227 to 0.1903 mg/kg and from 0.1090 to 0.2474 mg/kg, respectively. Similarly, Se concentrations in the grains and stems increased accordingly with augment of the amount of SOP applied, reaching highly significant correlations (r = 0.9945, p = 0.05 for grains and 0.9894, p = 0.05 for stems).

4 CONCLUSIONS

Selenium ore powder application increased the Se concentration in both gains and stems of the first season rice variety Ling Liangyou 211 and the second season rice variety Ziyou 218. At the same time, the Se treatment changes crop economical values. Selenium ore powders can be used as an important practice to improve Se-rich grain production.

REFERENCES

Liu, Peidi. 1993. *Selenium resources and their comprehensive exploitation and utilization*, Beijing: China Science and Technology Press.
Su, Xiaoyun. 1998. *China selenium resources and its exploitation and utilization*, Beijing: China Meteorological Press.
Xing, Danying. 2003. *The research and application of crop selenium-rich technology*, Wuhan: Huazhong Agricultural University.
Zhong, Na., Wang, Xiaoru & Chen, Dengyun. 2007. The research progress of organic selenium. *Chinese Pharmaceutical Affairs,* 21(4):268–271

Agronomic biofortification of Brachiaria with selenium along with urea

L.A. Faria*, M.C. Machado, A.L. Abdalla, P.P. Righeto & L.L. Campos
CENA/USP, Piracicaba/SP, Brazil

F.H.S. Karp & M.Y. Kamogawa
ESALQ/USP, Piracicaba/SP, Brazil

1 SELENIUM IN PASTURES OF BRAZIL

1.1 Selenium in Brazilian soils

Brazil has low levels of selenium (Se) in soil according to reported research in areas of South, Southeast and Midwest, which confirmed the occurrence of marginal Se deficiency in grazing cattle in Rio Grande do Sul (Valle et al., 2003), and low content of Se in grass in Southeast and Midwest (Lucci et al., 1984).

1.2 Pastures fertilized with selenium

Selenium fertilization has been an option to attain good results in areas with low bioavailability of Se. The addition of Se can supply the plants and animals with Se and consequently improve animal performance and nutritional quality of the food, milk and meat produced. These positive results already were verified in the absence of any symptoms of deficiency (Whelan et al., 1994).

2 MATERIALS AND METHODS

2.1 Experimental design

Experiment was carried out in a greenhouse in CENA/USP in randomized block design with four replicates per each respective measurement of sodium selenate equivalent to 0, 10, 20, 40, 80 and 160 g/ha Se in topdressing of the *Brachiaria brizantha* cv Marandu with two cuts.

2.2 Preparation of fertilizers and soil

Urea was coated by mixture of boric acid, copper sulfate and Se application. Fertilization was applied in topdressing at a rate of 100 kg/ha N. Soil was collected of 0–20 cm layer from a sandy Oxisol with low Se (<0.5 mg/dm^3) characterized for 18% clay, 2% silt and 80% sand, with pH (CaCl$_2$) = 4.6; P = 12 and S = 5 mg/dm^3; K = 0.3; Ca = 10; Mg = 8; Al = 0; H+Al = 38; SB = 18.3 and CTC = 56.3 mmol/dm^3; V = 33% e m = 5%; B = 0.24; Cu = 0.6; Fe = 97; Mn = 7.9; Zn = 1.4, Si = 7 e Na < 4 mg/dm^3.

The liming was applied with CaCO$_3$ to reach base saturation of 70%. Application of a nutrient solution to raise the P to 40 mg/dm^3 and the K to 3 mmol/dm^3 with KH$_2$PO$_4$ and KCl was applied after transplanting seedlings.

2.3 Evaluations

Two cuts were performed at 5 cm from the ground level at the beginning of the aging process (30 and 60 days after the treatments). The indirect determination of chlorophyll content with the SPAD-502 device took place a day before cutting. The shoot material was separated into two portions, one for the analysis of gases production and bromatological and chemical analysis. In other portion, the material was separated in fractions of leaves and stem+sheath. Materials were dried at 60°C to determinate dry matter production (DMP) and the proportions of leaves and stem + sheath.

The Se concentration in the leaf fraction was performed by digestion with HNO$_3$ and H$_2$O$_2$ in closed vessels and High Performance Microwave Digestion, and its determination was made by generator of hydrides coupled to the Atomic Absorption Spectrometer (FI-HG-AAS).

Gas production (GP) was realized on degradability assay *in vitro* (PG) in a semi-automatic system (Mauricio et al., 1999). GP technique and quantifying of gases species were applied during the *in vitro* fermentation (Theodorou et al., 1994) and adapted to semi-automatic system (Bueno et al., 2005).

3 RESULTS AND DISCUSSION

Concentrations of Se in leaves showed increasing linear effect of Se application in both cuts (Fig. 1).

The application of 10 g/ha in the first cutting was enough to supply the requirement of 0.1 to 0.3 mg/kg Se in dry matter for cattle diet (NRC, 2001), however, in the second cutting, the requirement was achieved with application next to treatment of 40 g/ha. The fibrous fraction as ADF, NDF, cellulose and hemicellulose were not influence by Se levels, except lignin; it was greater (122.1 g/kg) than control (73.9 g/kg)

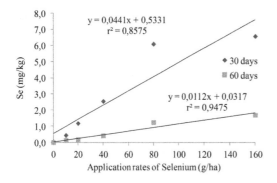

Figure 1. Selenium foliar (mg/kg DM) for Se application in fertilization.

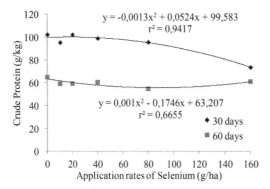

Figure 2. Crude protein (g/kg DM) in *Brachiaria brizantha* for Se doses in fertilization.

with application of 40 g/ha of Se. The organic matter degradability (OMD) was adjusted in a quadratic curve with peak in 59.5 g/ha Se.

Gas production/DM decreased linearly with Se levels. The gas production/OMD showed a significant interaction Se levels x cuts, effect, which was lower with increasing Se levels in the second cut to 60 days. The methane production, gas production/NDF and NDF degradation had no influence from Se treatments. The ammonia-N content was adjusted in a quadratic curve with lower values in highest Se levels. Positive effects was observed with Se levels up to 40 g/ha.

The forage plant showed no visual influence or dry matter production of shoots and roots, for Se application. The crude protein (CP) resulted in quadratic adjustment for interaction Se levels x cuts with opposite effects into the cuts (Fig. 2).

4 CONCLUSIONS

Selenium treatment levels of up to 40 g/ha were favorable for production and quality of forage considering to parameters quantitative and qualitative for animal feed.

REFERENCES

Bueno, I.C.S., Cabral Filho, S.L.S., Gobbo, S.P., Louvandini, H., Vitti, D.M.S.S. & Abdalla, A.L. 2005. Influence of inoculum source in gás production method. Animal feed science and technology, 123–124: 95–105.

Lucci, C.S., Moxon, A.L., Zanetti, M.A., Franzolin Neto, R. & Maracomini, D.G. 1984. Selênio em bovinos leiteiros do Estado de São Paulo. II. Níveis de selênio nas forragens e concentrados. Revista da Faculdade de Medicina Veterinaria e Zootecnia da Universidade de São Paulo, 21(1): 71–76.

Mauricio, R.M., Mould, F.L., Dhanoa, M.S., Owen, E., Channa, K.S. & Theodorou, M.K. 1999. A semi-automated in vitro gas production technique for ruminant feedstuff evaluation. Animal Feed Science and Technology. 79: 321–330.

NRC. 2001. NUTRIENT Requirements of Dairy Cattle. 7 ed. Washington: National Academy.

Theodorou, M.K., Williams, B.A., Dhanoa, M.S., McAllan, A.B. & France, J. 1994. A simple gas production method using a pressure transducer to determine the fermentation kinetics of ruminant feeds. Animal Feed Science and Technology. 48: 185–197.

Valle, S.F., González, F.D., Rocha, D., Scalzilli, H.B., Campo, R. & Larosa, V.L. 2003. Mineral deficiencies in beef cattle from southern Brazil. Brazilian Journal of Veterinary Research and Animal Science, 40: 47–53.

Whelan, B.R., Barrow, N.J. & Peter, D.W. 1994. Selenium fertilizers for pastures grazed by sheep. II.* Wool and Liveweight Responses to Selenium. Australian Journal Agriculture Research, 45: 877–887.

Effects of different foliar selenium-enriched fertilizers on selenium accumulation in rice (*Oryza sativa*)

Q. Wang, X.F. Wang, J.X. Li, Y.B. Guo & H.F. Li*
College of Resources and Environmental Sciences, China Agricultural University, Beijing, China

1 INTRODUCTION

Selenium (Se) is an essential trace element for humans and livestocks. Inadaquate Se intake can cause health disorders, while the window between Se deficiency and toxicity is rather narrow. The referenced Se intake dose was specified by World Health Organization (WHO) to be 40–200 μg/day. However, it was estimated that approximately 72% of the land in China was Se deficient and the average dietary intake of Se was only 26–32 μg/day for Chinese population (Chen et al., 2002). Thus, it is important to seek a scientific and feasible method to safely enhance the Se dietary intake.

Dietary Se intake can be supplemented through dietary diversification, medicine supplementation, food fortification, and crop biofortification, including fertilization and genetic approaches (Yang *et al.*, 1981; White & Broadley, 2009). Biofortification of Se through foliar fertilizer has been practiced in Finland, UK, Malawi and China (Eurola *et al.*, 1991; Rayma, 2008; Fang *et al.*, 2008; Chilimba *et al.*, 2012). In those studies, Se contents of grain were significantly increased by application of foliar selenite or selenate compared with the control. However, there have been relatively few studies that investigated the application of other forms of Se fertilizers, such as SeMet or chemical nano-Se. Therefore, the aim of this study was to determine different effects of four foliar Se fertilizers, including selenite, selenite, SeMet and chemical nano-Se, on the Se concentrations in rice.

2 MATERIALS AND METHODS

The field trial was conducted at the Fangzheng county (longitude 128°83′E, latitude 45°83′N), Harbin, Heilongjiang Province, China. The soil type was black soil of pH 5.96. The major physical and chemical properties of the original soil include: soil organic matter (OM) = 34.0 mg/kg, total N = 2.3 g/kg, available P = 69.8 mg/kg, available K = 169.0 mg/kg, total Se = 0.2 mg/kg. A common japonica rice cultivar (Daohuaxiang 2) was selected in this study. Rice plants were transplanted in May 2014, and harvested in September 2014. Selenium fertilizers were applied to leaves during the flowering stage in June. The five treatments were (1) control (without Se addition), (2) Na_2SeO_3 (SeIV), (3) Na_2SeO_4 (SeVI), (4) SeMet, and (5) nano-Se; the Se application rate was 7.5 g Se per hm^2 (or ha). The field layout was a randomized block design with three replicates per treatment. The plot size was $5 \times 5\,m^2$ with a 2 m buffer zone at each end. Selenium chemicals (Na_2SeO_3, Na_2SeO_4, and SeMet) were obtained from Sigma-Aldrich (St Louis, MO, USA). Chemical nano-Se was prepared by the method described by Lin and Wang (2005). The average particle size was 140 nm in diameter.

For the analysis of total Se content, plant samples of straw, grain and husk were acid-digested in 8 mL of HNO_3 (GR) using a microwave sample preparation system (CEM, MARS 5, CEM Corp., Matthews, NC, USA). Concentrations of Se were determined by AFS (AFS-920 Dual-channel Atomic Fluorescence Spectrometer, Beijing Jitian Instruments Co. Ltd., Beijing, China). During the chemical analysis process, certified reference materials (GBW10014) and blanks were included for quality assurance. The recovery of Se in GBW10014 was 85–110%.

3 RESULTS AND DISCUSSION

Concentrations of Se in rice grain, husk and straw were significantly influenced by the forms of applied Se (Fig. 1). At harvest, the ranges of Se concentrations in root, husk and straw were 0.06–0.84 mg/kg, 0.09–1.09 mg/kg and 0.13–1.04 mg/kg, respectively. Under +SeMet treatment, Se concentrations in different parts of rice were the highest, 0.72 mg/kg in grain, 0.95 mg/kg in husk and 0.91 mg/kg in straw, which were approximately 5-, 7-, and 5-fold higher than those of the control. Even with the high Se concentration, no phytotoxic symptoms were observed in any of SeMet-treated plots.

Selenium concentrations in grain were significantly ($p < 0.05$) increased by foliar application of Se fertilizers compared with the control, except the nano-Se treatment (Fig. 1a). When supplied with chemical nano-Se, Se concentrations in the different parts of

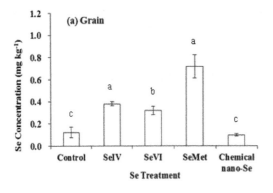
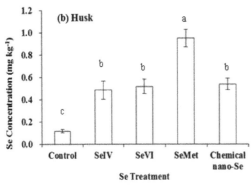
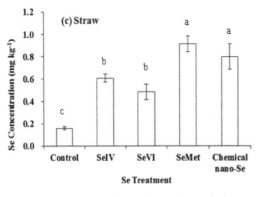

Figure 1. Concentrations of Se in rice grain (a), husk (b) and straw (c) after treated wih different forms of Se. The different letters indicate significant differences ($P < 0.05$) among the same part of rice. Data are presented as mean ± standard error (n = 3).

rice were as follows: straw > husk > grain, indicating that it was difficult for Se to translocate to grains. Elemental Se has long been considered as the least toxic due to its solid state and low bioavailability (Chapman et al., 2010), which is consistent with our result.

4 CONCLUSIONS

Different forms of Se foliar fertilizers influenced Se accumulation in grain significantly. With the application of SeMet, Se concentration in rice was the highest, followed by selenite, selenite, and nano-Se.

REFERENCES

Chapman, P.M., Adams, W.J., Brooks, M., Delos C.G., Luoma, S.N., Maher, W.A., et al. 2010. Ecological assessment of selenium in the aquatic environments. Pensacola: SETAC Ecological Press.

Chen, L.C., Yang, F.M., Xu, J., Hu, Y., Hu, Q.H., Zhang, Y.L., et al. 2002. Determination of selenium concentration of rice in China and effect of fertilization of selenite and selenate on selenium content of rice. Journal of Agricultural and Food Chemistry, 50(18): 5128–5130.

Chilimba, A.D.C., Young, S.D., Black, C.R., Meacham, M.C., Lammel, J., & Broadley, M.R. 2012. Agronomic biofortification of maize with selenium (Se) in Malawi. Field Crops Research, 125: 118–128.

Eurola, M.H., Ekholm, P.I., Ylinen, M.E., Varo, P.T., & Koivistoinen, P.E. 1991. Selenium in Finnish foods after beginning the use of selenate-supplemented fertilisers. Journal of the Science of Food and Agriculture, 56(1): 57–70.

Fang, Y., Wang, L., Xin, Z.H., Zhao, L.Y., An X.X., & Hu, Q.H. 2008. Effect of foliar application of zinc, selenium, and iron fertilizers on nutrients concentration and yield of rice grain in China. Journal of Agricultural and Food Chemistry, 56(6): 2079–2084.

Golubkina, N.A., Folmanis, G.E. & Tananaev, I.G. 2012. Comparative evaluation of selenium accumulation by allium species after foliar application of selenium nanoparticles, sodium selenite and sodium selenate. Doklady Biological Sciences. 444(1): 176–179.

Lin, Z.H. & Wang, C.R.C. 2005. Evidence on the size-dependent absorption spectral evolution of selenium nanoparticles. Materials Chemistry and Physics, 178(1): 92–102.

Rayman, M.P. 2008. Food-chain selenium and human health: emphasis on intake. British Journal of Nutrition, 100(2): 254–268.

Yang, G.Q., Yin, T.A., Liu, S.J., Wang, G.Y., Zhou, R.H., Gu, L.Z., et al. 1982. Approaches to the supplementation of selenium in the prevention of Keshan disease. Acta Nutriments Sinica, 4(1): 1–7.

Using agronomic biofortification to reduce micronutrient deficiency in food crops on loess soil in China

H. Mao
College of Resources and Environment, Northwest Agriculture and Forestry University, Shaanxi, China; Key Laboratory of Plant Nutrition and the Agri-environment in Northwest China, Ministry of Agriculture, China

G.H. Lyons
School of Agriculture, Food & Wine, University of Adelaide, Waite Campus, Glen Osmond, Australia

Z.H. Wang*
College of Resources and Environment, Northwest Agriculture & Forestry University, Shaanxi, China

1 INTRODUCTION

Micronutrient malnutrition in humans is derived from micronutrient deficiency in soils and thus crops and foods derived from them. Agronomic biofortification of food crops is proposed to alleviate micronutrient deficiencies, for example selenium (Se), zinc (Zn) and iodine (I), in an area with high prevalence of the osteoarthropathy, Kashin-Beck disease. The Loess Plateau of central China is notably Se deficient, and deficiencies of Zn and I are also common. Loess is a deep, windblown, high pH, calcareous soil. Studies by our group have previously shown that this soil type is relatively high in most nutrients and micronutrients, but these are generally of low availability to crops. Total Se in loess is around 0.086 mg/kg, but water soluble Se in soil is usually only around 2% of total Se, hence around 0.002 mg/kg (Tan et al., 2002).

2 MATERIALS AND METHODS

A pilot trial conducted in a glasshouse at NWAFU found promising results of Se, Zn and I biofortification, and this was followed by field trials at Yangmazhuang village, Yongshou County (N34°49′ E108°11′, elevation 1127.76 m), located around 100 km north of NWAFU, a typical Kashin-Beck disease area on the Loess Plateau. The village has around 260 inhabitants with subsistence agriculture predominating and apples as a cash crop. The field trials were designed to evaluate the feasibility of agronomic biofortification of food crops with Se, Zn and I. The crops were spring maize (Zhengdan 958), popular local varieties of soybean, potato and cabbage (April to Sept 2008), and winter wheat (Jinmai 47) and canola (Ganza 1) (Sept 2008 to June 2009).

Field trial 1: The size of each plot was 16 m^2 (4 m × 4 m), a randomized block design was used and usual basal NP fertilizer applied. The plots were sown with micronutrients at planting as follows: 0.2 kg Se per ha as sodium selenate, 23 kg Zn per ha as ZnSO$_4$.7H$_2$O, 0.5 kg I per ha as KIO$_3$ (as well as soil B, Cu, Co and foliar Fe, Mn). As this was the first trial conducted in this region using Se, Zn and I, and represented a concept/feasibility trial rather than a rate study, relatively high single rates of these micronutrients were used.

Field trial 2: To assess the effect of foliar application of Se and Zn to winter wheat (Jinmai 47), another field trial was conducted (2008–2009) at the same site as Trial 1. Foliar Se and Zn were applied 3 times in 3 weeks from late booting stage (Zadoks scale 48–49) (Zadoks et al., 1974). The Se rates were 0.00, 0.004, 0.02, 0.03, 0.05 and 0.06 kg/ha as sodium selenite. Zinc rates were 0.00, 0.23, 0.45, 0.68, 0.91 and 1.14 kg/ha as ZnSO$_4$. Analyses of variance were conducted using PASW statistics 17.0 edition. Reported yields are means of 3 or 4 replications for crops other than maize (8 replications).

After harvest in each field trial, the biomass and yield of each plot were measured. Samples were oven-dried at 70°C for 48 hours and were analysed using ICP-OES or ICP-MS for nitrogen (Kjeldahl), macro- and micro nutrient elements.

3 RESULTS AND DISCUSSION

Field trial 1: Concentrations of Se in grain, tuber and leaf increased significantly, showing that biofortification through Se fertilizer applied to the soil was successful. The largest effect was for winter wheat: a 230-fold increase to reach a grain Se concentration of 4.5 mg/kg (Table 1); this concentration would need to be diluted with low-Se wheat to a level of <1 mg/kg to be safe as a consumable food product. A Se application of 18 g/ha (as selenate) would likely produce ~0.3 mg/kg, a suitable grain Se target level for human health. Similar findings were reported by Chilimba et al. (2012) (maize) and Lyons et al. (2004) (wheat).

Table 1. Micronutrient concentration in edible crop parts after micronutrient application to soil at planting, compared to control. Data shown are means and standard errors in parentheses ($n = 4$, but 8 for maize).

Crop	Applied micronutrient					
	Se (µg/kg)		Zn (mg/kg)		I (µg/kg)	
	CK	SeO$_4$	CK	Zn	CK	I
Wheat	16	3667	18	18	11	21
	(3)	(270)	(2)	(1)	(1)	(2)
Maize	11	1093	17	19	11	20
	(1)	(82)	(4)	(1)	(1)	(4)
Soybean	32	1850	53	60	13	19
	(6)	(50)	(3)	(2)	(1)	(2.5)
Potato	14	1823	25	23	63	58
	(3)	(160)	(7)	(6)	(1)	(6)
Canola	11	865	30	36	975	890
	(1)	(35)	(1.5)	(1)	(45)	(50)
Cabbage	97	16234	24	59	580	1150
	(7)	(1600)	(4)	(5)	(40)	(35)

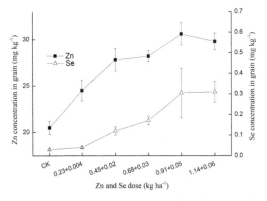

Figure 1. The effect of different foliar application rate Zn, Se application rate on Zn and Se concentration in wheat grain.

Soil-applied Zn and I, however, were effective in increasing levels in cabbage only, by 3- and 2-fold, respectively (Table 1). A high proportion of applied Zn was adsorbed on soil colloids, becoming unavailable. Moreover, Zn translocation to edible parts was inefficient, as Zn levels were considerably higher in leaves of wheat, maize, soybean and potato (data not shown). Other studies on different soils report more efficient Zn biofortification from soil-applied Zn than we found, and a combination of soil/foliar Zn is generally most effective (Yilmaz et al., 1997; Cakmak, 2008). Iodine was taken up by plants into leaves (see cabbage, Table 1) but did not reach grain or tuber, as it is transported mostly in the xylem tissue (Mackowiak & Grossl, 1999). Concentrations of I in control cabbage and canola in the present study (0.5–1 mg/kg) indicate that this is not an I-deficient area, and human I status could be increased by eating more leafy vegetables.

There was high variability in yield within treatments and no overall effect of micronutrient application on yield (of grain or tubers) was found, in contrast to the pilot glasshouse trial, where large increases were found, likely related to minimized insect/disease infestation. Apparently, soil available Zn level in this area was not deficient for these crops. However, Se was associated with higher cabbage yield. Macronutrient concentrations in plant tissues were not affected by micronutrient application.

Field trial 2: Foliar application of Se at rates of 0.02–0.06 kg/ha significantly increased the grain Se concentration linearly, with a 10-fold increase over control at the 0.05 and 0.06 kg/ha rates (Fig. 1). Comparison of foliar and soil-applied Se (using extrapolation) suggests that soil-selenate is around 3 times as effective as foliar-selenite at this site.

Grain Zn concentrations were increased by all the foliar Zn applications above 0.23 kg/ha by a mean 43% (Fig. 1). This observation showed that a grain Zn concentration increase of around 40% can be obtained by foliar application of just 0.5 kg/ha; a very efficient example of biofortification, in contrast to soil-applied Zn on this soil. As with Trial 1, there was no effect on biomass and grain yield of wheat with any of the Zn and Se application.

4 CONCLUSIONS

Using a range of food crops in field trials, soil-applied selenate and foliar-applied selenite were found to be effective for biofortification on the Loess Plateau. Foliar zinc sulphate was effective in biofortifying winter wheat, and would be likely to be effective for other crops as well. Soil-applied Zn and I (as potassium iodate) were only effective in increasing leaf levels of these micronutrients. Further research on optimum timing for foliar Se and Zn application and comparison of foliar application of selenate and selenite is needed.

REFERENCES

Cakmak, I. 2008. Enrichment of cereal grains with zinc: agronomic or genetic biofortification? Plant and Soil 302: 1–17.

Chilimba, A.D.C., Young, S.D., Black, C.R., Meacham, M.C., Lammel, J., & Broadlet, M.R. 2012. Agronomic biofortification of maize with selenium (Se) in Malawi. Field Crops Research 125: 118–128.

Lyons, G., Lewis, J., Lorimer, M.F., Holloway, R.E., Brace, D.M., Stangoulis, J.C.R., et al. 2004. High-selenium wheat: agronomic biofortification strategies to improve human nutrition. Food, Agriculture and Environment. 2: 171–178.

Mackowiak, C.L. & Grossl, P.R. 1999. Iodate and iodide effects on iodine uptake and partitioning in rice (Oryza sativa L.) grown in solution culture. Plant Soil, 212: 135–143.

Tan, J.A., Zhu, W.Y., Wang, W.Y., Li, R.B., Hou, S.F., Wang, D.C., et al. 2002. Selenium in soil and endemic diseases in China. Science of the Total Environment, 284: 227–235.

Effect of selenium treatment on biomass production and mineral content in common bean varieties

M.A. de Figueiredo*, D.P. Oliveira, M.J.B. de Andrade & L.R.G. Guilherme
Federal University of Lavras, Lavras, Minas Gerais, Brazil

L. Li
Cornell University – USDA-Agricultural Research Service, Ithaca, New York, USA

1 INTRODUCTION

Beans are important part of basic diet for the Brazilian population, as well as for most of the world's population. Selenium (Se) is important to human health, and Se deficiency affects a large number of people worldwide. Agronomic biofortification, together with genetic biofortification, is a crucial strategy to improve nutritional quality in plants and seeds. The aim of this study was to evaluate common bean cultivars for their responses to Se nutrition for ultimately improving Se concentrations in beans.

2 MATERIALS AND METHODS

The seeds of common beans, carioca (BRS Notável – B, Pérola-C, IPR Tangará-D, BRS Requinte – E, BRS Estilo – F, BRS Pontal-G, BRSMG Madrepérola-H, BRS Cometa – J cultivars) and black (BRS Supremo – A, BRS Grafite – I cultivars), were obtained from EMBRAPA (Rice and Beans, Goiânia-GO, Brazil). They were selected due to their good agronomic and genetic characteristics as well as being preferably planted in different regions in Brazil.

The experiments were conducted in a greenhouse at Cornell University in the U.S. The seeds were surface-sterilized in 10% NaOCl solution for 10 minutes, rinsed in Milli-Q deionized water, and germinated on moistened filter paper for 3 days. After that, the seedlings were transferred to pots containing modified Hoagland and Arnon's solution. Five days later the treatment of $5\,\mu M$ Na_2SeO_4 was applied, along with control (modified Hoagland and Arnon's solution). The experiment was conducted using completely randomized design, and each treatment had three replicates. After 22 days, the plants were harvested, separated into shoot and roots, oven-dried at 65°C for 3 days, and analyzed using ICP in triplicate.

The data were analyzed for ANOVA using the Statistical Analysis Software Sisvar®. The significant treatment effects were determined by comparing the means using Scott-Knott test at $\alpha = 0.05$.

3 RESULTS AND DISCUSSION

The biomass production of the common bean cultivars was measured as an indication of plant growth. Selenate supplement at $5\,\mu M$ did not show any significantly negative effects on common bean shoot growth (Fig. 1). While the shoot biomass of nearly all cultivars was found to be similar between control and the

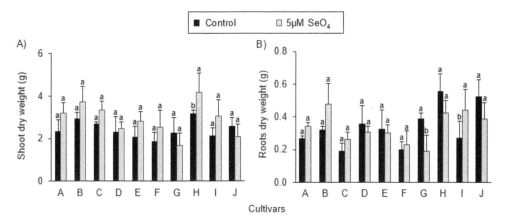

Figure 1. Shoot and root dry weight of common beans with and without $5\,\mu M$ Na_2SeO_4 treatment. Error bars indicate standard error of the mean (SEM) (n = 3).

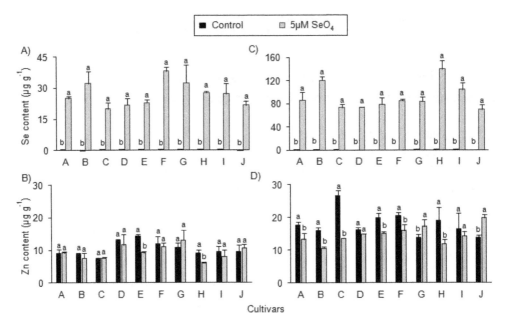

Figure 2. Total Se and Zn concentrations in shoot (A, B) and roots (C, D) of ten cultivars of common beans treated with and without 5 μM Na$_2$SeO$_4$. Error bars indicate standard error of the mean (SEM) (n = 3).

Se treatment, one variety showed significant increase of dry biomass (Fig. 1a). This is consistent with other previous studies that reported the genetic variation of plant growth in response to Se treatment and the beneficial effect of low Se concentration on stimulating plant growth (Ramos et al., 2011; Souza et al., 2013).

The selenate treatment of 5 μM also exhibited limited effects on common bean root growth (Fig. 1b). The root biomasses of eight cultivars showed no significant difference with and without 5 μM Na$_2$SeO$_4$ supply. A reduction of root growth was observed in one line and an increase was observed in another line, showing significant variation among the selected common bean cultivars.

Selenium concentrations in shoots and roots of these common bean cultivars were also determined. These cultivars showed different levels of Se accumulation when the growth substrate was treated with 5 μM Na$_2$SeO$_4$ (Fig. 2a, c). Approximately 2-fold difference in total Se content between cultivars was observed in both shoots and roots, indicating different capacity of the common bean cultivars in accumulating Se. Genetic variation for Se content in other crops was also reported in other studies (Ávila et al., 2013; Ramos et al., 2011; Souza et al., 2013).

Zinc (Zn) is also important to plant growth and human health. Concentrations of Zn in the bean cultivars were also studied with the selenate treatment. The 5 μM selenate treatment exhibited minimal effects on Zn accumulation in shoots of most cultivars (Fig. 2), but showed negative effects on Zn content in roots of majority varieties (Fig. 2d). A previous study reports no significant correlation between Zn and Se (Hacisalihoglu et al., 2005). These results reinforce genetic variation in response to Se treatment and provide information to breed cultivars with high capacity to accumulate Se.

REFERENCES

Ávila, F.W., Faquin, V., Yang, Y., Ramos, S.J., Guilherme, L.R., Thannhauser, T.W. et al. 2013. Assessment of the anticancer compounds se-methylselenocysteine and glucosinolates in se-biofortified broccoli (*Brassica oleracea* L. var. italica) Sprouts and Florets. *J. Agric. Food Chem.* 61: 6216–6223.

Cakmak, I. 2008. Enrichment of cereal grains with zinc: agronomic or genetic biofortification? *Plant Soil* 302: 1–17.

Hacisalihoglu, G., Kochian, L.V. & Vallejos, C.E. 2005. Distribution of seed mineral nutrients and their correlation in *Phaseolus vulgaris*. *Proc. Fla. State Hort. Soc.* 118: 102–105.

Ramos, S.J., Rutzke, M.A., Hayes, R.J., Faquin, V., Guilherme, L.R.G. & Li, L. 2011. Selenium accumulation in lettuce germplasm. *Planta* 233: 649–660.

Souza, G. A., Carvalho, J.C., Rutzke, M., Albrecht, J.C., Guilherme, L.R.G. & Li, L. 2013. Evaluation of germplasm effect on Fe, Zn and Se content in wheat seedlings. *Plant Science* 210: 206–213.

Biofortification of irrigated wheat with Se fertilizer: Timing, rate, method and type of wheat

I. Ortiz-Monasterio* & M.E. Cárdenas
Global Conservation Agriculture Program, International Maize and Wheat Improvement Center, Mexico D.F., Mexico

G.H. Lyons
School of Agriculture, Food & Wine, University of Adelaide, Waite Campus, Glen Osmond, SA, Australia

1 INTRODUCTION

Selenium (Se) is an essential micronutrient for humans and animals, but is deficient in at least a billion people worldwide. Wheat (*Triticum aestivum L.*) is an important source of Se in many countries. Biofortification is a strategy for increasing micronutrient density in staple crops. There are two main strategies for increasing Se grain concentration plant breeding and bio-technology, and Se fertilization. No significant genotypic variability for grain Se concentration was observed in modern wheat cultivars in an earlier study (Lyons et al., 2005). On the other hand Se fertilization, whether soil or foliar, can result in large increases in grain Se concentration (Lyons et al., 2004).

2 MATERIALS AND METHODS

Six field experiments were conducted during 2006 to 2011 in the Yaqui Valley near Ciudad Obregon, Sonora, Mexico (27 N 109 W, 40 masl). The first three years' experiments were established at CENEB (Centro Experimental Norman E. Borlaug, Experimental Field Norman E. Borlaug) using a split-plot design with main plots arranged as a randomized complete block design (RCBD) with three replications. The varieties (Rayon F89 and Jupare C2001) were main plots and sub-plots a factorial combination of two application times × selenium rates (0, 10, 30, 100 and 300 g Na_2SeO_4/ha). In 05–06, the subplots were two application times [soil and foliar application at 45 days after sowing (DAS)] × Se rates. In 06–07, subplots were two application times [foliar application at 45 DAS and 7 days after anthesis (DAA)] × Se rates. In 07–08, the subplots were two application times (foliar application at 45DAS and at booting) × Se rates. The last three years' experiments were established in farmers' fields in Yaqui Valley, and only foliar applications at 7DAA were tested. In 08–09, a split-plot design was used with main plots arranged as a RCBD with three replications. The varieties (Cevy Oro C2008, Cirno C2008, Tacupeto F2001, Roelfs F2007 and Navojoa M2007) were main plot and sub-plots selenium rates (0, 150, 300, 450 and 600 g Na_2SeO_4/ha). During 09–10 and 10–11 the experiments were planted using a RCBD with four replications. The rates tested in 2009–2010 were 0, 500, 1000 and 1500 g Na_2SeO_4/ha in varieties Atil C200 and Kronstand F2004, and 0, 1000, 1500, 2000, and 2500 g Na_2SeO_4/ha in 10–11 with varieties Navojoa M2007 and Cirno C2008. The source of Se from 2005–2006 to 2008–2009 was sodium selenate (SS) (Na_2SeO_4) as reagent grade. In 09–10, two sources of Na_2SeO_4 were used; reagent grade and fertilizer grade (Pacific Rare Specialty Metals & Chemicals, Inc (Philippines)). In 10–11 only fertilizer grade was used. All experiments were run under irrigated conditions with good agronomic management practices. All samples were digested as described by Zarcinas et al. (1987) and analyzed for Se at the Waite Analytical Services using ICP mass spectrometry.

3 RESULTS

3.1 Method

During 05–06, we studied the effect of soil applied Se, compared to foliar applications of Se at 45 DAS. Figure 1, shows how the response to a foliar application

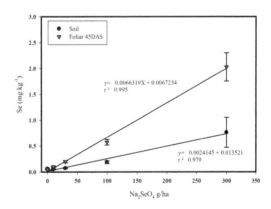

Figure 1. Comparison of soil application of sodium selenate at sowing and foliar application at 45 days after sowing (DAS).

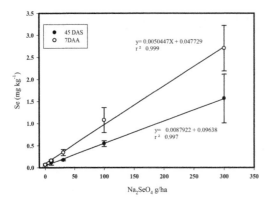

Figure 2. Foliar application of sodium selenate at 45 days after sowing (DAS) and seven days after anthesis (DAA).

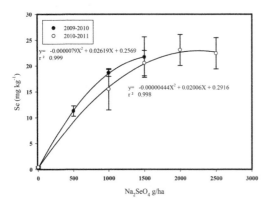

Figure 3. Foliar application of sodium selenate at 7 days after anthesis (DAA).

at 45 DAS increased grain Se concentration per unit of Se applied. This observation is in contrast with other research reported in Australia, which showed that soil applied Se was more effective than foliar applications (Lyons et al., 2004).

3.2 Timing

During 06–07, we studied the foliar application time of Se at 45 DAS and at 7 DAA. Results showed that 7DAA increased grain Se significantly and more rapidly per unit of Se applied (Fig. 2). During 07–08, we studied the timing of Se application comparing foliar application at 45 DAS vs the booting stage at different application rates of Se. The results showed that only at some rates applied at booting were more effective than 45 DAS at increasing grain Se.

3.3 Rate

During the first three years we found that the response to foliar Se was linear between 0 and 300 g of SS. In 09–10 and 10–11, we explored higher rates of Se and were finally able to identify the Se rate that maximized grain Se concentration, namely 2259 g selenate/ha. This Se rate resulted in 23 mg/kg Se in the grain (Fig. 3).

3.4 Crop

During the first three cycles the rates evaluated were 0 to 300 g selenate/ha, and in those three years there was no significant differences in grain Se concetrations between bread wheat and durum wheat. In 08–09, there were significant differences between bread wheat and durum but only at the higher rates of 450 and 600 g selenate/ha. In 09–10, there were again significant differences between bread wheat and durum but only at the higher rates of 500 and 1000 g selenate/ha. In 10–11, there were, however, no significant differences between durum and bread wheat at any rate. This suggests that there is a tendency for bread wheat to respond more to Se foliar fertilization, but this is not consistent across years and tends to be mainly at the higher rates.

4 CONCLUSIONS

The maximum efficiency of Se fertilization is obtained with foliar applications at 7 days after anthesis, then at booting stage, then at 45 days after sowing and last with soil applications. The foliar rate that maximized response to Se was 2259 g sodium selenate/ha, which increased grain Se to 23 mg kg^{-1}. Bread wheat tended to have a better response to foliar Se than durum wheat, but only at the higher rates of Se (>300 g selenate/ha) and it was not consistent in all years.

ACKNOWLEDGMENTS

We would like to thank Fundación Produce Sonora and HarvestPlus for providing funding to carry out this research Neither Fundación nor HarvestPlus were involved at any step of the research, writing or interpretation of results.

REFERENCES

Lyons, G., J. Lewis, M.F. Lorimer, R.E. Holloway, D.M. Brace, J.C.R. Stangoulis, et al. 2004. High-selenium wheat: agronomic biofortification strategies to improve human nutrition. Food, Agriculture & Environment, 2(1): 171–178.

Lyons, G., I. Ortiz-Monasterio, J. Stangoulis & R. Graham. 2005. Selenium concentration in wheat grain: sufficient genotypic variation for selection? *Plant and Soil* 269: 369–380.

Zarcinas, B.A., Cartwright, D., & Spouncer, L.R. 1987. Nitric acid digestion and multi-element analysis of plant material by inductively coupled plasma spectrometry. Commun Soil Sci Plant Anal 18: 131–146.

Egg and poultry meat enrichment of selenium

A.G. Bertechini*, V.A. Silva, F.M. Figueiredo & T.F.B. Oliveira
Department of Animal Science, Federal University of Lavras, Lavras, Minas Gerais, Brazil

1 INTRODUCTION

Brazilian soils generally have low concentrations of selenium (Se), particularly in southeastern and southern Brazil (Bertechini, 2012). Previous studies suggested that those regions with low levels of Se in soil have higher rates of human diseases such as prostate and breast cancer (Brooks et al., 2001; Klein et al., 2003). According to the National Cancer Institute, the prostate cancer rate (per 100,000 residents) in the U.S. was estimated: 91 in the South, 88 in the Southeast, 63 for the Midwest, 47 in the Northeast, and 30 to the North, while the breast cancer rate: 71 in the South, 71 in the Southeast, 51 in the Midwest, 37 in Northeast, and 21 in the North (INCA, 2014). Because the majority of Brazilian grains are produced in the areas with low concentrations of Se in soil, Se supplementation to the animal feed becomes necessary. In comparing with inorganic Se, organic Se has become popular due to its higher levels of Se accumulation in muscle tissue and eggs (Zhou & Wang, 2011; Oliveira et al., 2014; Pappas et al., 2005; Invernizzi et al., 2013).

Although organic Se has advantages over inorganic Se, it is not widely used due to its high cost in comparison to sodium selenite. Thus, the use of organic sources and strategic inclusions in the production periods would be a viable option to enrich animal products without increasing the costs of the feed. The enrichment of food products with Se is an alternative to improve the consumption of this mineral by the population. Products like chicken meat and eggs have low cost, a high nutritional value, and it is acceptable by the consumer market. Thus, the aim of this study was to evaluate the dietary supplementation of different levels and sources of Se for laying hens and broiler chickens on the deposition of this trace mineral in their respective products.

2 MATERIALS AND METHODS

The study was conducted two experimental in trials. In the first trial used 720 De Kalb-White hens (40 wks old). The hens were randomly distributed among 9 treatments in a factorial design of $4 \times 2 + 1$. These include; Se levels (0.15, 0.30, 0.45 and 0.60 mg/kg of feed) $\times$ 2 sources (organic – Se yeast and inorganic – sodium selenite) + control, with 8 replications per treatment. A basal diet was formulated with corn and soybean meal, according to Rostagno et al. (2011). Observations were collected every 3 days for 25 days. Eggs were weighed and 4 eggs were sampled from each treatment to determine the Se concentration.

In the second trial, we used 1,440 one d old male Ross 308 broilers. The birds were randomly distributed among 48 floor pens. There were 6 treatments with 8 replicates each and 30 birds in each pen. The Se sources used were Se yeast (SY-0.2% Se) and sodium selenite (SS: 45.6% of Se). The treatments were arranged as follows: T1 (1-42 d SS); T2 (1-42 d SY); T3 (1-14 d SS and 15-42 d SY); T4 (1-21 d SS and 22-42 d SY); T5 (1-28 d SS and 29-42 d SY) and T6 (1-35 d SS and 36-42 d SY). The formulation of the diets was based on corn and soybean meal (Bertechini, 2012). After 42 d, three chicken per replicate were euthanized, and their breasts were collected for chemical analysis.

Samples were acid digested in a digester oven (Merck Darmstadt, Germany). Concentrations of Se were measured using atomic absorption spectrophotometer (Varian SpectrAA 2000) equipped with hydride generation (VGA 77). The data was submitted to analysis of variance, using the SAS statistical software (2000).

3 RESULTS AND DISCUSSION

We observed an increasing linear ($p < 0.05$) concentration of Se as the levels of Se in the diets increased in the yolk and in the edible part of the egg (yolk + albumen), regardless of the Se source (Fig. 1). The correlation equations with the SS treatment: $y = 0.2745 + 0.0726x$, $r^2 = 0.69$, while with the SY treatment: $y = 0.328 + 0.1023x$, $r2 = 0.95$. The organic Se treatment resulted in 40% more Se accumulation than the inorganic Se. Ivernizzi et al. (2013) and Pappas et al. (2005) also observed that there was a positive correlation between the diet Se level and the yolk Se concentration.

The chicken treated with organic Se from Days 1 to 42 had higher Se concentrations in breast tissue ($p < 0.05$). The birds that received organic Se in the last 3 and 2 weeks of age had similar concentrations ($p > 0.05$). The lowest ($p < 0.05$) concentration of Se

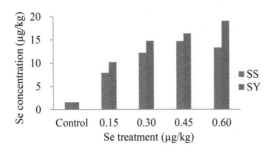

Figure 1. Concentrations of Se in eggs. SS: Se-enriched yeast; SS: sodium selenite.

Table 1. Selenium concentrations in the broiler chicken breast. Means with different letters are different ($p > 0.05$).

Treatments	Breast Se (μg/kg)
1-42 d SS	41.25^d
1-42 d SY	70.86^a
1-14 d SS and 15-42 d SY	51.47^c
1-21 d SS and 22-42 d SY	57.31^{bc}
1-28 d SS and 29-42 d SY	61.24^b
1-35 d SS and 36-42 d SY	62.39^b
Mean; CV(%):	57.42; 8.69%

in the breast was observed in birds receiving inorganic Se in the feed (Table 1).

The concentration of Se in the breast of birds receiving SY in the diet from one d until 42 d of age resulted in Se concentrations of 70.86 μg/kg of meat, compared to 41.25 μg/kg in the breasts of birds fed inorganic Se (with a difference of >70%). These results are consistent with those found by Dlouhá et al. (2008) and Oliveira et al. (2014), who reported significant differences in the concentrations of Se in chicken breasts where organic sources of Se were used in feed. Southem and Payne (2005) also observed increased tissue Se concentration in chicken fed with organic of Se, compared to those receiving inorganic Se.

4 CONCLUSIONS

The use of Se yeast is able to improve the concentration of Se in the yolk and broiler breast meat. The daily consumption of a 150 g of chicken breast and one egg (63 g) enriched with Se can achieve 19 and 35% of the daily intake recommendations for adults (55 μg), respectively.

REFERENCES

Bertechini, A.G. 2012. *Nutrição de monogástricos*. 2 ed. UFLA, Lavras, MG 373p.

Brooks, J.D., Metter, E.J., Chan, D.W., Sokoll, L.J., Landis, P., Nelson, W.G. et al., 2001. Plasma selenium level before diagnosis and the risk of prostate cancer development. *The Journal of Urology*. 166:2034–2038.

Combs, G.F. Jr. 2004. Status of selenium in prostate cancer prevention. *British Journal of Cancer* 91:195–199.

Dlouhá, G., Ševčíková, S., Dokoupilová A., Zita, L., Heindl, J., & Skøivanevciková, M. 2008. Effect of dietary selenium sources on growth performance, breast muscle selenium, glutathione peroxidase activity and oxidative stability in broilers. Czech *Journal of Animal Science*, 53(6): 265–269.

INCA (Instituto Nacional do Câncer). 2014. Disponível em:http//www2.inca.gov.br/wps/wcm/connect/agencianoticias/site/home/noticias/2013/inca_ministerio_saude_apresentam_estimativas_cancer_2014, acesso em 03/03/2015.

Invernizzi, G., Agazzi, A. & Ferroni, M. 2013. Effects of inclusion of selenium-enriched yeast in the diet of laying hens on performance, eggshell quality, and selenium tissue deposition. Italian *Journal of Animal Science*, 12(1): 1–8.

Klein, F.A., Thompson I.M., Lippman S.M., Goodman P.J., Albanes D., Taylor P.R., et al., 2003. The selenium and vitamin E cancer prevention trial. *World Journal Urology*. 21: 21–27.

Oliveira, T.F.B., Rivera, D.F.R., Mesquita, F.R., Braga, H., Ramos, E.M., & Bertechini, A.G. 2014. Effect of different sources and levels of selenium on performance, meat quality, and tissue characteristics of broilers. *Journal Applied of Poultry Research* 23:1522.

Pappas, A.C., Acamovict, T., Sparks, N.H.C., Surai, P.F., & McDevitt, R.M. 2005 Effects of supplementing broiler breeder diets with organic selenium and polyunsaturated fatty acids on egg quality during storage. *Poultry Science* 84:865–874.

Payne, R.L. & Southern, L.L. 2005. Comparison of inorganic and organic selenium sources for broilers. *Poultry Science*, 84:898–902.

Rostagno, H.S., Albino, L.F.T., Donzele, J.L., Gomes, P.C., Oliveira, R.F., Lopes, D.C. et al. 2011. *Tabelas brasileiras para aves e suínos. Composição de Alimentos e exigências nutricionais*. 3^a Ed. Viçosa, MG, 252p.

Statistical Analysis System (SAS). 2000. User's guide: statistics. V. 10. 14 ed. Cary: SC.

Zhou, X.E. & Wang, Y. 2011. Influence of dietary nanoelemental selenium on growth performance, tissue selenium distribution, meat quality, and glutathione peroxidase activity in Guangxi yellow chicken. *Poultry Science*, 90: 680–686.

Strategies for selenium supplementation in cattle: Se-yeast or agronomic biofortification

J.A. Hall & G. Bobe
Oregon State University, Corvallis, Oregon, USA

1 INTRODUCTION

Selenium (Se) is an essential trace mineral and is important for immune function and overall health for cattle. Clinical Se deficiency can cause nutritional myodegeneration, also known as white muscle disease, characterized by muscle weakness, heart failure, unthriftiness, and eventually death. Inadequate Se intake also causes subclinical disease resulting in poor livestock performance. The role of Se in animal health is based primarily on the functions of selenocysteine-containing proteins, many of which have antioxidant activities (Fairweather-Tait et al., 2010). Although reactive oxygen species (ROS) are a natural result of the body's normal metabolic activity, excessive stress, e.g., at parturition, can lead to over-production of ROS or ROS accumulation because of lack of antioxidants. Oxidative stress is a primary mechanism by which ROS influence biological processes and may contribute to the pathogenesis of disease by modifying the expression of proinflammatory genes. The beneficial effects of Se supplementation may result from selenoproteins converting harmful ROS to less reactive molecules, as well as influencing the expression of redox-regulated genes as reviewed by Hugejiletu et al. (2013).

In Se-deficient areas, several means of Se supplementation for livestock are available, e.g., organic Se-yeast can be added to feed, or inorganic Na selenite can be added to mineral mixes for free-choice consumption. Agronomic biofortification is an alternative approach, whereby the Se concentration of forage is increased through the use of Se-containing fertilizer amendments (Hall et al., 2013a). We have demonstrated in a series of studies that Se-fertilization of pastures or drenching with Se-yeast are both effective strategies for supranutritional Se supplementation in sheep (Hall et al., 2009; Hall et al., 2012). The purpose of this report was to evaluate the effect of these strategies on WB and colostral Se concentrations in dairy and beef cattle. Selenium concentrations were determined at parturition in dairy cattle fed Se-yeast supplement (Hall et al., 2014a,b) and in beef cattle fed Se-biofortified alfalfa hay.

2 MATERIALS AND METHODS

Se-Yeast Supplementation: During the last 8 wk before calving, Jersey dairy cows were fed either 0 (control cows, $n=24$) or 105 mg of Se-yeast once weekly (supranutritional Se-yeast-supplemented cows; $n=36$), in addition to Na selenite in their ration at 0.3 mg of Se/kg of dry matter (DM). The organic Se source (Prince Se Yeast 2000, Prince Agri Products Inc., Quincy, IL) had a guaranteed analysis of 2 g organically bound Se (78% as Se-Met) per kg supplement. The Se-yeast dosage was calculated to provide 15 mg of Se/d, which is equal to 5 × the maximal FDA-permitted level of 3 mg of Se/d. The amount of Se-yeast fed to each cow was 52.5 g once weekly. Blood samples were collected from cows after parturition (at the time of first milking) to measure WB-Se concentrations. Colostrum was collected within 2 h of calving from control and Se-yeast-supplemented cows and pooled.

Agronomic Biofortification: Inorganic Na selenate was mixed with water and sprayed onto the soil surface of alfalfa fields at application rates of 0, 45.0, or 89.9 g Se/ha after the second cutting of hay. A Penn State forage sampler was used to take 25 cores from random bales in each hay source. During the last 8 wk before calving, 45 Angus and Angus-cross cows were fed alfalfa hay at a rate of 2.5% body weight/d (3 pens of 5 cows/group). Based on nutrient analysis of hay samples (17% CP; 58.7% TDN), cows were fed 70% alfalfa hay and 30% grass hay (6.8% CP) to achieve a ration CP of 13.94%. Group 1 (control): three pens of cows ($n=15$) fed non-Se fortified alfalfa hay as a major portion of the ration plus a mineral supplement containing 120 mg/kg Se (US FDA regulations) from Na selenite. Group 2: three pens of cows ($n=15$) fed alfalfa hay harvested from a field fertilized with 45.0 g Se/ha and a mineral supplement without added Se. Group 3: three pens of cows ($n=15$) fed alfalfa hay harvested from fields fertilized with 89.9 g Se/ha and fed mineral supplement without added Se. Blood samples were collected from cows at baseline, after 4 wk of hay consumption, and at parturition to measure WB-Se concentrations. Colostrum was collected before calves suckled.

Selenium Analysis: Selenium concentrations were determined using an inductively coupled argon plasma emission spectrometry method by commercial laboratories (Center for Nutrition, Diagnostic Center for Population and Animal Health, Michigan State University, East Lansing, MI, for the WB dairy cow samples and Utah Veterinary Diagnostic Laboratory, Logan, UT, for all other samples). *Statistical Analysis:* Data were analyzed in SAS and reported as means ± standard error of the mean for averages of individual animals and means ± standard deviation for pooled samples. All tests were two-sided and significance was declared at $p \leq 0.05$.

3 RESULTS AND DISCUSSION

Se-Yeast Supplementation: Prior to Se-yeast supplementation, cows were fed Na selenite in their ration and had WB-Se concentrations (mean, 245 ng/mL; range, 200–271 ng/mL), which were within the reference interval of adult cows (120–300 ng/mL). At calving, cows supplemented with supranutritional Se-yeast had higher WB-Se concentrations (371 ± 7 ng/mL) compared with control cows (231 ± 7 ng/mL), $p < 0.0001$, and higher pooled colostrum Se-concentrations (384 ± 8 ng/mL, $n = 3$, vs. 219 ± 4 ng/mL, $n = 2$); $p < 0.0001$, respectively. *Agronomic Biofortification:* Fertilizing fields with increasing amounts of Na selenate increased the Se-concentration of third cutting alfalfa hay from 0.36 mg Se/kg DM (non-fertilized control) to 2.42 and 5.17 mg Se/kg DM for Na selenate application rates of 45.0 and 89.9 g Se/ha, respectively. The relationship between amount of Se applied by fertilization (g Se/ha) and observed forage Se concentration (mg Se/kg DM) was $y = 0.054x + 0.244$, $R^2 = 0.9931$. Calculated Se intake from dietary sources was 9.0, 32.9, and 70.3 mg Se/head/d for cows consuming hay with Se concentrations of 0.36 to 2.42 and 5.17 mg Se/kg DM, respectively. Prior to agronomic biofortification cows had WB-Se concentrations, (mean, 149 ng/mL; range, 106–243 ng/mL), which were within the reference interval of adult cows (120–300 ng/mL). WB-Se concentrations increased after 4 wk of feeding Se-fertilized alfalfa hay depending upon the Se-application rate (0, 45.0, or 89.9 g Se/ha) to 164 ± 6, 253 ± 8, and 339 ± 5 ng/mL ($P_{Linear} < 0.001$) and colostral Se concentration to 121 ± 6, 504 ± 9, and $1,339 \pm 19$ ng/mL ($P_{Linear} < 0.001$), respectively.

Health Benefits: Current FDA regulations limit the amount of dietary Se supplementation in ruminants to 0.3 mg/kg (as fed) of organic Se-yeast, which is equivalent to 3 mg per cow per day. Our previous studies suggest that Se-supplementation above current recommendation may have short- and long-term beneficial effects on immune function and production of the cow as well as their calf (Hall et al., 2014a; Hall et al., 2013b). The primary route by which maternal Se supplementation increases WB-Se concentration of calves is *in utero* transfer, as colostral Se transfer is rather ineffective at increasing WB-Se concentrations because the amount of Se that can be transferred is limited (Hall et al., 2014a).

Comparison: Both supranutritional Se supplementation strategies were effective for increasing WB-Se and colostral Se concentrations in Se-replete cows. Feeding Se-yeast is difficult when animals graze exclusively on pastures. In addition there are added costs for purchasing comparable dosages of Se-yeast as well as labor costs associated with feeding animals individually with some animals not consuming the supplement. A more efficient and economical alternative is to fertilize forages with Na selenate to increase organic Se content in forage.

4 CONCLUSIONS

Both supranutritional Se supplementation strategies, i.e., feeding Se-yeast and agronomic biofortification of forage hay, are effective for increasing WB-Se and colostral Se concentrations. The more economical alternative is agronomic biofortification because it involves less labor and product costs.

REFERENCES

Fairweather-Tait, S.J., Collings, R. & Hurst, R. 2010. Selenium bioavailability: Current knowledge and future research requirements. *Am J Clin Nutr* 91: 1484S–1491S.

Hall, J.A., Bobe, G., Hunter, J.K., Vorachek, W.R., Stewart, W.C., Vanegas, J.A., et al. 2013a. Effect of feeding selenium-fertilized alfalfa hay on performance of weaned beef calves. *PLoS One* 8(3):e58188.

Hall, J.A., Bobe, G., Vorachek, W.R., Estill, C.T., Mosher, W.D., Pirelli, G.J., et al. 2014a. Effect of supranutritional maternal and colostral selenium supplementation on passive absorption of immunoglobulin G in selenium-replete dairy calves. *J Dairy Sci* 97: 4379–91.

Hall, J.A., Bobe, G., Vorachek, W.R., Hugejiletu, Gorman, M.E., Mosher, W.D., et al. 2013b. Effects of feeding selenium-enriched alfalfa hay on immunity and health of weaned beef calves. *Biol Trace Elem Res* 156: 96–110.

Hall, J.A., Bobe, G., Vorachek, W.R., Kasper, K., Traber, M.G., Mosher, W.D., et al. 2014b. Effect of supranutritional organic selenium supplementation on postpartum blood micronutrients, antioxidants, metabolites, and inflammation biomarkers in selenium-replete dairy cows. *Biol Trace Elem Res* 161: 272–87.

Hall, J.A., Van Saun, R.J., Bobe, G., Stewart, W.C., Vorachek, W.R., Mosher, W.D., et al. 2012. Organic and inorganic selenium: I. Oral bioavailability in ewes. *J Anim Sci* 90: 568–76.

Hall, J.A., Van Saun, R.J., Nichols, T., Mosher, W., Pirelli, G., et al. 2009. Comparison of selenium status in sheep after short-term exposure to high-selenium-fertilized forage or mineral supplement. *Small Ruminant Research* 82: 40–45.

Hugejiletu, H., Bobe, G., Vorachek, W.R., Gorman, M.E., Mosher, W.D., Pirelli, G.J., et al. 2013. Selenium supplementation alters gene expression profiles associated with innate immunity in whole-blood neutrophils of sheep. *Biol Trace Elem Res* 154: 28–44.

Selenoneine content of traditional marine foods consumed by the Inuit in Nunavik, Northern Canada

P. Ayotte, A. Achouba, P. Dumas, N. Ouellet & M. Lemire*
Axe santé des populations et pratiques optimales en santé, Centre de recherche du CHU de Québec, Québec, Canada, G1V 2M2; Institut national de santé publique du Québec, Québec, G1V 5B3, Canada

L. Gautrin
LCABIE, Université de Pau et des Pays de l'Ardour, Pau, France

L. Chan
Department of Biology, University of Ottawa, Ottawa, Ontario, Canada

B. Laird
School of Public Health and Health Systems, University of Waterloo, Waterloo, Ontario, Canada

M. Kwan
Nunavik Research Center, Kuujjuaq, Québec, Canada

1 INTRODUCTION

Selenium (Se) is an essential element highly present in traditional marine foods consumed by Inuit (indigenous population inhabiting Artic regions of Greenland and North America), who exhibit one of the highest Se intake in the world (Lemire et al., 2015). In fish and marine mammal eating populations, there is increasing evidence suggesting that the high Se intake may play a role in offsetting some deleterious effects of methylmercury (MeHg) exposure (Fillion et al., 2013; Lemire et al., 2010, 2011; Valera & Dewailly, 2009; Ayotte et al., 2011). In 2010, Yamashita and Yamashita (2010) reported the isolation from the blood of bluefin tuna a novel selenocompound – selenoneine – which was later identified in the blood cell fraction of Japanese fishermen and in some marine species.

This project aimed at developing an analytical method for the quantification of selenoneine in marine foods consumed by the Inuit of Nunavik, starting with mattaaq (beluga skin with the underlying layer of fat), a delicacy highly praised in this population and a major source of Se in the Inuit diet (Lemire et al., 2015).

2 METHODS

The selenoneine standard was isolated from genetically modified *S. pombe* yeast cells overexpressing the egt+ gene (kindly donated by M. Yanagida (Pluskal et al., 2014), Dept. of Molecular Biotechnology, Hiroshima University), which were grown in the presence of selenate. After cell lysis, selenoneine was extracted and purified on a weak-cation-exchange solid phase extraction column. Beluga mattaaq samples were processed as follows. After removing the fat layer from the mattaaq sample, the rest of the sample (skin) was ground in liquid nitrogen with a mortar and pestle followed by cell disruption in a methanol-water mixture using a Bead Beater. After filtration on a 10 kDa filter and solvent evaporation, the resulting concentrated extract was analysed for selenoneine and other selenocompounds by anion-exchange liquid chromatography-inductively coupled tandem mass spectrometry (LC-ICP-MS/MS).

3 RESULTS

Using selenoneine biosynthesis, we obtained 10 mL of a standard solution containing 0.912 µg/L. The concentration was determined by total Se isotope dilution analysis and LC-ICP-MS/MS analysis, by comparison with the area under the peak obtained for a Se + 6 certified solution.

The total Se content in one beluga mattaaq sample analysed to date was 5.3 µg/g wet weight. We estimated the selenoneine concentration at 2.6 µg Se/g, representing nearly 50% of the total Se concentration present in the sample, while none of the other selenocompounds targeted by the method were detected (selenite, selenite, selenomethionine, selenocystine).

4 CONCLUSIONS

Our results to date suggest that selenoneine is the major form of Se in beluga mattaaq, a traditional marine food highly consumed by the Inuit in Nunavik. We are currently analysing other food items that are major sources of Se including; beluga meat, seal meat, seal liver and arctic char. The production of an isotope-labelled selenoneine standard is underway to allow for selenoneine quantification, and to study the role of selenoneine in the Se cycle and eventual interactions with MeHg. These data will improve our capacity to assess the benefits and risks of the traditional marine diet in this population.

REFERENCES

Ayotte, P., Carrier, A., Ouellet, N., Boiteau, V., Abdous, B., Sidi, E.A., et al. 2011. Relation between methylmercury exposure and plasma paraoxonase activity in inuit adults from Nunavik. Environ Health Perspect, 119(8): 1077–83.

Fillion, M., Lemire, M., Philibert, A., Frenette, B., Weiler, H.A., Deguire, J.R., et al. 2013. Toxic risks and nutritional benefits of traditional diet on near visual contrast sensitivity and color vision in the Brazilian Amazon. Neurotoxicology, 37: 173–81.

Lemire, M., Fillion, M., Frenette, B., Mayer, A., Philibert, A., Passos, C.J., et al. 2010. Selenium and mercury in the Brazilian Amazon: opposing influences on age-related cataracts. Environ Health Perspect., 118(11): 1584–9.

Lemire, M., Fillion, M., Frenette, B., Passos, C.J., Guimaraes, J.R., Barbosa, F., JR. et al. 2011. Selenium from dietary sources and motor functions in the Brazilian Amazon. Neurotoxicology, 32(6): 944–53.

Lemire, M., Kwan, M., Laouan-Sidi, A.E., Muckle, G., Pirkle, C., Ayotte, P., et al. 2015. Local country food sources of methylmercury, selenium and omega-3 fatty acids in Nunavik, Northern Quebec. Sci Total Environ., 509–510: 248–259.

Pluskal, T., Ueno, M., & Yanagida, M. 2014. Genetic and metabolomic dissection of the ergothioneine and selenoneine biosynthetic pathway in the fission yeast, S. pombe, and construction of an overproduction system. *PLoS One,* 9(5): e97774.

Valera, B., Dewailly, E., & Poirier, P. 2009. Environmental mercury exposure and blood pressure among Nunavik Inuit Adults. Hypertension, 54(5): 981–6.

Yamashita, Y. & Yamashita, M. 2010. Identification of a novel selenium-containing compound, selenoneine, as the predominant chemical form of organic selenium in the blood of bluefin tuna. *J Biol Chem.,* 285(24): 18134–8.

Yamashita, M., Yamashita, Y., Ando, T., Wakamiya, J., & Akiba, S. 2013. Identification and determination of selenoneine, 2-selenyl-N alpha, N alpha, N alpha-trimethyl-L-histidine, as the major organic selenium in blood cells in a fish-eating population on remote Japanese Islands. Biol Trace Elem Res., 156(1–3): 36–44.

Genotypic variation and agronomic biofortication of upland rice with selenium

H.P.G. Reis & J.P.Q. Barcelos
UNESP-Univ Estadual Paulista, Ilha Solteira-SP, Brazil

A.R. dos Reis*
UNESP-Univ Estadual Paulista, Ilha Solteira-SP, Brazil; UNESP-Univ Estadual Paulista, Tupã-SP, Brazil

M.F. Moraes
Federal University of Mato Grosso, Barra do Garças-MT, Brazil

1 INTRODUCTION

Cereal production has maintained the same growth rate as that of human population. It is anticipated that world demand for food will double in the time period of 1999 to 2030, and up to three and half times in developing countries (Daily et al., 1998; Garvin et al., 2006). On the other hand, malnutrition has increased, reaching almost half of mankind, particularly pregnant women, infants and children (Welch, 2001; Graham et al., 2007). This development is partly due to the focus of plant breeding, which aims on improving gains productivity. These gains are inversely related to the mineral content of the grains (Garvin et al., 2006).

Malnutrition is a consequence of large intake of cereals (rice, maize, wheat) poor in vitamins and minerals (Graham et al., 2007). Deficiencies of iron, selenium (Se) and zinc are the major nutritional concerns in relation to human health today, especially in developing countries. Among the strategies applied for reducing the prevalence of Se deficiency problem in human populations, enrichment (biofortification) of food crops with Se through agricultural approaches is a widely applied strategy. Agronomic biofortification (e.g., via fertilizer application) represents a complementary and cost-effective agricultural approach to the problem (White & Broadley, 2009).

This study is aimed to evaluate Brazilian upland rice genotypic variation on the accumulation of Se in edible parts, and to establish an agronomic biofortification strategy to enhance Se content in rice grains for better human nutrition.

2 MATERIALS AND METHODS

Experiment 1: A pot experiment was carried in a greenhouse at University of Sao Paulo, Brazil. One month before sowing, recommended amounts of both lime and nutrients were applied using pots with 3 dm^3 of an acid Oxisol soil. In all pots, 0.1 mg/kg of Se was applied as sodium selenate. Thirty five rice upland cultivars (Table 1) were selected according to the following characteristics: old and new cultivars, duration of the cycle, high and low content of minerals in grains, yield, type, and color of grain (red, black or white).

After maturation, plants were separated into root, stem, leaf, husk and grain. Total Se content in the samples was determined in a PerkinElmer Analyst 800 atomic absorption spectrophotometer (PerkinElmer Inc., San Jose, USA) with electrothermal atomization by (pyrolytic) graphite furnace. A certified reference material BCR402 "White Clover" was used for certification of the quality of analysis.

Experiment 2: The experiment was carried out in a greenhouse at University of Sao Paulo, Brazil. After evaluation of genotypic variation on Se accumulation, eight genotypes of rice were selected based on Se accumulation efficiency. The experiment was conducted in completely randomized design in a 8 × 2 factorial scheme, with three replicates by means of eight upland rice genotypes (e.g., CNA 10929, BRS Carisma, Piaui, BRS Relampago, IAC 202, Dourado Precoce, IAC 1246 and IAC 4), with or without Se (0.50 and 0.00 mg/kg, respectively) applied as sodium selenate. Total Se content was analyzed as described in Experiment 1. Data shown are mean ± standard deviations.

3 RESULTS AND DISCUSSION

Rice is one of major dietary source of Se. In addition, combining technologies such as breeding crops that acquire and/or use Se more efficiently and agronomic biofortification is one important strategy to improve human health.

Concentrations of Se were identified in grains of 35 efficient upland rice varieties under greenhouse conditions and genotypic variation that ranged from 15 to 122 μg/kg DW (Table 1).

Table 1. Upland rice genotypic variation on Se concentrations and accumulation in grains.

	Rice Genotypes	Concentration of Se (μg/kg, DW)	Accumulation of Se (ng/pot)
1	IAC 4440	57	1370
2	PB 11	35	479
3	PB 05	17	368
4	BRSMG Relâmpago	16	178
5	BRSMG Caravera	15	236
6	BRSMG Curinga	74	1274
7	Pérola	64	100
8	Bonança	42	542
9	Jaguari	48	82
10	Primavera	53	286
11	IAC 201	61	592
12	IAC 600	47	76
13	Guarani	56	669
14	Caiapó	63	595
15	Bico ganga	27	225
16	IAC 435	63	116
17	BRS Talento	55	714
18	Canastra	73	526
19	PB 01	17	264
20	Arroz preto	47	422
21	IAC 165	58	409
22	BRSMG Conai	40	528
23	Carajás	56	698
24	IAC 47	44	297
25	IAC 202	63	707
26	Pratão	39	29
27	Beira campo	36	487
28	Maravilha	42	270
29	IAC 25	48	450
30	Batatais	49	49
31	Dourado precoce	85	216
32	IAC 4	122	286
33	Cateto	40	99
34	IAC 1246	115	941
35	Cateto seda	71	484

Table 2. Agronomic biofortication of upland rice genotypes with selenium.

Treatments	Grain Se (μg/kg) DW	Productivity (g/pot)
BRSMG Relampago −Se	16.3 ± 3.4	27.4 ± 2.1
BRSMG Relampago +Se	6764.6 ± 317.4	27.9 ± 1.1
Piaui −Se	16.7 ± 3.6	10.4 ± 0.6
Piaui +Se	6344.2 ± 487.5	10.9 ± 3.9
IAC 202 −Se	70.2 ± 7.4	17.5 ± 4.3
IAC 202 +Se	9237.3 ± 305.5	23.6 ± 0.7
Dourado Precoce −Se	83.1 ± 2.1	10.1 ± 3.1
Dourado Precoce +Se	8889.3 ± 191.8	13.9 ± 3.9
IAC 4 −Se	89.4 ± 7.9	17.5 ± 4.3
IAC 4 +Se	8515.1 ± 356.7	17.1 ± 1.7
IAC 1246 −Se	91.1 ± 6.5	15.9 ± 4.7
IAC 1246 +Se	8446.7 ± 79.	13.3 ± 3.1
Carisma −Se	88.26 ± 7.31	18.9 ± 2.1
Carisma +Se	7735.53 ± 655.	19.2 ± 4.6
CNA 10929 −Se	85.8 ± 5.5	4.2 ± 1.3
CNA 10929 +Se	8715.7 ± 145.9	1.8 ± 1.1

After screening the rice genotypes based on Se accumulation in rice grains, eight cultivars were selected for agronomic biofortification studies (Table 2). Upland rice genotypic variation for Se concentrations in grains ranged from 16.3 to 91 μg/kg DW in the absence of Se fertilizer. In this study, Se application significantly increased the Se concentrations in grains of all varieties, showing that agronomic biofortification of rice genotypes with Se can be an important strategy for improving Se human nutrition. The use of Se fertilizers offers an effective means of increasing Se in upland rice grains. Therefore, new approaches involving genetic breeding to enhance Se accumulation potential are required for further studies.

4 CONCLUSIONS

Significant varietal differences in the Se concentration in Brazilian upland rice grains were observed, ranging from 15 to 122 μg/kg DW. Selenium application significantly increased the Se concentration in rice grains, showing that agronomic biofortification can be an important effort to improve Se human nutrition.

REFERENCES

Daily, G., Dasgupta, P., Bolin, B., Cross, P., Guerny, J., Ehrlich, P., et al. 1998. Global food supply: food production, population growth, and the environment. Science 281 (5381): 1291–1292.

Garvin, D.F., Welch, R.M. & Finley, J.W. 2006. Historical shifts in the seed mineral micronutrient concentrations of US hard red winter wheat germplasm. Journal of the Science of Food and Agriculture 86 (13): 2213–2220.

Graham, R.D., Welch, R.M., Saunders, D.A., Ortiz-Monasterio, I., Bouis, H.E., Bonierbale, M., et al. 2007. Nutritious subsistence food systems. Advances in Agronomy 92: 1–74.

Welch, R.M. 2001. Micronutrients, agriculture and nutrition: linkages for improved health and well being. In: K. Singh, S. Mori, & R.M. Welch (eds). Perspectives on the Micronutrient Nutrition of Crops. 247–289, Jodhpur: Scientific Publishers.

White, P.J. & Broadley, M.R. 2009. Biofortification of crops with seven mineral elements often lacking in human diets – iron, zinc, copper, calcium, magnesium, selenium and iodine. New Phytologist 182: 49–84.

Effect of selenium fertilization on nitrogen assimilation enzymes in rice plants

J.P.Q. Barcelos & H.P.G. Reis
UNESP-Univ Estadual Paulista, Ilha Solteira-SP, Brazil

A.R. dos Reis*
UNESP-Univ Estadual Paulista, Ilha Solteira-SP, Brazil; UNESP-Univ Estadual Paulista, Tupã-SP, Brazil

M.F. Moraes
Federal University of Mato Grosso, Barra do Garças-MT, Brazil

1 INTRODUCTION

Selenium (Se) has been deemed essential to animal nutrition since 1957. Selenium has low bioavailability in most crop soils, especially in parts of China, the UK, Eastern Europe, and Australia (Lyons et al., 2005). Thus, the concentration of this element is usually low in food crops grown in these areas leading to dietary Se deficiency in humans (Hartikainen et al., 2000). Therefore, a low bioavailability of Se and the fact that plants are one main dietary source of this element, recent studies have reported on ways to increase Se content in plant food (Cartes et al., 2005).

Combs (2001) reported that Se level in a population is highly correlated with Se content in agricultural crops. Thus, increasing Se content in food crops can have a positive impact on reducing Se deficiency throughout parts of the world. Selenium biofortification in agricultural crops by means of Se fertilization or selection of crop genotypes with high capacity of Se accumulation, can be useful approaches to increase the consumption of Se by animals and humans (Lyons et al., 2005). In contrast to its essentiality in humans and animals, Se is not considered to be essential for higher plants (Pilon-Smits and Quinn, 2010). Studies have, however, reported beneficial effects of Se because it increases the antioxidant activity in plants, leading to better plant yield, mainly under stress conditions (Hartikainen et al., 2000; Gomes-Junior et al., 2007; Rios et al., 2009).

There are few reports of Se application on nitrogen (N) metabolism. Nowak et al. (2002) demonstrated that Se influences the enzymes nitrate reductase (NR) and nitrite reductase (NiR), diminishing their activity and possibly affecting negatively N metabolism in general. Due to the new importance of Se biofortification in agricultural crops, and based upon a few reports of Se affecting N metabolism, this study aimed to evaluate the effects of Se on net photosynthesis, chlorophyll, and the N metabolism enzymes, such as NR and urease, in Brazilian upland rice genotypes.

2 MATERIALS AND METHODS

The experiment was carried out in a greenhouse at University of Sao Paulo, Brazil. Soil samples were taken from the 0–20 cm layer of a soil low in Se. The experiment was conducted in completely randomized design in a 8 × 2 factorial scheme, with three replicates for eight upland rice genotypes (i.e. CNA 10928, BRS Carisma, Piaui, BRS Relampago, IAC 202, Dourado Precoce, IAC 1246 and IAC 4), with or without Se (0.50 and 0.00 mg/kg, respectively) applied as sodium selenate. During the full tillering stage, net photosynthesis and chlorophyll were measured and afterwards rice leaves were collected for NR and urease assay (Reis et al., 2009).

3 RESULTS AND DISCUSSION

Net photosynthesis and N metabolism enzymes showed significant differences depending of rice genotype and Se application (Table 1). Except for the genotypes IAC 202 and IAC 1246, Se application increased the net photosynthesis in all other genotypes tested, with the highest values observed in Carisma and CNA 10928. The genotype IAC 202 showed the lowest net photosynthesis, but highest values of NR activity in leaves. Rios et al. (2010) found similar results regarding the influence of Se on increasing NR activity in lettuce plants. The first step in nitrate assimilation is its reduction to nitrite by the enzyme NR, which is regulated mainly by the nitrate concentration in leaves. This stage of nitrogen assimilation is defined as the most important and limiting step in this physiological process. Net photosynthesis might be regulated by the higher activity of NR in IAC 202 leaves. In the genotypes IAC 4, IAC 1246 and CNA 10928, the NR activity decreased due to Se application. The SPAD units showed different responses among the genotypes compared to net photosynthesis.

Table 1. Photosynthesis, chlorophylls and nitrogen metabolism enzymes in rice genotypes exposed to Se fertilizers.

Treatments	Photosynthesis μmol CO_2/m/s	SPAD unit	Nitrate Reductase μmol NO_2^- g/FW/h	Urease μmol NH_4^+ g/FW/h	Se (grains) Mg/kg DW
BRSMG Relampago −Se	18.34 ± 1.21	43.33 ± 2.54	30.33 ± 8.79	33.77 ± 5.46	16.3 ± 3.4
BRSMG Relampago +Se	18.94 ± 3.20	41.07 ± 2.06	34.73 ± 4.62	35.55 ± 8.44	6764.6 ± 317.4
Piaui −Se	16.94 ± 1.82	39.93 ± 5.00	21.27 ± 4.06	31.89 ± 4.73	16.7 ± 3.6
Piaui +Se	17.13 ± 2.48	40.23 ± 2.86	21.10 ± 1.90	37.43 ± 1.56	6344.2 ± 487.5
IAC 202 −Se	20.80 ± 4.40	40.60 ± 1.06	20.86 ± 6.32	31.39 ± 3.67	70.2 ± 7.4
IAC 202 +Se	17.76 ± 1.15	40.27 ± 1.91	47.45 ± 8.11	42.49 ± 13.13	9237.3 ± 305.5
Dourado Precoce −Se	15.67 ± 2.70	39.13 ± 2.55	31.62 ± 13.35	34.61 ± 5.96	83.1 ± 2.1
Dourado Precoce +Se	16.27 ± 1.64	39.97 ± 1.14	22.46 ± 6.24	25.09 ± 2.24	8889.3 ± 191.8
IAC 4 −Se	16.72 ± 3.13	41.93 ± 0.40	24.43 ± 3.51	35.59 ± 6.85	89.4 ± 7.9
IAC 4 +Se	19.44 ± 0.88	41.53 ± 1.32	17.79 ± 4.17	38.68 ± 4.45	8515.1 ± 356.7
IAC 1246 −Se	18.49 ± 3.41	38.03 ± 1.62	27.44 ± 5.48	32.59 ± 6.66	91.1 ± 6.5
IAC 1246 +Se	17.74 ± 3.29	40.80 ± 0.66	17.24 ± 2.61	41.54 ± 1.71	8446.7 ± 79.9
Carisma −Se	20.16 ± 1.71	39.07 ± 2.24	35.05 ± 12.44	38.21 ± 4.57	88.26 ± 7.31
Carisma +Se	22.53 ± 3.85	43.87 ± 1.79	46.94 ± 1.70	44.28 ± 12.64	7735.53 ± 655.1
CNA 10928 −Se	21.35 ± 3.21	42.07 ± 1.33	47.73 ± 8.58	44.28 ± 2.74	85.8 ± 5.5
CNA 10928 +Se	22.08 ± 1.40	41.23 ± 1.42	43.84 ± 6.91	47.65 ± 7.56	8715.7 ± 145.9

Regarding urease activity, the genotypes Piaui, IAC 202, IAC 1246 and CNA 10928 exhibited moderate high increases after Se application. There is little information in the literature regarding the role of Se on the regulation of urease activity. Nowak et al. (2002) observed that low levels of Se can stimulate the urease activity in soils.

4 CONCLUSIONS

Selenium fertilization significantly increased the nitrate reductase and urease activity in BRSMG Relampago, IAC 202 and Carisma rice genotypes. For better understanding the role of Se on nitrogen metabolism, further studies are needed, however, our findings confirm the significant influence of Se on N metabolism.

REFERENCES

Cartes, P., Gianfreda, L., & Mora, L.M. 2005. Uptake of selenium and its antioxidant activity in ryegrass when applied as selenate and selenite form. *Plant and Soil* 256: 359–367.

Combs, G.F. 2001. Selenium in global food systems. *British Journal of Nutrition* 85: 517–547.

Hartikainen, H., Xue, T.L., & Piironen, V. 2000. Selenium as an antioxidant and pro-oxidant in ryegrass. *Plant and Soil*, 225: 193–200.

Lyons, G., Ortiz-Monasterio, I., Stangoulis, J., & Graham, R. 2005. Selenium concentration in wheat grain. Is there sufficient genotypic variation to use in breeding? *Plant and Soil*, 269: 369–380.

Pilon-Smits, E. & Quinn, C. 2010. Selenium metabolism in plants. In: Hell. R. & Mendel, R.R. (eds), *Cell Biology of Metals and Nutrients*. Springer, Berlin, 225–241.

Nowak, J., Kaklewski, K., & Klodka, D. 2002. Influence of various concentrations of selenic acid (IV) in the activity of soil enzymes. The *Science of the Total Environment*, 291: 105–110.

Reis, A.R., Favarin, J.L., Malavolta, E., Júniorc, J.L., & Moraes, M.F. 2009. Photosynthesis, chlorophyll and SPAD readings in coffee leaves in relation to nitrogen supply. *Communication in Soil Science and Plant Analysis* 40: 1512–1528.

Rios, J.J., Blasco, B., Cervilla, L.M., Rosales, M.A., Sanchez-Rodriguez, E., Romero, L., et al. 2009. Production and detoxification of H_2O_2 in lettuce plants exposed to selenium. *Annals of Applied Biology*, 154: 107–116.

Rios, J.J., Blasco, B., Rosales, M.A., Sanchez-Rodriguez, E., Leyva, R., Cervilla, L.M., et al. 2010. Response of N metabolism in lettuce plants subjected to different doses and forms of selenium. *Journal of Science and Food Agriculture*, 90: 1914–1919.

Selenium status in Brazilian soils and crops: Agronomic biofortification as a strategy to improve food quality

André Rodrigues dos Reis
UNESP-Univ Estadual Paulista, Tupã-SP, Brazil

1 INTRODUCTION

Selenium (Se) is an essential mineral element for human and animal nutrition, but it is not required by plants. There is conclusive evidence of Se deficiencies in Brazilian soils, pasture grass and agriculture products. Selenium deficiency concentrations in soils are considered to range from 100 to 600 µg/kg. However, in this mini-review, the highest Se concentration shown in Brazilian agriculture soils was approximately 210 µg/kg (very low Se levels), and the foliar Se concentration in pasture grass (Brachiaria sp. and Stylosanthes sp.) ranged from 40 to 66 µg/kg. The amount of Se in plants is strongly related to available Se in soils. There is evidence of low Se intake in Brazilian population, but no accurate and conclusive study has been conducted. In addition, biofortification of crops with Se is not included in the HarvestPlus Program of Brazil. Therefore, more research on soil-Se levels and Se levels in agricultural products covering all Brazilian states is needed. The application of Se-containing fertilizers is efficacious at correcting low Se levels in human and animals in Finland, New Zealand and Australia, etc.

2 SELENIUM CONTENT IN BRAZILIAN SOILS

Various compounds and/or chemical elements are important to ensure the quality of food such as protein, carbohydrates, fats, vitamins, iron (Fe), iodine (I), zinc (Zn) and Se (Rayman et al., 2012). Genetic breeding programs act positively to the development of improved varieties, which reach higher productivity. However, problems of nutritional deficiencies are experienced by almost half the world's population, especially Fe, I, Se, vitamin A and Zn in developing countries (Rayman et al., 2012). These phenomena happen basically due to two reasons: (1) low concentration of micronutrients in the soils, which are affected by texture class or soil pH; (2) is a dilution effect of the essential micronutrients and vitamins for human health in the highest productive varieties or cultivars. Therefore, Se concentrations in agricultural products (food) depends on their concentrations in the soil. In addition, genotypic variation can also influence the absorption capacity of Se by plants. Although Brazilian agricultural products have low levels of Se, it is important to understand the behavior of Se in soils and assess its transfer to edible parts of plants, which are primary sources of food for the population and animals.

The two major inorganic forms of Se in soils are selenate and selenite. However, selenite forms usually are more strongly retained in the soil colloids in comparison to selenate, which depends on the environment characteristics, such as pH, ionic strength, ion concentration and other effects. Moreover, the range between essential and toxic level of Se in plants and animals is very narrow (Lyons et al., 2003). In this context, the study of Se levels in agricultural soils and their sorption behavior under different conditions of pH, ionic strength and concentration of competing ions becomes highly relevant to a better understanding of the dynamics of Se in Brazilian soils. These studies will greatly contribute to future research aiming to provide optimal levels of Se to be applied through fertilizer (agronomic biofortification) to increase the natural intake of Se by Brazilian population. There is a lack of information about the distribution of Se in Brazilian soils. Table 1 summarizes the Se and sulfur concentrations in soils collected from different regions of Brazil.

There is evidence of Se deficiency in Brazilian population; however, no extensive research on the subject is available (Moraes et al., 2009). Ferreira et al. (2002) observed that food consumed in Brazil has significantly low concentrations of Se. This observation might be due to low Se concentrations in Brazilian soils. Similar results were reported by Faria (2009), showing very low Se concentrations in pasture grass (*Brachiaria sp.* and *Stylosanthes sp.*) ranging from 40 to 66 µg/kg. On the other hand, Brazilian nuts growing in the North-West of Brazil is considered the richest food source for Se with concentrations ranging from 0.03 to 512 mg/kg (Abreu, 2010). Unfortunately, there is no information regarding Se levels in Amazon or Para State soils available in the literature.

3 AGRONOMIC BIOFORTIFICATION AS A STRATEGY TO IMPROVE FOOD QUALITY

In areas where the bioavailability of Se is low in the soil, the supplementation using commercial fertilizers

Table 1. Selenium and sulfur concentrations in Brazilian soils (Adapted from Moraes et al., 2009).

Region/City	S Concentration (g/kg)	Se Concentration (µg/kg)	References
Matão	6–10	2.3–15	Faria (2009)
São Paulo	–	0–800	Paiva Neto & Gargantini (1956)
São Paulo	–	38–212	Anno (2001)
Itirapina	6–8	0–0.26	Faria (2009)
Goiás	–	1–8	Fichtner et al. (1990)
Piracicaba	5–10	5–155	Faria (2009)
Cerrados Region	–	10–80	Carvalho (2011)
Piracicaba	10–16	90–206	Faria (2009)
Piracicaba	4–32	0–90	Faria (2009)
Analândia	5–13	0–25	Faria (2009)
Pirassununga	8–25	0	Faria (2009)
Pirassununga	6–15	0–59	Faria (2009)
Se deficiency range	–	100–600	Lyons et al. (2003)

can be an effective strategy to provide the micronutrient to the plants. In 1984, the Ministry of Agriculture and Forestry of Finland decided to initiate the supplementation of mineral fertilizers with sodium selenate. At first, supplementation was 16 mg/kg of Se for cereals and vegetables and 6 mg/kg of Se for pasture. After a few years and extensive research, the content of Se in fertilizers increased to 10 mg/kg in 1998 (Hartikainen, 2005). This supplementation with Se improved nutritional value of the food chain soil for plants, animals, and humans. The amount of Se used in this study was considered safe and adequate and also improved the production of plants (Hartikainen, 2005).

4 CONCLUSIONS

More research is needed related to the Se concentrations in soils from different States of Brazil. In areas where the bioavailability of Se is low, an efficient alternative is supplementation of commercial fertilizers with Se to the soils or foliarly applied, which corrects low levels of Se in the pastures, animals, and humans.

REFERENCES

Anno, R.M. 2001. Selenium in soils and in *Brachiaria brizantha* (Hochsr) Stapf. Cv. Marandú. Bachelor in Animal Science – Universidade de São Paulo.
Carvalho, G.S. 2011. Selenium and mercury in cerrados soils of Brazil. PhD Thesis in Soil Science. Universidade Federal de Lavras. 93p.
Faria, L.A. 2009. Levantamento sobre selênio em solos e plantas do Estado de São Paulo e sua aplicação em plantas forrageiras. Master of Science Thesis, Universidade de São Paulo. 75p.
Fichtner, S.S., Paula, A.N., Jardin, E.C., Silva, E.C., & Lopes, H.O.S. 1990. Estudo da composição mineral de solos, forragens e tecido animal de bovinos do município de Rio Verde, Goiás. IV. Cobre, molibdênio e selênio. Anais das Escolas de Agronomia e Veterinária, 20: 1–6.
Hartikainen, H. 2005. Biogeochemistry of selenium and its impact on food chain quality and human health. J Trace Elem Med Biol, 18: 309–318.
Lyons, G., Stangoulis, L., & Graham, R. 2003. High-selenium wheat: biofortification for better health. Nutr Res Rev., 16: 45–60.
Moraes, M.D., Welch, R.M., Nutti, M.R., Carvalho, J.L.V., & Watanabe, E. 2009. Evidences of selenium in Brazil: from soil to human nutrition. In: Banuelos, G.S., Lin, Z.Q., & Yin, X.B. (eds). Selenium: Deficiency, Toxicity and Biofortification for Human Health. Hefei: University of Science and Technology of China Press. 73–74.
Paiva Neto, J.E., & Gargantini, H. 1956. Dosagem de selênio no solo. Bragantia, 15: 13–16.
Rayman, M.P. 2012. Selenium and human health. Lancet, 379: 1256–1268.

Effects of selenium-enrichment on fruit ripening and senescence in mulberry trees

Jie Wu, Zhilin Wu & Miao Li*
Key Laboratory of Agri-Food Safety of Anhui Province, School of Plant Protection, Anhui Agriculture University, Hefei, Anhui, China

Yongjin Deng
Sericulture Research Institute, Anhui Academy of Agricultural Sciences, Hefei, Anhui, China

Gary S. Bañuelos
ARS-USDA, Parlier, California, USA

Zhi-Qing Lin
Environmental Sciences Program, Southern Illinois University Edwardsville, Edwardsville, Illinois, USA

1 INTRODUCTION

Selenium (Se) is an essential trace element, and a natural antioxidant agent both in humans and animals (Pezzarossa et al., 1999). In plants, the essentiality of Se as a micronutrient has not been conclusively demonstrated, however, it is believed that Se could delay plant senescence and decrease postharvest losses (Malorgio, 2009). Previous work has demonstrated that Se is effective in delaying the onset of tomato plant senescence and fruit ripening (Pezzarossa et al., 2014). Furthermore, Se has been shown to be effective in decreasing the production of ethylene and PAL activity, consequently improving the quality of leafy vegetables and the shelf life in lettuce and chicory species (Lei et al., 2014).

Little is known about the efficacy of Se on fruit ripening and senescence in mulberry trees. The current study was to evaluate the delaying effect of Se on fruit ripening and senescence in mulberry trees, and the possible mechanisms were also discussed by which Se delayed the fruit ripening and senescence in mulberry trees.

2 MATERIALS AND METHODS

Field experiments were conducted in Spring 2013 at the Taoxi Orchard in Hefei, including two mulberry species of *Morus alba* L. and *Morus nigra* L. The experimental mulberry trees were three years old. The soil nano-Se fertilizer was presented by Suzhou Setek Co. Ltd, and it was applied via soil/root at the application rate of 35 or 50 mg nano-Se fertilizer per mulberry tree. Mulberry trees that were not treated with nano-Se were used as control. Each mulberry variety (i.e. white mulberry or black mulberry) had two Se fertilizer treatments (*i.e.* 35 mg and 50 mg Se per tree). Each treatment had five replicates. A total of 10 trees of each variety were treated with Se fertilizer, along with 10 individual trees of each variety as control.

Mulberry fruits were harvested at commercial maturity (in late May). Plant leaves were also collected from each replicate tree of each treatment and the control. Mulberry fruit and foliar fresh weights were recorded, and their dry weights were determined after oven-dried at 45°C for 48 hours. Carotenoid (lutein, lycopene, β-carotene) extraction was performed according to the reported method (Pezzarossa et al., 2014). The extraction of flavonoids (quercetin, naringenin, and rutin) was carried out according to the reported method (Pezzarossa et al., 2014). The measurements of carotenoids and flavonoids were carried out by HPLC consisting of a low pressure gradient pump with four solvent model UV-Vis detector and a multichannel UV. The metabolites were identified based on retention times and quantified using calibration curves with pure standards of known concentration. Total Se content was determined at harvest in oven-dried ground leaves and fruit samples after digestion with nitric and perchloric acids and reduction by hydrochloric acid. The digests were analyzed by atomic absorption spectrophotometry. One way ANOVA was used to determine the treatment effects in a completely randomized experimental design. The level of significance (α) was 0.05.

3 RESULTS AND DISCUSSION

Selenium added to the soil was absorbed by roots and accumulated in fruits and leaves. The greater the amount of Se added, the greater the amount of Se accumulated in fruits and leaves. Concentrations of Se, on a dry weight basis, was higher in leaves than in

Table 1. Concentrations of total Se (µg/g, DW) in leaves and fruits from black mulberry trees treated with 0, 35, and 50 mg Se per tree. Values with different letters in the same column are statistically significantly different ($p < 0.05$).

Se add	Se content (µg/g) DW	
(mg Se per tree)	Leaf	Fruit
0	0.5a	0.3a
35	46.3b	35.2b
50	51.6b	38.9b

Table 2. Production and qualitative characteristics of fruits from black mulberry trees treated at 0 and 35 mg, as well as 50 mg Se per tree. Data are the means (n = 3). Values with different letters in the same column are statistically significantly different ($p < 0.05$).

	Se Application Rate (mg Se per tree)		
Parameters	0	35	50
Total yield (g/tree)	1432a	1456a	1428a
Fruit weight (g)	0.6a	0.7a	0.6a
Solid content (%)	1.3a	1.2a	1.1a
Lycopene (µg/g DW)	4.4a	5.6a	5.4a
Lutein (µg/g DW)	0.87a	1.11a	0.99a
Maturity Index	9.2a	8.6a	8.9a
Harvest Index	0.81a	0.73a	0.75a

fruit. Fruit yield was not statistically influenced by Se addition (Table 1). Selenium addition influenced the carotenoid concentration in fruits at harvest. Lycopene was higher in the treated samples, particularly at 35 mg Se per tree, whereas β-carotene and lutein contents significantly decreased when Se was added at 50 but not at 35 mg Se per tree. The addition of Se, either at 35 or 50 mg Se per tree, decreased the content of naringenin in fruits, whereas the highest dose of Se induced a dramatic increase in the quercetin concentration (Table 2). The content of rutin was not influenced by the Se addition (data not shown).

The data indicate that Se accumulating in mulberry fruits may affect fruit secondary metabolism and composition at ripening. The significant decrease of β-carotene and the slight increase in lycopene in the 50 mg Se per tree treatment indicated that the carotenoid biosynthetic pathway is affected by Se (Table 2). The ripening delay and anti-senescence effects of Se are evident when evaluating the evolution of mulberry fruit pigmentation after detachment over a 6-day storage period at 4°C (data not shown). In fact, Se-enriched fruit showed a reduced rate in color change, confirming that ripening-related processes, such as the degradation of chlorophyll and the synthesis of carotenoids, are affected by this element, with potential benefits in terms of storage and shelf-life (Table 2).

4 CONCLUSIONS

The results of this study support the hypothesis that Se (at the tested concentrations) may modulate mulberry fruits development, in particular the ripening and senescence process (with beneficial effects in terms of post-harvest commercial life) possibly through its anti-oxidant and anti-senescence properties. Fully understanding how the molecular and biochemical mechanisms are directly or indirectly affected by Se in fruit tissues at ripening and senescence, as well as during the postharvest phase should help considerably in optimizing the treatment procedures and protocols.

REFERENCES

Lei, C., Ma, Q., Tang, Q.Y., Ai, X.R., Zhou, Z., Yao, L., et al. 2014. Sodium selenite regulates phenolics accumulation and tuber development of purple potatoes. *Scientia Horticulturae*, 165(22): 142–147.

Malorgio, F., Diaz, K., Ferrante, A., Mensuali, A., & Pezzarossa, B. 2009. Effects of selenium addition on minimally processed leafy vegetables grown in floating system. Journal of the Science of Food and Agriculture, 89(13): 2243–2251.

Pezzarossa, B., Malorgio, F., & Tonutti, P. 1999. Effects of selenium uptake by tomato plants on senescense, fruit ripening and ethylene evolution. 275–276. In: A.K. Kanellis, C. Chang, H. Klee, A.B. Bleecker, J.C. Pech, & D. Grierson (eds.), *Biology and Biotechnology of the Plant Hormone Ethylene II*. Kluwer Academic Publishers, Dordrecht, Netherlands.

Pezzarossa, B., Rosellini, I., Borghesi, E., Tonutti, P., & Malorgio, F. 2014. Effects of Se-enrichment on yield, fruit composition and ripening of tomato (*Solanum lycopersicum*) plants grown in hydroponics. *Scientia Horticulturae*. 165(22):106–110.

Wheat biofortification: Genotypic variation and selenium fertilization in Brazil

C.R.S. Domingues & J.A.L. Pascoalino
Federal University of Parana, Graduate Program of Soil Science, Curitiba, PR, Brazil

M.F. Moraes
Federal University of Mato Grosso, Graduate Program of Tropical Agriculture, Barra do Garças, MT, Brazil

C.L.R. Santos
Federal University of Parana, Graduate Program of Soil Science, Curitiba, PR, Brazil

A.R. dos Reis
Sao Paulo State University, Campus of Tupa, Tupa, SP, Brazil

F.A. Franco & A. Evangelista
Center for Agricultural Research, COODETEC, Cascavel, PR, Brazil

P.L. Scheeren
Embrapa Wheat, Passo Fundo, RS, Brazil

1 INTRODUCTION

Biofortification with selenium (Se) can be done through genetic methods or by application of fertilizers. Both approaches can be used together to increase the levels of nutrients and vitamins in food crops, but fertilization stands out for its practicality, agility and viability, making it a promising way to biofortify crops with Se.

Studies have indicated that agronomic biofortification is an efficient and safe strategy for the natural enrichment of nutrients such as iron (Fe), zinc (Zn) and Se in target crops (Gissel-Nielsen & Bisbjerg, 1970). Research in Brazil has shown that agricultural soils typically have low levels of Se, ranging from <0.001 to 0.80 mg/kg (Moraes et al., 2009). Consequently, Se levels in food crops are low compared to international standards. This situation has attracted the attention of researchers to pursue biofortification strategies with Se.

The objective of this study was to assess the yield and nutritional performance of wheat cultivars exposed to application of Se to the soil.

2 MATERIALS AND METHODS

The experiment was conducted in a greenhouse located on the Federal University of Parana in Curitiba, PR, Brazil. The soil used was a typic dystrophic Red-yellow Latosol (Oxisol), obtained at a depth of 0 to 20 cm, in the municipality of Ponta Grossa, PR. According to Neal (1993), this soil is sandy and has low Se content (<0.01 to 0.05 mg/kg).

The experiment was fully randomized in an 8 × 2 factorial design with three replications. The treatments were combinations of eight wheat genotypes (Abalone, BRS Guamirim, BRS Parrudo, BRS 210, CD 150, Embrapa 21, Londrina, Mentana), with and without application of Se (0.5 mg/kg) in the form of sodium selenate.

Plants were grown in pots containing 3 kg of soil. Ten seeds were planted in each pot, and 15 days after shoot emergence, plants were thinned to three plants per pot. At flowering, flag leaves were collected from each pot for nutritional diagnosis.

Selenium was determined in plant tissue by GF-AAS. Certified samples (White Clover – BCR 402 by Institute for Reference Materials) and Measurements were used for quality control in Se analyses. The data were submitted to analysis of variance (F-test) and when significant, means were compared by the Tukey test at 5% probability.

3 RESULTS AND DISCUSSION

The Se fertilization did not significantly affect the grain yield. Yields ranged from 26.1 to 32.1 g/pot (with Se application) to 24.7 to 30.9 g/pot (without Se application).

This result corroborates that reported by Lyons et al. (2004) who did not observe an increase in wheat yields with Se application.

In the cultivars that did not receive Se application, concentrations of Se in the grains were below the limits of detection. However, for the cultivars that received an application of Se, the Se concentration in the grains

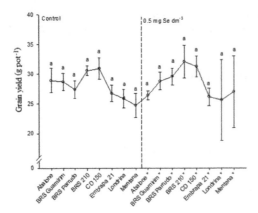

Figure 1. Grain yield of wheat cultivars with and without application of Se to the soil.

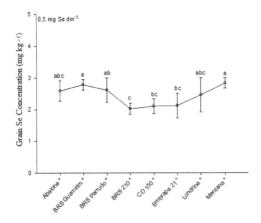

Figure 2. Concentration of Se in the grains of wheat cultivars subjected to Se fertilization.

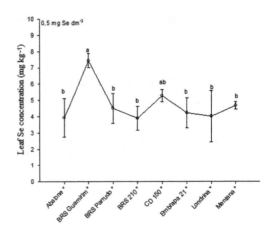

Figure 3. Leaf Se concentration of wheat cultivars subjected to Se fertilization.

4 CONCLUSIONS

The application of Se to the soil did not significantly influence the grain yield of the wheat cultivars studied, but it did increase the concentration of Se in the grains and leaves. There was low genotypic variation among the cultivars on their ability to accumulate Se in grains and leaves.

ACKNOWLEDGMENTS

For Coordination for the Improvement of Higher Education Personnel (CAPES) to provide fellowship to first author and to the Brazilian National Council for Scientific and Technological Development (CNPq) for the grant # 486150/2011-9.

varied from 2.0 to 2.8 mg/kg (Fig. 2), characterizing low genotypic variation on Se accumulation among the cultivars. In our study, cultivars BRS Guamirim and Mentana accumulated the greatest Se concentrations, while the lowest Se concentration was found in BRS 210. According to Poblaciones et al. (2014), Se fertilization in the soil was associated with higher concentrations of the element in the grains of durum wheat.

The leaf concentration of Se increased with Se application to the soil, although there was little difference on Se content in grains among cultivars. The highest Se concentration was in BRS Guamirim (7.4 mg/kg) while the lowest concentrations were in Abalone (3.9 mg/kg) and BRS 210 (3.8 mg/kg) (Fig. 3). Since the cultivar with the highest leaf Se concentration was also the one with the highest grain Se concentration, we inferred that Se is transported from the leaves to the grains. Also, the leaf Se concentrations tended to be higher than in the grains, suggesting that biofortification with Se may also be efficient through foliar application.

REFERENCES

Gissel-Nielsen, G., & Bisbjerg, B. 1970. The uptake of applied selenium by agricultural plants. Plant Soil, 32(1–3): 382–396.

Lyons, G.H., Judson, G.J., Stangoulis, J.C., Palmer L.T., Jones J.A., & Graham, R.D. 2004. Trends in selenium status of South Australians. Med J Aust. 180(8): 383–386

Moraes, M.F., Welch, R.M., Nutti, M.R., Carvalho, J.L.V. & Watanabe, E. 2009. Evidences of selenium deficiency in Brazil: from soil to human nutrition. In: G.S. Banuelos, Z.-Q. Lin, & X.B. Yin (eds.), Selenium: Deficiency Toxicity and Biofortification for Human Health. USTC Press. Hefei, China. 73–74.

Neal, R.H. 1993. Selenium. 237–260. In: B.J. Alloway (ed.), Heavy Metals in Soil. Blackie & John Wiley & Sons, Inc., Glasgow.

Poblaciones, M.J., Rodrigo, S., Santamarío, O., Chen, Y., & Mcgrath, S.P. 2014. Agronomic selenium biofortification in Triticum durum under Mediterranean conditions: from grain to cooked pasta. Food Chem., 146: 378–384.

Improving selenium nutritional value of major crops

L. Li*
Robert W. Holley Center for Agriculture and Health, USDA-ARS, Cornell University, Ithaca, NY, USA

F.W. Avila, G.A. Souza, P.F. Boldrin, M.A. de Figueiredo, V. Faquin,
M.J.B. Andrade & L.R.G. Guilherme
Federal University of Lavras, Lavras, Minas Gerais, Brazil

S.J. Ramos
Vale Institute of Technology, Ouro Preto, Minas Gerais, Brazil

1 INTRODUCTION

Micronutrient malnutrition, including selenium (Se) deficiency, affects a large number of people worldwide because many crops contain low levels of these essential nutrients. Biofortification of crops with Se by means of fertilization and/or breeding varieties with a high capacity of accumulation provides an effective approach to obtain adequate amounts of Se in our diet (Malagoli et al., 2015; White & Broadley, 2009; Zhu et al., 2009). This not only helps reduce Se deficiency problem, but also promotes better health with enhanced chemopreventive activity. A comprehensive understanding of Se nutrition and metabolism in plants is critical to improve Se nutritional value and health-promoting properties in food crops. Recently, we have evaluated the genotypic variation of multiple major crops response to Se application and investigated the effects of Se supplements on plant growth, Se and metabolite accumulation, mineral interaction, as well as Se effects at molecular/biochemical levels.

2 MAJOR CROPS EXHIBIT CONSIDERABLE GENOTYPIC VARIATION IN SE LEVELS

Exploration of genetic variation for micronutrient content is increasingly seen as an effective strategy to breed nutrient-dense crops. Genotypic variation of Se content was observed in a number of germplasms including broccoli and lettuce with a broad range of market types, as well as wheat and common beans of Brazilian varieties. Broccoli, as a Se secondary accumulator, accumulated large amounts of total Se and produced bioactive forms of Se such as Se-methylselenocysteine (SeMSCys) when exposed to 20 μM Na_2SeO_4. There was more than a 2-fold difference between broccoli accessions (Ramos et al., 2011b). Additionally, broccoli florets contained significantly more total Se and SeMSCys than broccoli sprouts (Avila et al., 2013). Lettuce, as Se non-accumulator, accumulated low levels of total Se. There was more than a 2-fold difference between lettuce lines when lettuce was treated with either 15 μM Na_2SeO_4 or 15 μM Na_2SeO_3 (Ramos et al., 2011a). Wheat grain is an important dietary source of Se. In contrast to some other regions of modern wheat cultivars, a 1.5-fold difference in Se content was noted in grains of Brazilian wheat lines (Souza et al., 2014). Moreover, the shoot tissue showed a 4-fold difference in total Se levels (Souza et al., 2013). Similarly, genetic variation of Se content in seeds of common beans was observed (Figueiredo et al., unpublished). Variation in Se content among genotypes of these crops suggests the opportunity for breeding better cultivars that have a higher capacity to accumulate Se.

3 SELENIUM FORMS DIFFERENTLY AFFECT PLANT GROWTH AND SE ACCUMULATION

Selenium is acquired by plants mainly as selenate and selenite through active transport and passive uptake, respectively (Terry et al., 2000). Selenate is known to be less toxic to plants than selenite. Under low dosages of selenate supply for biofortification, plant growth (in term of biomass) was not affected in broccoli (Ramos et al., 2011b), lettuce (Ramos et al., 2011a), wheat seedling (Souza et al., 2013), or in common beans (Figueiredo et al., unpublished). Reduced wheat biomass was, however, observed in nearly all wheat lines when treated with 10 μM Na_2SeO_3 (Souza et al., 2013).

Selenium supplementation is effective in promoting total Se accumulation in shoots and sprouts. A much larger increase of total Se levels was observed when treated with the same dosages of selenate than selenite, especially in wheat shoots (Avila et al., 2013; Ramos et al., 2011a; Souza et al., 2013). However, in wheat grains, selenite was as effective as selenate in enhancing Se content in some germplasm (Souza et al., unpublished data), suggesting tissue specific accumulation responses to selenate and selenite.

Selenate is reportedly more effective in stimulating anticancer SeMSCys biosynthesis than selenite in broccoli florets and leaves (Lyi et al., 2005). Interestingly, SeMSCys content in broccoli sprouts was found not to be affected by the Se forms supplied. Both selenate and selenite promote SeMSCys synthesis to similar levels in broccoli sprouts when treated with the same dosages of Se (Avila et al., 2013).

4 INTERACTIONS BETWEEN SE AND OTHER NUTRIENTS

Selenium as an analog of sulfur (S) is known to have a synergistic or antagonistic relationship with sulfate depending on the selenate dosages supplied. Under low dosages, selenate was found to enhance S level in all lines of lettuce (Ramos et al., 2011a) and seedlings of wheat (Souza et al., 2013). Sulfur levels increased in wheat shoots as selenate was applied up to $10\,\mu M$ (Boldrin et al., unpublished). However, selenate treatment did not change the S level in wheat seeds, indicating a relative consistent ability of grains to accumulate S or there is a tissue-specific effect of selenate on S accumulation. In contrast, selenite had a minimal effect on S levels in these crops.

Selenium fertilization was also found to affect other nutrient levels. In wheat seedlings, synergistic effects of selenate in enhancing Ca, Mg, Fe, Zn and Mo levels were observed, while selenite reduced these nutrient contents (Souza et al., 2013). Both Fe and Zn are major targets for crop biofortification. Selenate supplement enhanced Zn level and improved Fe content in grains of some wheat lines, suggesting the feasibility of simultaneous enhancement of Zn and Se along with Fe (Souza et al., 2014).

5 SIMULTANEOUS ACCUMULATION OF ANTICANCER SE AND S COMPOUNDS

Brassica crops accumulate chemopreventive glucosinolates. In Se-biofortified broccoli, an antagonistic relationship between SeMSCys and glucosinolate accumulation is often observed in florets (Avila et al., 2013), although genotypic variation without negative correlation was also noted (Ramos et al., 2011b). By contrast, the accumulation of SeMSCys by Se treatment exerted minimal effects on both the total and composition of glucosinolates in sprouts from the six most consumed Brassica vegetables. This observation indicates that Se can simultaneously accumulate both groups of anticancer compounds (Avila et al., 2013, 2014). While a comparable capacity to synthesize SeMSCys was observed among sprouts of these crops, Se-biofortified broccoli sprouts had the highest glucoraphanin content that makes it as an excellent source of Se with high chemopreventive activity (Avila et al., 2014).

6 BASIS OF SE NUTRITION IN CROPS

The selenate-stimulated growth effects appeared to be associated with altered antioxidant enzyme activities (Ramos et al., 2011a). Our recent studies indicate that the selenate-promoted S accumulation was likely due to the combined action of upregulation of *SULTR1;1* for enhanced capacity of uptake and down regulation of *APS* expression for decrease of assimilation (Boldrin et al., unpublished).

REFERENCES

Avila, F.W., Faquin, V., Yang, Y., Ramos, S.J., Guilherme, L.R., Thannhauser, T.W., et al. 2013. Assessment of the anticancer compounds Se-methylselenocysteine and glucosinolates in Se-biofortified broccoli (*Brassica oleracea* L. var. italica) sprouts and florets. J. Agric. Food Chem. 61: 6216–6223.

Avila, F.W., Yang, Y., Faquin, V., Ramos, S.J., Guilherme, L.R., Thannhauser, T.W., et al. 2014. Impact of selenium supply on Se-methylselenocysteine and glucosinolate accumulation in selenium-biofortified Brassica sprouts. Food Chem. 165: 578–586.

Lyi, S.M., Heller, L.I., Rutzke, M., Welch, R.M., Kochian, L.V., and Li, L. 2005. Molecular and biochemical characterization of the selenocysteine Se-methyltransferase gene and Se-methylselenocysteine synthesis in broccoli. Plant Physiol. 138: 409–420.

Malagoli, M., Schiavon, M., dall'Acqua, S., and Pilon-Smits, E.A. 2015. Effects of selenium biofortification on crop nutritional quality. Front Plant Sci. 6: 280.

Ramos, S.J., Rutzke, M.A., Hayes, R.J., Faquin, V., Guilherme, L.R., and Li, L. 2011a. Selenium accumulation in lettuce germplasm. Planta. 233: 649–660.

Ramos, S.J., Yuan, Y., Faquin, V., Guilherme, L.R., and Li, L. 2011b. Evaluation of genotypic variation of broccoli (*Brassica oleracea* var. italic) in response to selenium treatment. J. Agric. Food Chem. 59: 3657–3665.

Souza, G.A., de Carvalho, J.G., Rutzke, M., Albrecht, J.C., Guilherme, L.R., and Li, L. 2013. Evaluation of germplasm effect on Fe, Zn and Se content in wheat seedlings. Plant Sci. 210: 206–213.

Souza, G.A., Hart, J.J., Carvalho, J.G., Rutzke, M.A., Albrecht, J.C., Guilherme, L.R., et al. 2014. Genotypic variation of zinc and selenium concentration in grains of Brazilian wheat lines. Plant Sci. 224: 27–35.

Terry, N., Zayed, A.M., De Souza, M.P., & Tarun, A.S. 2000. Selenium in higher plants. Annu Rev Plant Physiol Plant Mol Biol 51: 401–432.

White, P.J., & Broadley, M.R. 2009. Biofortification of crops with seven mineral elements often lacking in human diets–iron, zinc, copper, calcium, magnesium, selenium and iodine. New Phytol. 182: 49–84.

Zhu, Y.G., Pilon-Smits, E.A., Zhao, F.J., Williams, P.N., & Meharg, A.A. 2009. Selenium in higher plants: understanding mechanisms for biofortification and phytoremediation. Trends Plant Sci. 14: 436–442.

The changing selenium content in vegetables and nutritional status of Chinese residents

Weiming Shi*, Sumei Li, Ju Min & Longhua Wu
State Key Laboratory of Soil and Sustainable Agriculture, Institute of Soil Science, Chinese Academy of Sciences, Nanjing, China

Gary S. Bañuelos
USDA-Agricultural Research Service, San Joaquin Valley Agricultural Sciences Center, Parlier, CA, USA

1 INTRODUCTION

Selenium (Se) is an essential trace element in human nutrition. However, according to World Health Organization (WHO), there are more than 40 countries designated as low Se or Se deficient in the world. In China the Se deficient areas account for 72% of the country's total area, and this deficiency affects over 70 million people who face potential adverse health impacts due to Se deficiency. Overt Se deficiency has caused serious health consequences in low Se areas of China, such as endemic Keshan disease (endemic cardiomyopathy) and Kaschin-Beck disease (endemic osteoarthropathy).

Selenium enters the food chain through plants, and the amount of Se in foods is directly affected by Se levels in the soil in which they are grown. Intake of Se varies considerably between regions and countries largely due to the variability of the Se concentration in food products. The concentration of Se in hair is a commonly-used bioindicator to evaluate body Se load, as it reflects the long-term Se level in human body. Hair Se can also be used as an important indicator on endemic diseases and may play a critical role in the etiologic research on the Keshan disease, Kashin–Beck disease, and on local selenosis in China. Dietary Se provided from vegetables can play an important role in satisfying Se nutrition requirements of Chinese people. Selenium concentrations in plant foods are directly related to Se levels in the soil, while the soil Se concentration and bioavailability can be very much affected by fertilization, besides soil parent material and environmental conditions. The excessive use of nitrogen fertilizer leads to soil acidification and accumulation of nitrate that competes for Se uptake. In the past decades large amounts of fertilizers have been applied in soils to support the increasing demand for growing foods. Consequently, it is important to understand the relationship between the fertilizer nitrogen input and the Se content in vegetables. Previous studies also showed that even though soils containing adequate amounts of Se (or high concentrations of total Se), they produced crops with low levels of Se.

Moreover, Chinese residents' dietary changes greatly with rapid development of economics. Therefore, the relationship between the current Se concentrations in foods and Se nutritional status of Chinese residents has not been well explored. Our objective is to examine more carefully the Se nutritional status of the Chinese.

2 MATERIALS AND METHODS

The present study aims: (1) to evaluate the Se nutritional status of Chinese residents by conducting a systematic survey on hair Se concentrations of residents crossing 10 provinces extending from northeast to southeast China; (2) to compare the Se nutrition of current residents with data reported in the past, and to analyze if the changes in hair Se content is due to reduced Se intake via rice as a staple food over time; (3) to determine effect of continuous application of inorganic N-fertilizer on Se concentration in vegetables.

A total of 408 hair samples were collected from local healthy people of both genders ranging from 4 to 76 years old across the 10 provinces—there were 46 children (4–18 years old), 319 adults (19–60 years old) and 43 seniors (61–76 years old). Approximately 2.0 g of hair samples were cut between 1 and 3 cm from the nape of the neck. The effect of inorganic N-fertilizer application on the Se content in soils and plants was evaluated in vegetables produced in a poly-tunnel greenhouse vegetable system located in Yixing, Jiangsu province of China with typical subtropical region. The Soil was classified as a sandy clay loam, with soil pH 5.99 and extractable NO_3-N concentration, Olsen-P and extractable S were 32.2 mg/kg N, 18.6 mg/kg P, and 80.1 mg/kg S, respectively, and total Se averaged 0.30 mg/kg. The Se concentration in each sample was determined by Hydride Generation Atomic Fluorescence Spectrometry. The crop sequence was tomato (April to July), fallow (July to September), cucumber (September to December), and celery (December to next April). We applied annually a total N of 870 kg/ha from urea as inorganic N-fertilizer treatment, in addition to 234 kg ha N applied from manure. Besides, each crop received 150 kg/ha K_2O and 120 kg/ha P_2O_5 as baseline fertilizers for each growing season.

3 RESULTS

By measuring the hair Se concentrations in Chinese inhabitants across northeast to southeast China, the results indicated that generally 84% of all residents have normal hair Se content. Between the sexes, the average hair Se content of males was higher than that of females, irrespective of districts. When comparing geographical regions, the average hair Se content of southern residents was greater than that of northern residents, regardless of gender. Historically, the overall hair Se content of today's inhabitants decreased between 24% and 46% when compared with the inhabitants living in the same geographic region 20 years ago. The decrease of hair Se content may be related to the overall decrease of grain consumption and the lower Se content in the staple food rice.

Six years of continuous application of inorganic N-fertilizer resulted in a significant accumulation of NO_3-N, Olsen-P, and extractable S in the soil over the course of the experiment. Soil NO_3-N concentration accumulated up to 600 mg/kg with application of inorganic N-fertilizer, while soil phosphate extractable Se, and vegetable Se concentrations were significantly lower than that those grown in control treatments. The negative effect of continuous application of inorganic N-fertilizer on vegetable Se concentration may be a consequent of excessive accumulated soil NO_3-N. When NO_3-N concentration was greater than 300 mg/kg, vegetable Se concentration was negatively correlated with soil NO_3-N.

4 CONCLUSIONS

Our results show that the negative effect of continuous application of inorganic N-fertilizer on vegetable Se concentrations was likely the result of soil nitrate's competitive or antagonistic effect on Se uptake by vegetables grown in an intensive polytunnel vegetable cultivation system. Moreover, the eventual decrease of hair Se content of Chinese residents may also be related to the lower Se content in vegetables and rice, as well as an overall decrease of grain consumption.

REFERENCES

Chen, Y.Q., Xu, X.Y., & Gao, Q. 2011 The selenium daily intake of the residents in the southern Songnen plain of Heilongjiang province. *J. China West Norm. Univ.* 32: 198–200.

Cheng, L.C., Yang, F.M., Xu, J., Hu, H., Hu, Q.H., Zhang, Y.L., et al. 2002. Determination of selenium concentration of rice in China and effect of fertilization of selenite and selenate on selenium content of rice. *J. Agric. Food Chem.* 50: 5128–5130.

Gao, J., Liu, Y., Huang, Y., Lin, Z.Q., Bañuelos, G.S., Lam, M.H., et al. 2011. Daily selenium intake in a moderate selenium deficiency area of Suzhou, China. *Food Chem.* 126: 1088–1093.

Huang, Y., Wang, Q., Gao, J., Lin, Z., Bañuelos, G.S., Yuan, L. et al. 2013. Daily dietary selenium intake in a high selenium area of Enshi, China. *Nutrients* 5: 700–701.

Laker, M. 1982. On determining trace element levels in man: The uses of blood and hair. *Lancet* 2: 260–262.

Lyons, G.H., Judson, G.J., Stangoulis, J.C.R., Palmer, L.T., Jones, J.A., & Graham, R.D. 2004. Trends in selenium status of South Australians. *Med. J. Aus.* 180: 383–386.

Rayman, M.P. 2000. The importance of selenium to human health. *Lancet*, 356, 233–241.

Research Group of Environment and Endemic Disease, Chinese Academy of Sciences. 1982. Geographical distribution of selenium content in human hair in Keshan disease and non-disease zones in China. *Acta Geogr. Sin.* 37: 136–143 (in Chinese).

Sumei Li, Gary S. Bañuelos, Longhua Wu & Weiming Shi. 2014. The changing selenium nutritional status of Chinese residents. *Nutrients* 6: 1103–1114.

Sumei Li, Gary S. Bañuelos, Ju Min, & Weiming Shi. 2015. Effect of continuous application of inorganic nitrogen fertilizer on selenium concentration in vegetables grown in the Taihu Lake region of China. *Plant and Soil*, doi: 10.1007/s11104-015-2496-3.

Tan, J., Zhu, W., Wang, W., Li, R., Hou, S., Wang, D., et al. 2001. Selenium in soil and endemic diseases in China. *Sci. Total Environ.* 284: 227–235.

Yang, G.Q., Zhou, R.H., Yin, S.A., Pu, J.H., Zhu, L.Z., & Liu, S.J. 1985. Study of selenium requirements for people. *J. Hyg. Res.* 14: 24–28.

Selenium and nano-selenium biofortified sprouts using micro-farm systems

Hassan El-Ramady & Tarek Alshaal
Soil and Water Department, Faculty of Agriculture, Kafrelsheikh University, Egypt

Neama Abdalla
Plant Biotechnology Department, Genetic Engineering Division, National Research Center, Egypt

Jószef Prokisch & Attila Sztrik
Bio- and Environmental Enegetics Institute, Nano Food Lab, Debrecen University, Hungary

Miklós Fári & Éva Domokos-Szabolcsy
Agricultural Botanics, Plant Physiology and Biotechnology Department, Debrecen University, Hungary

1 INTRODUCTION

Selenium (Se), is an essential trace element for many organisms, including humans. The trace element is becoming more and more insufficient in food crops as a result of intensive plant production in many countries (El-Ramady et al., 2014a). Due to both deficiency and toxicity problems worldwide, producing Se biofortified food crops with Se is emerging issue.

The production of safe and nutritious foods, including biofortified crops or animals can occur using such supplements as folate, vitamins (A, C, E), pro-vitamin A, carotenoids or some essential elements such as Fe, Zn, I, and Se. Regarding Se, sodium selenite and selenate are generally considered the most suitable fortificants for margarine, table salt, infant foods and sports drinks (Bonsmann and Hurrell, 2009). Concerning Se-biofortification of food crops. This strategy has been practiced in some Se-deficient regions by adding inorganic-Se containing fertilizers to soils in several countries such as Finland, the UK, China, Australia and New Zealand (Bañuelos et al., 2015).

Nanotechnologies are one of the potentially effective options of significantly enhancing the global agricultural productions needed to meet the future demands of the growing population (Liu and Lal, 2015). Due to the low efficiency of the conventional fertilizers (30–50%) and its few management options to enhance the rates, nanotechnology strategies play an important role in fertilizer research and development (De Rosa et al., 2010).

Due to the consumption of sprouts at the beginning growing phase, their nutrient content remains very high (Marton et al., 2010). Therefore, cultivation of Se-enriched crop sprout under micro-farm system has been established (El-Ramady et al., 2014b). Thus, the objective of this work is to focus on nano-Se under micro-farm system as a promising candidate for producing of Se-enriched crop sprout.

2 MATERIALS AND METHODS

Nano-Se suspension was biologically produced by *Lactobacillus acidophilus*. These bacteria were grown in sterile MRS liquid medium, containing 20 mg/l sodium selenite for 36–48 h at 37°C in aerobic environment. Nano-Se was recovered by acidic hydrolysis with HCl and after it was purified by distilled water leaching as described by Eszenyi et al. (2011). About 50 g from 4 seeds including wheat, alfalfa, radish, water cress, as well as tubers of Jerusalem artichoke were selected. These seeds were soaked into bleach solution (30%) for 15 min. Then, the seeds as well as tubers were placed in replicates of 5 into different concentrations of Se (2 and 5 mg/kg) and nano-Se (100 mg/kg) were added using the micro-farm system. Sprouts were maintained for 10 days under white fluorescent lamps (41 μmol/m^2/s photon flux density), at 24°C and 8/16 h photoperiod and then harvested (Fig. 1). The biofortified and sprouted tubers were cultivated in the field at Experimental Demonstration Garden, Debrecen University from April to November 2014.

Figure 1. Biofortification of seeds (1) and tubers (2) with Se using micro-farm system. (Photos by El-Ramady).

Table 1. Total Se contents in sprouts of cultivated plants under micro-farm conditions for 10 days.

Plant species	Se-addition form (Rate, mg/L)	Total Se (μg/g)
Radish (whole plant)	Na_2SeO_4 (2)	69.2
	Na_2SeO_4 (5)	211.0
	Nano-Se (100)	916.5
Alfalfa (whole plant)	Na_2SeO_4 (2)	89.5
	Na_2SeO_4 (5)	169.1
	Nano-Se (100)	793.5
Water cress (whole plant)	Na_2SeO_4 (2)	41.3
	Na_2SeO_4 (5)	74.9
	Nano-Se (100)	492.7
Wheat (roots + grains)	Na_2SeO_4 (2)	13.3
	Na_2SeO_4 (5)	25.5
	Nano-Se (100)	71.6
Wheat (shoots)	Na_2SeO_4 (2)	18.9
	Na_2SeO_4 (5)	61.5
	Nano-Se (100)	46.9

Table 2. Concentrations of total Se in *L. acidophilus* sprouts cultivated under micro-farm conditions for 10 compared to total Se in cultivated plant leaves grown in the field for 6 months.

	Se-treatment (mg/L)	Total Se (μg/g)
Micro-farm (sprouts)	Control	0.06
	Na_2SeO_4 (1)	7.31
	Na_2SeO_4 (2)	12.74
	Na_2SeO_4 (5)	25.37
In field (harvested leaves)	Control	0.08
	Na_2SeO_4 (1)	0.15
	Na_2SeO_4 (2)	0.24
	Na_2SeO_4 (5)	0.80

Exactly 0.2 g lyophilized sample was digested in 5 ml HNO_3 at 100°C for 45 min and then 2.5 ml H_2O_2 were added at 120°C for 45 min. The digestion solution was diluted to 10 ml with 3 M HCl. Total Se content was measured using Hydride Generation Atomic Fluorescence Spectroscopy (Millennium System, England) according to the method reported by Cabanero et al. (2004).

3 RESULTS AND DISCUSSION

The addition of inorganic and nano Se increased the Se concentrations in all sprouts and tubers produced in the micro-farm system (Tables 1, 2). The level of Se concentration was measured in the following order: radish > alfalfa > water cress > wheat > Jerusalem artichoke. Due to several advantages of micro-farm system, including a constant flow rate of the nutrient solution at a given temperature and humidity we demonstrated that this system can be used at any time of the year (with or without soil) for Se biofortification with the aid of inorganic or nano-Se.

4 CONCLUSIONS

We concluded that micro-farm system can be used automatically to produce a large variety of Se-enriched spouts and baby greens under different climate conditions with or without soil. The total Se content measured in sprouts is an acceptable example of Se biofortification. The recommended Se-dose is less than 2 mg/kg and further research is needed to study the nutritional and hormonal status within these Se-enriched plants, including antioxidants and Se-speciation.

ACKNOWLEDGMENT

The authors are grateful for the funding support from Hungarian Scholarship Board (HSB), Balassi Institute, Hungary.

REFERENCES

Banuelos, G.S., I. Arroyo, I.J. Pickering, S.I. Yang & J.L. Freeman. 2015. Selenium biofortification of broccoli and carrots grown in soil amended with Se-enriched hyperaccumulator *Stanleya pinnata*. *Food Chemistry* 166: 603–608.

Bonsmann, S.S. & R.F. Hurrell. 2009. The impact of trace elements from plants on human nutrition a case for biofortification. pp. 1–15. In: G.S. Banuelos & Z.-Q. Lin (eds.), Development and Uses of Biofortified Agricultural Products. Taylor & Francis.

Cabanero, A.I., Y. Madrid & C. Camara 2004. Selenium and mercury bioaccessibility in fish samples: an in vitro digestion method. *Anal Chim Acta* 526: 51–61.

De Rosa, M.C., C. Monreal, M. Schnitzer, R. Walsh & Y. Sultan 2010. Nanotechnology in fertilizers. *Nat. Nanotechnol.* 5: 91.

El-Ramady, H., T. Alshaal, S.A. Shehata, É. Domokos-Szabolcsy, N. Elhawat, J. Prokisch, et al. 2014. Plant nutrition: from liquid medium to micro-farm. pp. 449–508. In: E. Lichtfouse (ed.), Sustainable Agriculture Reviews, **14**, Springer International Publishing Switzerland.

El-Ramady H., Domokos-Szabolcsy É., Abdalla, N.A., Alshaal T.A., Shalaby, T.A., Sztrik, A., et al. 2014a. Selenium and nano-selenium in agroecosystems. *Environmental Chemistry Letters*, 12(4): 495–510.

Eszenyi, P., A. Sztrik, B. Babka & J. Prokisch. 2011. Elemental, nanosized (100–500 nm) selenium production by probiotic lactic acid bacteria. *Inter J Biosci Biochem Bioinform* 1: 148–152.

Liu, R. & R. Lal. 2015. Potentials of engineered nanoparticles as fertilizers for increasing agronomic productions. *Science of the Total Environment* 514: 131–139.

Marton, M., Z. Mandoki, Z. Csapo-Kiss & J. Csapo. 2010. The role of sprouts in human nutrition: A review. *Acta Univ. Sapientiae, Alimentaria*, 3: 81–117.

The standardization of selenium biofortification in China

Xuebin Yin*
Advanced Lab for Selenium and Human Health, Suzhou Institute of USTC, Suzhou, Jiangsu, China
Jiangsu Bio-Engineering Research Centre of Selenium, Suzhou, Jiangsu, China

Fei Li
Jiangsu Bio-Engineering Research Centre of Selenium, Suzhou, Jiangsu, China

1 INTRODUCTION

Functional agriculture aims to improve the functional nutrients that benefit consumers' health in China (Zhao, 2011). Selenium (Se) biofortification, which is an important part of functional agriculture, can largely improve the consumers' Se intake with better Se accumulation and increased bioavailability in Se-enhanced food products. Standards, which are designed to be used consistently as a rule or guideline, can greatly help to increase the reliability and the effectiveness of the biofortification (Garcia-Maraver et al., 2011). In this paper, current Chinese standards related to Se biofortification are reviewed and the necessity of making national and international Se-related standards are discussed.

2 NATIONAL AND MINISTERIAL STANDARDS IN CHINA

In China, the standardization system is made up of national standards, industry standards, provincial standards, and enterprise standards which are promissory or referential national wide, in one specific industry, in a particular area or inside a company, respectively. Currently, there is only one national standard and two industry standards for selenium (Se)-enriched agricultural products. To define Se-enriched rice and its processing requests, *Selenium Enriched Paddy* (GB/T 22499) was carried out in 2008 to confirm that Se-enriched paddy has a naturally accumulated Se concentrations of 0.04 mg/kg to 0.3 mg/kg.
Selenium Enriched Tea (NY/T600) and *Selenium Fortified Salt* (QB2238.3) are two industry standards that aim to increase the reliability and effectiveness of Se-enriched tea and Se-fortified salt. The former clarified that tea with a naturally accumulated Se concentration of 0.25 to 4 mg/kg can be labeled as Se-enriched tea in China, while the latter was defined as a Se-fortified table salt having a Se concentration of 3–5 mg/kg.

3 PROVINCIAL STANDARDS IN CHINA

Many regions in China are suffering from Se deficiency, while there are also several seleniferous areas that produce natural Se-enriched food (Niu & Luo, 2011; Tian & Chen, 2011). Consequently, the provincial standards are set to clarity the definitions or requests of Se-enriched food products in seleniferous areas, and to promote the development of the natural Se resources.

Four provincial standards have been published to guide the Se biofortified food production especially in areas with high Se such as Enshi city in Hubei Pivince, Ziyang city in Shanxi Province, Yichun city in Jiangxi Province, and Bama City in Guangxi province (Table 1). However, in these four seleniferous areas, the provincial standards are usually set by the local government authorities who are more concerned about the concentrations of the soil Se and the Se-enriched agricultural products produced from the areas. As a result, the Se-enriched food standards vary around the country causing some food products to be considered Se-enriched in one area but not in the other.

4 THE NECESSITY OF SETTING NATIONAL AND INTERNATIONAL STANDARDS FOR SE BIOFORTIFICATION

The standard system for Se biofortification, as well as functional agriculture lacks national and industry standards. However, based on the current provincial standard, the Se dietary intake could range over 60 μg/d (Table 2).

Taking both Chinese recommended food consumption and the selenium content in the food into consideration, a Chinese national standard for selenium enriched products is suggested to guarantee that one Chinese adult can get enough selenium intake from daily diet with avoiding the risk of selenium toxicity. With the main food in Chinese dietary pagoda and the recommended selenium content, the selenium dietary

Table 1. Four provincial standards of selenium enriched product.

Standard No.	Issued time	Name	Content
DB42/211	2002	The label of Se enriched food (Hubei)	The definition and label requests of Se enriched food produced in Hubei province
DBS42/002	2014	Selenium content request of organic Se enriched food (Hubei)	The total Se and organic Se requests of Se enriched food produced in Hubei province
DB6124.01	2012	Standard for Se enriched food (Shanxi)	The definition and Se content requests for Se enriched food in Shanxi province
DBD36T566	2009	Standard for Se enriched food (Jiangxi)	The definition and Se content requests for Se enriched food in Jiangxi province
DB45T 1061	2014	Standard for Se enriched food (Guangxi)	The definition and Se content requests for Se enriched food in Guangxi province

Table 2. The selenium dietary intake under four provincial standards.

Food consumption	Hubei	Shanxi	Jiangxi	Guangxi
Se dietary intake (μg/d)	80–320	$\geq$60	51–310	61–460
Cereal (g)	300–500			
Meat & egg (g)	125–200			
Vegetable (g)	400–500			
Fruit (g)	100–200			

Table 3. The recommended selenium concentration in the selenium enriched food in this study.

Food	Recommended Se content (mg/kg)	Recommended consumption (g/d)	Se intake (μg/d)
Cereals	0.15–0.3	300–500	45–150
Meat & eggs	0.2–0.5	125–200	25–100
Vegetables	0.01–0.50	400–500	4–25
Fruits	0.01–0.50	100–200	2–10
Total Intake	76–285		

intake could range from 76 μg/d to 285 μg/d (Table 3) which is efficiency and safety.

Because of the large spatial variations of Se background, weathering, and soil composition, it is too complex to set international or worldwide standards for selenium biofortification process in agricultural strategies, however, it's easier and more feasible to set a national or international standard for Se biofortified products to help with the consumers' selenium dietary intake and benefit their health.

ACKNOWLEDGEMENTS

This work was supported by National Natural Science Foundation of China (NNSFC31400091), Natural Science Foundation of Jiangsu Province (BK2012195, BK2012202), and Applied Basic Research Project of Suzhou (SYN201306).

REFERENCES

García-Maraver, A., Popov, V. & Zamorano, M. 2011. A review of European standards for pellet quality. Renewable Energy 36(12): 3537–3540.

Niu, C. & Luo, K. 2011. Relationship of selenium, arsenic and sulfur in soil and plants in Enshi County, China. Journal of Food Agriculture & Environment, 9(2): 646–651.

Tian, Y. & Chen, F. 2011. Effect of soil properties on selenium availability in Kashin-Beck Disease-endemic area, Ruoergai county, Sichuan, China. 2011 5th International Conference on Bioinformatics and Biomedical Engineering, 10–12 May 2011, Wuhan, China.

Zhao, Q.G. 2011. Agricultural Science and Technology in China: A Roadmap to 2050. Science Press and Springer, Beijing and Berlin, 100–126.

The chemical form of selenium in dietary supplements

G.N. George*, S.I. Yang & I.J. Pickering
Department of Geological Sciences, University of Saskatchewan, Saskatoon, Saskatchewan, Canada

1 SELENIUM DIETARY SUPPLELMENTS

Selenium (Se) is an essential element for all six kingdoms of life. Selenium's sole biological manifestation with a defined catalytic function at the active sites of enzymes is as L-selenocysteine, the highly-reactive 21st amino acid, which has several well understood roles in humans. Two other Se containing amino acids are common; L-selenomethionine and Se-methyl-L-selenocysteine. The former has no known function, and from a molecular perspective is very difficult to distinguish from L-methionine, while the latter is a plant product that aside from nutrition, has no known functional role in humans.

A visit to almost any pharmacist or health food store in North America or Europe will confirm that Se supplements are widely available, and come in a variety of different chemical forms. The current boom in Se supplements has origins with the Nutritional Prevention of Cancer (NPC) clinical trial which was active in the 1990's and used selenized baker's yeast (*Saccharomyces cerevisiae*). This trial reported an unprecedented decrease in prostate cancer (Clark et al., 1998). Subsequent clinical trials have used L-selenomethionine, with the Selenium and Vitamin E Cancer Prevention Trial (SELECT) trial being among the most notable. In contrast to NPC, SELECT found no benefit of Se (or vitamin E) for prevention of prostate cancer (Dunn et al., 2010, Kristal et al., 2014). Indeed, SELECT was prematurely terminated due an *increased* risk of cancer in the vitamin E group (Kristal et al., 2014). In a complete about-face from the recommendations arising from NPC, SELECT has prompted recommendations that men using Se supplements should stop taking them.

One potentially important difference between the NPC and SELECT trials is that while the chemical form of Se in SELECT was precisely controlled and well-defined, that of the Se in the selenized yeast NPC trial was not. Moreover, other reports of detrimental effects of Se supplementation have come from the original NPC cohort, who continued taking selenized yeast or placebo after completion of the trial (Stranges et al., 2007; Rayman & Stranges, 2013). This study found an increased cumulative incidence of type 2 diabetes in the Se group.

Thus, there is conflicting information in the literature concerning the benefits or detriments of consuming dietary Se supplements. However, different Se compounds have differing biochemical fates (Weekley et al., 2014) and toxicological profiles, so that a knowledge of which chemical form of Se is present in the supplements might help understand these complexities and inform clinical recommendations.

2 X-RAY ABSORPTION SPECTROSCOPY

X-ray absorption spectroscopy (XAS) has unique capabilities for investigation of the chemical environment of elements such as Se. A major benefit of XAS is that it requires no pre-treatment or extraction, and thus provides a tool that can probe chemical species *in situ*. X-ray absorption spectra arise from photoexcitation of a core electron, a Se 1s electron for a Se K-edge. The XAS spectrum can be arbitrarily divided into two overlapping regions – the near-edge spectrum (often called the X-ray absorption near-edge structure or XANES), which is the structured region within about 50 eV of the absorption edge, and the extended X-ray absorption fine structure (EXAFS), which forms oscillations on the high-energy side of the absorption edge and can be accurately interpreted in terms of a local radial structure. Near-edge spectra effectively provide a fingerprint of chemical type. In general the exact chemical form of a compound cannot be obtained, but rather the category of compound. Thus, L-selenomethionine and Se-methyl-L-selenocysteine both contain CH_3-Se-CH_2-groups that have essentially indistinguishable near edge spectra, whereas L-selenocysteine and L-selenocysteineate show very distinct near-edge spectra (Pickering et al., 1999). The near-edge spectrum of an unknown is conveniently analyzed by comparison with a library of spectra from standards of known structure, and mixtures can be quantitated by fitting to linear combinations of reference spectra. The sensitivity of the method is good, and at the time of writing near-edge speciation is possible in aqueous solutions at around μM concentration. An example of an XAS spectrum (that of selenophene) is shown in Figure 1 with the different regions (near-edge and EXAFS) indicated.

3 SELENIUM SPECIATION OF SUPPLEMENTS

We have used Se K-edge XAS to investigate the Se speciation of a number of Se nutritional supplements.

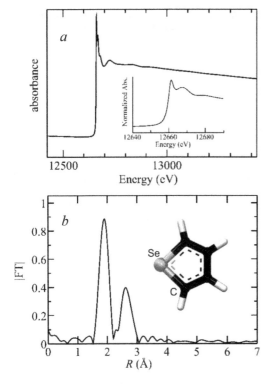

Figure 1. Selenium K-edge X-ray absorption spectroscopy (XAS) of a standard compound, selenophene. *a* shows the raw XAS spectrum with the inset showing the near edge. *b* shows the EXAFS Fourier transform, comprising a radial structure with peaks corresponding to the distances of nearby atoms, together with the known structure of selenophene in the inset.

These included a range of commercially available supplements, some of which were described as selenized yeast, some of which purportedly contained recognized Se compounds and even some that purportedly contained Se compounds that in fact do not exist. We have also examined the selenized yeast tablets used in the NPC clinical trial and the supplement tablets used in the SELECT clinical trial. The Se speciation of supplements was found to be highly variable with commercial brand and did not always correspond with the stated content. In particular, the NPC trial material contained components that have not been previously reported.

REFERENCES

Clark, L.C., Dalkin, B., Krongrad, A., Combs, G.F. Jr., Turnbull, B.W., Slate, E.H., Witherington, R., Herlong, J.H., Janosko, E., Carpenter, D., Borosso, C., Falk, S. & Rounder, J. 1998. Decreased incidence of prostate cancer with selenium supplementation: results of a double-blind cancer prevention trial. *British Journal of Urology* 81(5): 730–734.

Dunn, B.K., Richmond, E.S., Minasian, L.M., Ryan, A.M. & Ford, L.G. 2010. A nutrient approach to prostate cancer prevention: the selenium and vitamin E cancer prevention trial (SELECT). *Nutrition & Cancer* 62(7): 896–918.

Kristal, A.R., Darke, A.K., Morris, J.S., Tangen, C.M., Goodman, P.J., Thompson, I.M., Meyskens, F.L. Jr., Goodman, G.E., Minasian, L.M., Parnes, H.L., Lippman, S.M. & Klein, E.A. 2014. Baseline selenium status and effects of selenium and vitamin E supplementation on prostate cancer risk. *Journal of the National Cancer Institute* 106(3): 10.1093/jnci/djt456.

Pickering, I.J., George, G.N., van Fleet-Stalder, V., Chasteen, T.G. & Prince, R.C. 1999. X-ray absorption spectroscopy of selenium-containing amino acids. *Journal of Biological Inorganic Chemistry* 4(6): 791–794.

Rayman, M.P. & Stranges, S. 2013. Epidemiology of selenium and type 2 diabetes: can we make sense of it? *Free Radical Biology and Medicine* 65: 1557–1564.

Stranges, S., Marshall, J.R., Natarajan, R., Donahue, R.P., Trevisan, M., Combs, G.F., Cappuccio, F.P., Ceriello, A. & Reid, M.E. 2007. Effects of long-term selenium supplementation on the incidence of type 2 diabetes: a randomized trial. *Annals of Internal Medicine* 147(4): 217–223.

Weekley, C.M. & Harris, H.H. 2013. Which form is that? The importance of selenium speciation and metabolism in the prevention and treatment of disease. *Chemical Society Reviews* 42: 8870–8894.

Selenium pollution control

Phytoremediation of selenium-contaminated soil and water

Norman Terry
University of California, Berkeley, California, USA

1 INTRODUCTION

Selenium (Se) contamination of soil and water is a problem worldwide. It is encountered in arid and semiarid regions of the world that have seleniferous, alkaline soils derived from the weathering of Se-bearing rocks and shales. Selenium is also released into the environment by various industrial activities; for example, oil refineries and electric utilities generate Se-contaminated aqueous discharges. Because of the presence of excessive and potentially toxic levels of Se in the environment, it is important that we find ways of removing or detoxifying Se in Se-contaminated soil and water. Phytoremediation, using plants to remove, stabilize, or detoxify pollutants, is a promising technology to achieve this end (Terry et al., 2000).

Because Se and sulfur are chemically similar, plants are able to extract Se from soils and water into their tissues, which can be harvested and removed – a process known as *phytoextraction*. Some unique species, called Se hyperaccumulators, are naturally able to accumulate thousands of parts per million of Se. Although these hyperaccumulators are efficient Se extractors, their phytoremediation potential is often limited by their slow growth rate and low biomass. More effective Se phytoremediation has been achieved using fast-growing plant species with only moderate Se accumulation abilities, such as Indian mustard (*Brassica juncea*). An important goal of our research therefore was to combine the fast-growing ability of Indian mustard with the superior Se-accumulating ability of a Se hyperaccumulator.

Agricultural and industrial wastewaters contaminated with Se also represent a serious eco-toxic risk to humans and wildlife. Of all the technologies available for removing Se from water, constructed wetlands are the least expensive. A major interest of our laboratory therefore has been to test the efficacy of constructed wetland water treatment systems as a cost-effective technology for removing Se from large volumes of contaminated water.

2 MATERIALS AND METHODS

Indian mustard plants were genetically engineered by overexpressing ATP sulfurylase (APS) (Pilon-Smits et al., 1999), and then tested for their ability to phytoremediate Se under field conditions (Bañuelos et al., 2005). By overexpressing the gene for selenocysteine methyltransferase (SMT), as well as APS in Indian mustard, we were able to develop a double transgenic with the characteristics of a Se hyperaccumulator (LeDuc et al., 2006). In order to develop more efficient constructed wetland water treatment systems for Se removal we used wetland mesocosms to test different design parameters with respect to such factors as plant species selection, types of amendments, sediment structure and composition (Huang et al., 2012).

3 RESULTS AND DISCUSSION

3.1 *Genetic engineering of Indian mustard*

Normally, Indian mustard plants are secondary Se accumulators that accumulate Se to hundreds of parts per million. Research in our laboratory showed that transgenic Indian mustards simultaneously overexpressing APS and SMT were able to accumulate Se to ~4000 µg Se/g DW – without exhibiting toxicity – and at levels typical of Se hyperaccumulators such as *Astragalus bisulcatus* (Fig. 1). Furthermore, genetically enhanced Indian mustard plants were shown to be significantly more successful than wildtype even under real-world field conditions: transgenic Indian mustard plants overexpressing ATP sulfurylase alone enhanced Se accumulation 4- to 5-fold (Bañuelos et al., 2005).

3.2 *Use of constructed wetlands for Se removal*

Research in our laboratory has shown that constructed wetlands are reasonably efficient in the removal of selenite-Se from oil refinery wastewater (Hansen et al., 1998) and selenate-Se from agricultural irrigation drainage water (Lin & Terry, 2003). Wetlands filter Se from the inflow mostly by immobilization in the sediments as elemental Se or insoluble organic Se, but also by biologically volatilizing Se to the atmosphere. The latter pathway is particularly desirable because Se is removed from the wetland ecosystem, thereby minimizing Se buildup and potential Se ecotoxicity to local wildlife.

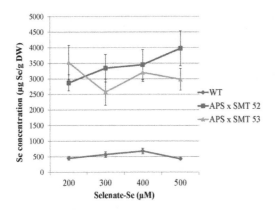

Figure 1. Two APS × SMT double transgenic lines accumulated up to 4000 μg Se/g DW in Indian mustard seedlings compared to ∼500 μg Se/g DW for wild type plants (WT).

Although constructed wetlands have been shown to remove from 60 to 90% of the Se from the inflow, depending on whether Se is present as selenate or selenite, Se concentration in the outflow remains at a stubborn 2 to 5 μg/L – a level that is potentially high enough to raise concerns of Se ecotoxicity. How then, can we improve wetland design to reduce Se levels in the outflow to say, less than 1 μg/L? Using wetland mesocosms to test different design parameters we were able to reduce Se concentration in the outflow from 15 to 0.1 μg Se/L in 72 h (Huang et al., 2012). To further enhance Se removal, especially through biological volatilization, we proposed adding an algal treatment cell to the wetland water treatment system (Huang et al., 2013). Thus, using this approach much of the inflow Se could potentially be volatilized by algae and wetland plants, thereby attenuating the risk of the buildup of potentially eco-toxic forms of Se within the wetland ecosystem.

4 CONCLUSIONS

The ability of plants to absorb, sequester, and volatilize Se has important implications in the management of environmental Se contamination by phytoremediation. Under dryland conditions, Indian mustard has been identified as a plant species that has a superior ability to accumulate and volatilize Se – characteristics that can be further enhanced through appropriate genetic manipulation. Indian mustard has other advantages in that it grows rapidly on Se-contaminated soil, tolerates salinity, and provides a safe source of forage for Se-deficient livestock at low Se concentrations. With respect to the treatment of Se contaminated wastewaters, constructed wetlands have been shown to be effective for the treatment of Se-contaminated wastewater. However, our studies with wetland mesocosms have shown that there is considerable potential to improve the efficiency of Se by the judicious selection of certain wetland design parameters, and possibly also by the inclusion of an algal pretreatment cell.

REFERENCES

Bañuelos, G., Terry, N., LeDuc, D., Pilon-Smits, E.A.H., & Mackey, B. 2005. Field trial of transgenic Indian mustard plants shows enhanced phytoremediation of selenium-contaminated soil. *Environ. Sci. Technol.,* 39(6): 1771–1777.

Gao, S., Tanji, K.K., Lin, Z.Q., Terry, N. & Peters, D.W. 2003. Selenium removal and mass balance in a constructed flow-through wetland system. *J. Environ. Quality* 32 (4): 1557–1570.

Hansen, D., Duda, P.J., Zayed, A. & Terry, N. 1998. Selenium removal by constructed wetlands: role of biological volatilization. *Environ. Sci. Technol.* 32 (3): 591–597.

Huang, J.C., Passeport, E. & Terry, N. 2012. Development of a constructed wetland water treatment system for selenium removal: use of mesocosms to evaluate design parameters. *Environ. Sci. Technol.* 46 (21): 12021–12029.

Huang, J.C., Suárez, M.C., Yang, S.I., Lin, Z.Q., & Terry, N. 2013. Development of a constructed wetland water treatment system for selenium removal: Incorporation of an algal treatment component. *Environ. Sci. & Technol.*, 47: 10518–10525.

LeDuc, D.L., AbdelSamie, M., Móntes-Bayon, M., Wu, C.P., Sarah, J. Reisinger, S.J. et al. 2006. Over-expressing both ATP sulfurylase and selenocysteine methyltransferase enhances selenium phytoremediation traits in Indian mustard. *Environmental Pollution*, 144: 70–76.

Lin, Z.-Q. & Terry, N. 2003. Selenium removal by constructed wetlands: quantitative importance of biological volatilization in the treatment of selenium-laden agricultural drainage water. *Environ. Sci. Technol.*, 37(3): 606–615.

Pilon-Smits, E.A.H., Hwang, S., Lytle, C.M., Zhu, Y.L., Tai, J.C., Bravo, R.C., et al. 1999. Over-expression of ATP sulfurylase in *Brassica juncea* leads to increased selenate uptake, reduction, and tolerance. *Plant Physiol.,* 119: 1–10.

Terry, N., A.M. Zayed, M.P. de Souza & A.S. Tarun. 2000. Selenium in higher plants, *Ann. Rev. Plant Physiol. Plant Mol. Biol.*, 51: 401–32.

Microbe-assisted selenium phytoremediation and phytomanagement of natural seleniferous areas

M. Yasin
Department of Microbiology and Molecular Genetics, University of the Punjab, Quaid-e-Azam Campus, Lahore, Pakistan
Colorado State University, Biology Department, Fort Collins, CO, USA

M. Faisal
Department of Microbiology and Molecular Genetics, University of the Punjab, Quaid-e-Azam Campus, Lahore, Pakistan

A.F. El Mehdawi & E.A.H. Pilon-Smits
Colorado State University, Biology Department, Fort Collins, CO, USA

1 INTRODUCTION

Natural seleniferous areas are found worldwide and are generally considered unsafe to use as agricultural land to grow crops because of the high concentration of selenium (Se) in the soil. Selenium is an important environmental contaminant when present at high concentration in food and it could be toxic for animals, humans and plants. Seleniferous areas (containing > 0.5 mg/kg) are particularly prevalent in China, India, Australia, Ireland and the U.S.A. (Dhillon et al., 2003). To avoid Se toxicity and reduce the impact of naturally occurring Se on the biological environment, extensive research has been conducted in the western United States and a vast amount of resources have been allocated to develop management strategies and remediation technologies (Banuelos et al., 2002). Selenium toxicity and deficiency can exist side by side, due to soil Se heterogeneity. Thus, Se removed from seleniferous areas may be beneficial for adjacent low-Se areas.

All plant species can take up and accumulate various levels of Se in their above and below ground parts. Moreover, excessively high Se levels in soil can be toxic for plant species and may result in reduced plant growth and biomass production. Most plant species are non-accumulator (e.g. grasses) and seldom accumulate >50 mg/kg (DW). However, some plant species can accumulate very high levels of Se (>4 g/kg DW) in their above ground parts (shoots, leaves, stems, flowers, seeds and pollens) without showing any visible signs of Se toxicity. These plants are called Se hyperaccumulators (e.g., *Stanleya pinnata* and *Astragalus bisulcatus*) and are usually found in natural seleniferous areas. Members of Brassicaceae family such as Indian mustard (*Brassica juncea*) are also well known for their accumulation potential as secondary accumulators (10–1000 mg/kg DW) but are not hyperaccumulators.

Phytoremediation (a plant-based technology) has received much attention as a low cost, environmentally friendly approach for managing the toxic effects of Se in the natural and artificially Se contaminated environments. To improve Se phytoremediation efficiency, an active area of research is the application of genetic engineering for the production of transgenic plants with enhanced phytoremediation capacity. For Se, promising transgenics are those overexpressing the enzymes ATP sulfurylase (APS) and selenocysteine methyltransferase (SMT), which are involved in Se assimilation and selenocysteine methylation, respectively (LeDuc et al., 2006). A drawback of these approaches is that the release of new transgenics in the natural environment is tightly regulated. In this study we explore an alternative approach: the application of Se accumulator plants in combination with naturally occurring Se-resistant microbes that may promote Se phytoremediation by improving plant growth and Se uptake. This approach is not only simple and economical but also an environmentally friendly approach to clean up natural seleniferous areas. In this regard, *Brassica juncea* was tested for its capacity to detoxify and extract Se from natural seleniferous shale rock derived soil, both with and without microbial consortia.

2 MATERIALS AND METHODS

Indian mustard plants (*B. juncea*, a secondary Se accumulator) were used in combination with several Se resistant bacterial strains *Bacillus* sp. to extract Se from seleniferous soil collected adjacent to Se hyperaccumulator plants (*Astragalus bisulcatus*) from a natural seleniferous area (Pine Ridge Natural Area) in West Fort Collins, CO. The *B. juncea* plants were grown in this natural seleniferous soil under greenhouse conditions both in the presence and absence of

Figure 1. Comparison of growth of three-week old *B. juncea* plants grown in seleniferous and non-seleniferous potting soil rich in peat moss (A). In natural seleniferous soil (B) comparison of growth of bacterial inoculated *B. juncea* compared to un-inoculated control plants.

Table 1. The concentrations (mg/kg DW) of Se, Ca and S in leaves of inoculated *B. juncea* plants compared to control grown in shale rock derived natural seleniferous soil. Mean followed by different letters in a column for the same parameter are significantly different ($p < 0.05$, $n = 5$). Data are mean ± SE.

Treatments	Selenium	Calcium	Sulfur
Control	780 ± 75^a	33835 ± 4174^a	21490 ± 1560^a
Inoculated plants	805 ± 30^a	51755 ± 2070^b	25707 ± 447^b

Se resistant microbes collected from a polluted area in Lahore, Pakistan. Analysis for Se and other nutrient elements levels in soil and various plants parts was carried out by ICP-OES, after acid digestion. The effects of Se on plant growth and non-protein thiol levels were recorded during early growth and at maturity.

3 RESULTS AND DISCUSSION

In shale rock-derived natural seleniferous soil, *B. juncea* plants exhibited reduced growth and delayed onset of reproductive phase compared to plants grown in non-seleniferous potting soil rich in peat moss (Fig. 1). This observation may result from a combination of generally less favorable soil properties in the seleniferous soil and Se-specific effects. Selenium concentration was 8.2 mg/kg in the tested soil. The Se levels in the *B. juncea* plants grown on this seleniferous soil were around 800 mg/kg (DW), which may be enough to cause toxicity and reduce growth, as reported by Prins et al. (2011). For non-accumulators, e.g., *Zea mays*, Se can already be toxic around 50 mg/kg DW, while Se hyperaccumulator plants (e.g., *Astragalus bisulcatus*) can accumulate 1,000–15,000 mg/kg DW without experiencing toxicity. These plants circumvent Se toxicity by methylating SeCys via the enzyme SeCys methyl-transferase (SMT), which does not get incorporated in proteins (Neuhierl & Bock, 1996).

The reduced growth of *B. juncea* in seleniferous soil could be ameliorated when the plants were inoculated with Se-resistant bacterial consortia. The plants exhibited enhanced growth as measured from increased seed weight, shoot length, shoot fresh and dry biomass and earlier onset of reproductive phase compared to un-inoculated control plants (Fig. 1B). Inoculation with bacterial consortia resulted in a remarkable increase in non-protein thiol content in plant leaves. Plant growth promoting rhizobacteria (PGPR) can stimulate plant growth through various direct and indirect mechanisms e.g. by synthesis of phytohormones (auxins), nitrogen fixation, phosphorous solubilization, synthesis of antioxidant substances and enhanced mineral uptake through improved plant root growth.

Bacterial inoculation may also enhance uptake of certain elements by stimulation of plant root growth. In seleniferous soil, the Indian mustard plants accumulated very high concentration of Se in leaves and seeds (>800 mg/kg); inoculation with bacterial consortium did not cause affect Se uptake and accumulation in leaves. However, bacterial inoculation may still enhance the capacity of *B. juncea* accumulate Se, by promoting biomass production. Moreover, inoculated *B. juncea* plants showed significant increases in calcium (Ca) and sulfur (S) concentration in leaves compared with un-inoculated plants (Table 1).

4 CONCLUSIONS

B. juncea and other Se accumulating crops, in association with Se tolerant bacteria, can be used to farm seleniferous soils that would otherwise be considered undesirable for farming. Over time, the seleniferous soil may become more amenable to farming other crop species. The high concentration of Se in *B. juncea* leaves (bioconcentration factor = 100) and other parts, show the potential of this species for Se phytoremediation as well as Se biofortification.

REFERENCES

Bañuelos, G.S., Lin, Z.Q., Wu, L., & Terry, N. 2002. Phytoremediation of selenium-contaminated soils and waters: fundamentals and future prospects. *Rev Environ Health*, 17: 291–306.

Dhillon, K.S., & Dhillon, S.K. 2003. Distribution and management of seleniferous soils. *Advances in Agronomy*, 79: 119–184.

LeDuc, D.L., AbdelSamie, M., Móntes-Bayon, M., Wu, C.P., Reisinger, S.J., & Terry, N. 2006. Overexpressing both ATP sulfurylase and selenocysteine methyltransferase enhances selenium phytoremediation traits in Indian mustard. *Environmental Pollution*, 144: 70–76.

Neuhierl, B., & Bock, A. 1996. On the mechanism of selenium tolerance in selenium accumulating plants: purification and characterization of a specific selenocysteine methyltransferase from cultured cells of *Astragalus bisulcatus*. *European Journal of Biochemistry*, 239: 235–238.

Prins, C.N., Hantzis, L.J., Quinn, C.F., & Pilon-Smits, E.A.H. 2011. Effects of selenium accumulation on reproductive functions in *Brassica juncea* and *Stanleya pinnata*. *Journal of Experimental Botany*, 62: 5633–5640.

Phytoremediation of selenium contaminated soils: strategies and limitations

Karaj S. Dhillon* & Surjit K. Dhillon
Department of Soil Science, Punjab Agricultural University, Ludhiana, India

1 INTRODUCTION

Selenium (Se) poses a unique environmental problem. In trace amounts, it is essential to human and other animal health but can be harmful at slightly greater concentrations. Strong relationships exist between high concentrations of Se in soils and Se toxicity to plants, animals and humans, as reported in several countries, including USA, China and India (Dhillon & Dhillon, 2003). For reducing Se in soils to safer levels, a phytoremediation strategy was advocated by Parker & Page (1994). Recently, plant biotechnology has created several new avenues for determining more precise and accurate process of phytoremediation of pollutants (Adki et al., 2014). During the last two decades, a number of phytoremediation alternatives have been proposed, including phytoextraction/phytoaccumulation (plant uptake and assimilation of Se), phytofiltration (plant roots removing contaminants), phytovolatilization (release of volatile Se species by plants) and phytodegradation (degradation of contaminants by plants). Among these strategies, phytoextraction has been successfully employed in a number of pilot and full-scale field demonstrations for decontaminating Se polluted soils. Reduction in total soil Se has been recorded up to 40% under greenhouse conditions and up to 20% under field conditions after one growing season. Under field situations, Se removal from the soil is generally more than that removed through the process of phytoextraction because microbiology leaching and volatilization (through soil and plants) also contribute to the process of phytoremediation. Phytoextraction is essentially a process of concentrating Se from one phase (soil) to another phase (plant). Although it has developed rapidly into an efficient, environment friendly and low cost technology, the process has some limitations.

2 LIMITATIONS

(1) In highly contaminated soils, the time involved in reducing Se to safe levels is the most limiting factor. If a phytoremediation crop is able to remove 1 kg Se/ha/year from the contaminated site having 50 kg Se/ha in the soil profile up to 120 cm, the strategy may take more than 50 years to reclaim that site. Also during the phytoremediation process, appreciable quantity of Se may continue to leach beyond the rooting zone, leading to contamination of underground water bodies.

(2) In some situations, the farmer simply cannot sacrifice a crop or growing season just for the process of phytoremediation. In addition, due to senescence, Se-rich leaves fall on the ground and cannot be removed along with harvested biomass at maturity. The amount of Se contained in the leaves (200–350 g Se/ha) thus gets redeposited into the soil and thereby lowering the efficiency of phytoextraction.

(3) Safe disposal of Se-rich biomass is one of the important limitations associated with phytoremediation of seleniferous soils. Harvesting of the phytoremediation crops results in the accumulation of Se-rich plant material ranging from 5 to 10 tons/ha/year. Disposal of Se-contaminated material without proper treatment may cause increased selenium's exposure to human, animals and the environment. Careful blending of Se-rich plant biomass with Se-deficient material and its utilization for animal and human consumption may serve as an important alternative.

(4) Plants while growing are known to transfer Se across the media i.e. from soil to air, which is known as phytovolatilization and thus complementing phytoextraction process in removing Se from the contaminated sites (Terry & Zayed, 1994). In most of the studies reported, this factor has not been considered.

(5) Phytoremediation of Se contaminated soil may be influenced by higher trophic levels including insects. Presence of insects may improve phytoremediation by increasing biotransformation of inorganic Se and release of volatile Se species.

(6) Phytoremediation appears to be low cost technology when considering the initial implementation requirements because the agronomic technology for the cultivation of trees/crops is already well established. The cost of technology must include the expenditure on growing crops on a long-term basis and the disposal of toxic phytoremediation material.

(7) Phytoremediation may not become fully acceptable to the actual user until the crop attains economic value. In order to off-set a part or whole

of the costs involved in cleaning up the toxic sites, the Se-rich biomass must be valued according to its Se content.

Due to several limitations involved in phytoremediation of Se-rich soil, there are limited investigations on reclaiming large contaminated areas. More field research is ultimately necessary on a large scale to help validate the phytoremediation of Se as a viable strategy.

3 CONCLUSIONS

In spite of several limitations, phytoremediation continues to gain acceptance as a tool for managing metal-contaminated sites. Among the available alternatives, phytoextraction has been successfully employed for decontaminating Se polluted soils. Although so far it work well in soils having light and shallow contamination, researchers need to work on decreasing the length of time for phytoremediation, protecting wildlife from feeding on remedial plants, increasing plant uptake capacity, tolerance and biomass through conventional/genetic manipulations.

REFERENCES

Adki, V.S., Jadhav, J.P. & Bapat, V.A. 2014. At the cross roads of environmental pollutants and phytoremediation: a promising bio remedial approach. Journal of Plant Biochemistry and Biotechnology 23: 125–140.

Dhillon, K.S. & Dhillon, S.K. 2003. Distribution and management of seleniferous soils. Advances in Agronomy 79: 119–184. Academic Press, London.

Parker, D.R. & Page, A.L. 1994. Vegetation management strategies for remediation of selenium contaminated soils. In: W.T. Frankenberger, Jr. & S. Benson (eds) Selenium in the Environment. pp. 327–342. Marcel Dekker Inc., New York.

Terry, N. & Zayed, M. 1994. Selenium volatilization by plants. pp. 343–367. In: W.T. Frankenberger & S. Benson (eds) Selenium in the Environment. Marcel Dekker Inc., New York.

Enhanced selenium removal and lipid production from wastewater by microalgae with low-energy ion implantation

Zhilin Wu & Miao Li*
Key Laboratory of Agri-Food Safety of Anhui Province, School of Resources and Environment & School of Plant Protection, Anhui Agriculture University, Hefei, Anhui, China

Mei Zhu & Lei Qiu
School of Engineering, Anhui Agriculture University, Hefei, Anhui, China

1 INTRODUCTION

The release of large amounts of Selenium (Se) into the environment from industrial and agricultural activities can lead to bioaccumulation Se into the food chain. High concentrations of Se are known to be toxic to organisms. Hence, remediation is important and some approaches have been proposed to deal with Se contamination in soil and water (Huang et al., 2013). Recent investigations have shown that phytoremediation, which utilizes algae for the removal of Se compounds from the soil and water could be a viable option (Huang et al., 2013).

The use of microalgae in bioremediation of heavy metal pollution has attracted great interest due to their central role in carbon dioxide fixation recently (Park et al., 2011). In addition, the algae biomass has great potential as feedstock for biofuel production. Phytoremediation of domestic wastewater and algae biomass production for sustainable biofuel production are useful for environmental sustainability (Park et al., 2011). The integration of microalgae-based biofuel and bioproducts production with wastewater treatment has major advantages for both processes (Park et al., 2011). However, there are major challenges to the breeding of microalgae in a way that allows for upstream processing to produce biofuels and other bioproducts of value. In this study, the feasibility of application of ion implantation for enhancing Se removal efficiency and lipid production from wastewater by microalgae is explored. Low-energy ion implantation is a technology used for effective production of genetic mutants (Li et al., 2014). This study demonstrates that algal breeding using a green colonial microalgae, *Botryococcus braunii*, by ion beam mutagenesis can improve Se removal efficiency and lipid yield in laboratory experiments.

2 MATERIALS AND METHODS

B. braunii strain was isolated in Chaohu Lake, China. The wild type (WT) and mutant seed cultures were grown in BG-11 culture medium in a column photobioreactor (60 cm high, diameter 4 cm, 300 mL culture volume) under continuous illumination of a low light intensity. Culture mixing was provided by aeration with compressed air containing 2% CO_2. After 4 days, the seed culture were collected by centrifugation (3200 × g, 5 min) and transferred into 300 mL or 10 L nitrogen-limited BG-11 medium (4.25 mM $NaNO_3$) in column or panel photobioreactors, respectively. The initial OD_{750} of algal culture was about 0.2. WT and mutant cultures were grown under continuous illumination with aeration containing 2% CO_2 for two weeks. All strains were cultured in 3 replicates at room temperature.

B. braunii strain was maintained in BG-11 medium in 100 mL Erlenmeyer flask under low light illumination at room temperature for 3 days. Algal cells at the exponential growth phase were collected by centrifugation (3200 × g, 3 min) and washed with sterile water, then re-suspended in fresh BG-11 medium. The cell concentration was adjusted to 1×10^6 cells/mL, and exposed to N^+ ion beam provided by the Low Energy Ion Implantation Facility at Hefei, Institute of Plasma Physics, Chinese Academy of Sciences. Irradiation treatments were conducted at 10 keV, 0–9 × 10^{16} N^+ ions/cm^2, and there were at least three algae samples for every dose treatment.

After irradiation, algae cells were plated on BG-11 agar plates in triplicate and cultured at 25°C under low light until algal colonies occurred on the plates. The colonies derived from the irradiated cells were selected and transferred to BG-11 agar plates several times to obtain purified monoclonal strains, which

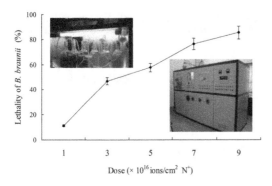

Figure 1. Effects of irradiation on the cells viability of *B. braunii*.

Table 1. Composition of total lipid in *B. braunii*.

Lipid contents	WT	IPP-AA68
Neutral lipid (%)	69.31 ± 1.46	78.86 ± 1.18
Glycolipid (%)	11.62 ± 0.87	8.78 ± 0.66
Phospholipid (%)	5.42 ± 0.58	1.98 ± 0.32

were regarded as putative mutants and constituted the mutant library. Total lipids were extracted and quantified according to the Bigogno's method with minor modifications (4). A t-test was applied to ascertain significant differences using SPSS version 10.0 (SPSS Inc., Shanghai, China) and the level of statistical significance was $p < 0.05$.

3 RESULTS AND DISCUSSION

The lethality of *B. braunii* after N^+ ion beam irradiations was shown in Figure 1. The relationship between the lethality of *B. braunii* and the irradiation dose of N^+ beam was fitted to a logistic curve equation, which indicated the death rate of cells increased with increasing the radiation dosage. After ion beam treatment, the algae colonies that appeared on agar plates were considered putative mutants. A preliminary screening of the putative mutants by light microscopy showed that the morphological characteristics of the putative mutants were indistinguishable from the wild type cells. Then photosynthetic characteristics were used as alternative parameters to further characterize the putative mutants by using a chlorophyll fluorescence technique.

All colonies were also subjected to screening for lipid-overproduction mutants using a modified Nile red method in conjunction with the determination of growth rate, as indicated by optical density of algal culture measured at 750 nm. The mutants obtained through the first round screening with Imaging-PAM and Nile red fluorescence method would be subjected to the next rounds of screening. Among these mutants, a wide range of phenotypic distribution of PEMs was observed and approximately in line with the normal distribution. In the following screening procedures, the lipid contents and photosynthetic efficiency of mutants were assayed when they were cultured in a column bioreactor. Lipid contents of many PEMs was significantly different with WT under stress conditions, which indicated that ion irradiation induced mutations of unicellular microalgae had a wide spectrum and a high frequency as observed in plants and mammalian cells (Table 1). After 5–6 rounds of the screening process mentioned above, a mutant named IPP-AA68 with higher photosynthetic efficiency and lipid contents than WT was obtained. In a batch culture mode, once inoculated in a BG-11 medium and exposed to light illumination, *B. braunii* grew rapidly for 4–6 days after a temporary lag phase (0–24 hours), and reached a stationary phase thereafter. During the first 3 days, almost all nitrates in the medium were consumed by algal cells. Under low light conditions, the mutant also showed the same biomass profile as wild type.

4 CONCLUSIONS

Our results showed that after irradiation by N^+ ion beam of 10 keV, 5×10^{16} N^+ ions/cm^2, followed by screening of resulting mutants on 24-well microplates, more than 100 mutants were obtained. One of those, named IPP-AA68, exhibited higher nitrogen and phosphorus removal efficiencies and lipid productivity of 0.259 g/L per day; 18.9% higher than wild type, respectively. This work demonstrated that the low-energy ion implantation combined with high-throughput screening is an effective means for microalgae could be considered as a trait improvement approach for simultaneous Se removal and lipid production for biofuel feedstock from wastewater.

REFERENCES

Huang, J.C., Suárez, M.C., Yang, S.I., Lin, Z.Q. & Terry, N. 2013. Development of a constructed wetland water treatment system for selenium removal: incorporation of an algal treatment component. *Environmental Science & Technology* 47(18): 10518–10525.

Li, M., Sheng, G.P., Wu, Y.J., Yu, Z.L., Bañuelos, G.S. & Yu, H.Q. 2014. Enhancement of nitrogen and phosphorus removal from eutrophic water by economic plant annual ryegrass (*Lolium multiflorum*) with ion implantation. *Environmental Science and Pollution Research* 21(16): 9617–9625.

Park, J.B.K., Craggs, R.J. & Shilton, A.N. 2011. Wastewater treatment high rate algal ponds for biofuel production. *Bioresource Technology* 102(1): 35–42.

Simultaneous production of biofuels and treatment of Se-laden wastewater using duckweed

Zhilin Wu & Miao Li*
Key Laboratory of Agri-Food Safety of Anhui Province, School of Resources and Environment–School of Plant Protection, Anhui Agriculture University, Hefei, Anhui, China

Lei Qiu & Mei Zhu
School of Engineering, Anhui Agriculture University, Hefei, Anhui, China

1 INTRODUCTION

Selenium (Se) is important trace element with key physiological roles in oxidative stress and immune function. However, the release of large amounts of Se into the environment from industrial and agricultural activities can lead to bioaccumulation Se into the food chain (Bañuelos *et al.*, 2010). High concentrations of Se are known to be toxic to many forms of living organisms (Bañuelos *et al.*, 2010). Hence, remediation is important and a variety of approaches have been proposed to deal with Se contamination in soil and water (Bañuelos *et al.*, 2010). Recent investigations have shown that phytoremediation, which utilizes plant for the removal of Se compounds from the soil and water could be a viable option (Bañuelos *et al.*, 2010).

Duckweed plants are widely used for the phytoremediation of water pollution and represent an attractive feedstock for biofuel production (Cheng & Stomp, 2009). The substantial biomass of duckweed plant together with the relatively high levels of starch, cell wall carbohydrates and lipids make them the feedstock of choice for bioethanol production (Xu & Shen, 2011). Pyrolysis of *S. oligorrhiza* plant biomass produces a different range of bio-oil components that can potentially be used for the production of green gasoline and diesel fuel supplement or as a glycerine-free component of biodiesel. Other identified bio-oil components from pyrolysis of *S. oligorrhiza* plant biomass can be converted into petrochemicals using catalytic hydrodeoxygenation. In the present study, the feasibility of using duckweed plants for Se-laden wastewater treatment and production of biofuel is explored.Results may show an attractive, ecologically friendly, and cost-effective solution for efficient treatment of Se-laden wastewater, renewable energy, and petrochemicals production from generated duckweed plant biomass.

2 MATERIALS AND METHODS

Duckweed was acquired from the campus of Anhui Agricultural University, China. Plants were collected, rinsed in deionized water and any unwanted debris removed. Experiments were carried out in plastic containers (250 mL) filled with 200 mL of Se-laden wastewater (SLW). Control experiments were conducted in sterile 1/2 Hoagland's medium. The samples were placed in growth chambers ($23 \pm 2°C$) with 16 h photoperiod provided by three fluorescent tubes. Three replicates were included for each treatment. Destructive sampling was conducted to evaluate the Se concentrations in plants at day 6. Growth was monitored every two days test period by weighting fresh weight.

Water samples from different dilutions of SLW were collected at the beginning and at the end of the experimental period. Plants from each treatment were blotted on filter paper and dried at 70°C overnight. Plant extracts and wastewater were analyzed for total selenium concentration by inductively coupled plasma mass spectrometry (ICP-MS).

Pyrolysis experiments were carried out in a quartz tube reactor. Samples were dried at 110°C overnight before the experiment was initiated. Heating of the quartz tube reactor was carried out using a tube furnace. Two thermocouples were used to monitor pyrolysis temperature. The furnace was vertically aligned, so that the liquid products dripped into the condenser. After condensation of the liquid product, the gas passed through a glass wool filter before being collected in a gas-sampling bag. The condenser was weighed before and after the reaction to obtain the weight of bio-oil collected. The solid product biochar was dislodged from the pyrolysis reactor after the experiment and weighed. The volume of the pyrolysis gas was measured and the gas sample was analyzed by GC with thermal conductivity detector for permanent

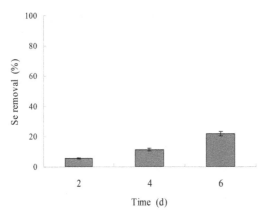

Figure 1. Se removal efficiency from SLW by duckweed plant.

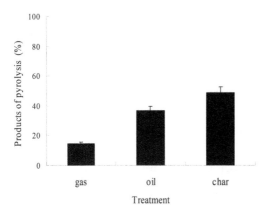

Figure 2. Production of bio-oil, bio-gas and biochar after pyrolysis of duckweed plant.

gases and flame ionization detector for hydrocarbon gases while the bio-oil of pyrolysis was analyzed by GC-MS.

All treatments in this study were conducted in triplicate. The experimental data were subjected to the one-way analysis of variance (ANOVA). Tukey simultaneous tests were conducted to determine the statistical differences between treatments ($p < 0.05$).

3 RESULTS AND DISCUSSION

The ability of duckweed plant to remove Se from the SLW is shown in Figure 1. Duckweed plant showed the capacity to reduce the concentration of Se in wastewater and accumulate Se in their tissues. Growing in SLW duckweed showed up to 20% of Se removal after 6 days. This high efficiency of phytoextraction can be explained by higher tolerance of duckweed to SLW and also by its higher capacity for Se uptake.

The distribution of biomass pyrolysis products: gas, liquid and solids (char) is shown in Figure 2. No significant differences were observed in the distribution of gas, liquid and solid products between duckweed. These plants showed good mass balances of the major pyrolysis products: gas, liquid and solids (char) and their distribution. Char was the main product of duckweed and elodea pyrolysis, whereas, water clover produced higher yield of the liquid fraction with correspondingly lower yield of gaseous and solid products.

4 CONCLUSIONS

Duckweed is effective at biofiltration of Se from Se-laden wastewater via accumulating Se in its fast growing biomass. Along with production of bio-gas and bio-solid components, pyrolysis of duckweed plant produced a range of liquid petrochemicals, which can be directly used as a diesel fuel supplement or as a glycerine-free component of biodiesel. Other identified bio-oil components can be converted into petrochemicals using existing techniques. A dual application of duckweed plant for wastewater treatment and production of biofuel and value-added chemicals offers an ecologically friendly and cost-effective solution for water pollution control and bioenergy production.

REFERENCES

Bañuelos, G..S., Roche, J.D. & Robinson, J. 2010. Developing selenium-enriched animal feed and biofuel from canola planted for managing Se-laden drainage waters in the westside of Central California. *International Journal of Phytoremediation* 12(3): 243–254.

Cheng, J.J. & Stomp, A.M. 2009. Growing duckweed to recover nutrients from wastewaters and for production of fuel ethanol and animal feed. *Clean–Soil, Air, Water* 37(1): 17–26.

Xu, J.L. & Shen, G.X. 2011. Growing duckweed in swine wastewater for nutrient recovery and biomass production. *Bioresource Technology* 102: 848–853.

Author index

Abdalla, A.L. 159
Abdalla, N. 189
Abdur, R. 47
Abreu, L.B. 15
Achouba, A. 65, 173
Alfthan, G. 155
Alshaal, T. 189
Ander, E.L. 151
Andrade, M.J.B. 185
Anouar, Y. 61
Arai, K. 49
Araujo, A.M. 23, 25, 27
Arnér, E.S.J. 59
Atarodi, B. 91
Avila, F.W. 185
Ayotte, P. 65, 173

Bai, G. 113
Bamberg, S.M. 137
Bao, Z.Y. 29, 37, 139, 141
Barcelos, J.P.Q. 175, 177
Batista, K.D. 17
Bauer, J. 129
Bañuelos, G.S. 103, 147, 181, 187
Behra, R. 9
Bertechini, A.G. 169
Bissardon, C. 77
Biu Ngigi, P. 101
Blazina, T. 7
Bobe, G. 171
Bohic, S. 77
Boldrin, P.F. 185
Boukhalfa, I. 61
Broadley, M.R. 19, 151

Campos, L.L. 159
Cappa, J.J. 115, 125
Cárdenas, M.E. 167
Carneiro, M.A.C. 137
Carvalho, G.S. 15
Chan, L. 65, 173
Charlet, L. 19, 77
Chaudhary, R.J. 93
Chawla, R. 93
Chen, H. 157
Chen, S. 39
Chen, X.L. 29, 37, 141
Chi, Feng-qin 21
Chilimba, A.D.C. 19, 149
Cochran, A.T. 125, 129
Correa, L.B. 97
Couture, R.M. 19

Cozzolino, S.M.F. 97
Cui, Z.W. 99
Cunha, J.A. 97

da Silva Júnior, E.C. 133
da Silva, K.E. 17, 133
da Silva, L.M. 17
de Andrade, M.J.B. 165
de Figueiredo, M.A. 165, 185
de Lima, R.M.B. 133
de Oliveira Júnior, R.C. 17
de Souza, G.A. 133
Deng, Y. 181
Dhanjal, N.I. 123
Dhillon, K.S. 93, 201
Dhillon, S.K. 201
Domingues, C.R.S. 183
Domokos-Szabolcsy, É. 189
Du Laing, G. 79, 101
Du, B. 157
Dumas, P. 173
Dumas, P.Y. 65
Dumesnil, A. 61

Ekholm, P. 155
El Mehdawi, A.F. 115, 125, 199
El-Ramady, H. 189
Eurola, M. 155
Evangelista, A. 183

Faisal, M. 199
Fan, S.X. 107
Faquin, V. 185
Fári, M. 189
Faria, L.A. 159
Fedirko, V. 53
Feng, P.Y. 31
Fernandes, J.B.K. 87
Fernandez Martinez, A. 19
Figueiredo, F.M. 169
Fotovat, A. 91
Franco, F.A. 183

Gautrin, L. 173
George, G.N. 43, 193
Goda, J. 51
Gota, V. 51
Gu, A.Q. 141
Guedes, M.C. 17, 133
Guilherme, L.R.G. 15, 23, 25, 27, 133, 165, 185

Guo, Y.B. 161
Gupta, R.C. 83

Hall, J.A. 171
Han, D. 33, 41
Han, Y.Y. 107
Harouki, N. 61
He, W. 157
Henry, J.-P. 61
Hesketh, J.E. 53
Hoffmann, P.R. 67
Hosseinkhani, B. 79
Hou, Q.Y. 5
Hou, Y. 147
Huang, J. 99
Huang, Z. 47
Hughes, D.J. 53
Hybsier, S. 53

Imtiaz, M. 41
Iwaoka, M. 49

Jain, V.K. 51
Jamil, Z. 87
Jayne, T.S. 117
Jenab, M. 53
Jiang, H. 105
Johnson, T.M. 89
Jones, G.D. 13
Jones, J.S. 53
Jones, L. 11
Joy, E.J.M. 151
Joy, E.M. 149
Júnior, N.J.M. 17

Kamogawa, M.Y. 159
Kaniz, F. 35
Karp, F.H.S. 159
Khan, I. 77
Khan, K.U. 87
Kirchner, J. 7
Korbas, M. 43
Krone, P.H. 43
Kuang, En-jun 21
Kumssa, D.B. 151
Kunwar, A. 49, 51
Kwan, M. 65, 173

Lachat, C. 101
Laird, B. 65, 173
Lavu, R.V.S. 79
Lemire, M. 65, 173

Lessa, J.H.L. 23, 25, 27
Li, F. 191
Li, H. 41
Li, H.F. 143, 161
Li, J. 31
Li, J.X. 161
Li, L. 165, 185
Li, M. 5, 69, 103, 105, 181, 203, 205
Li, S. 85, 187
Li, T. 113
Li, Z. 127, 131, 135
Liang, D.L. 31, 99, 127, 131
Lin, Z.-Q. 11, 35, 89, 103, 181
Liu, D. 95
Liu, H. 157
Liu, L. 39
Liu, Y. 95
Loomba, R. 93
Lopes, G. 23, 25, 27, 133
Lui, M. 29
Lungu, M. 153
Luo, Y. 135, 157
Luo, Y.Y. 141
Luo, Z. 143
Luxem, K.E. 9
Lv, Y.Y. 5
Lyons, G.H. 163, 167
Läderach, A. 7
Lăcătuşu, A.-R. 153
Lăcătuşu, R. 153

Ma, B. 19
MacDonald, T.C. 43
Machado, M.C. 159
Magalhães, C.A.S. 17
Mahajan, R. 11
Man, N. 31
Mao, H. 163
Marques, J.J. 15
Martinez, M. 65
Martins, G.C. 17
Min, J. 187
Minich, W.B. 81
Moraes, M.F. 175, 177, 183
Mulder, P. 61
Musilova, L. 129
Muyanga, M. 117
Méplan, C. 53

Nazir, S. 87
Nicol, L. 61

Oliveira, D.P. 165
Oliveira, T.F.B. 169
Ortiz-Monasterio, I. 167
Otieno, S.B. 117
Ouellet, N. 65, 173
Ouvrard-Pascaud, A. 61

Pascoalino, J.A.L. 183
Pavelka, S. 55, 57
Peng, F. 121
Peng, Q. 127
Pickering, I.J. 43, 193
Pilon, M. 115
Pilon-Smits, E.A.H. 115, 125, 129, 199
Pratti, V.L. 79
Priyadarsini, K.I. 49, 51
Prokisch, J. 189

Qian, W. 135
Qiao, Y. 85
Qin, H.-B. 89
Qin, S.Y. 31, 131
Qiu, L. 203, 205
Qiu, Z. 69
Qu, Z. 121

Ralston, N.V.C. 71, 73
Ramos, S.J. 137, 185
Rashid, M.M. 35
Raymond, L.J. 71, 73
Reis, A.R. dos 133, 175, 177, 179, 183
Reis, H.P.G. 175, 177
Ren, W. 39
Renko, K. 81
Reynolds, R.J.B. 125, 129
Riboli, E. 53
Richard, V. 61
Richterova, K. 129
Righeto, P.P. 159
Rémy-Jouet, I. 61

Santana, R.S.S. 97
Santos, C.L.R. 183
Saran Netto, A. 97
Sarwar, H. 87
Scheeren, P.L. 183
Schomburg, L. 53, 63, 81
Schuette, A. 81
Schwiebert, C. 81
Sharma, S. 123
Shi, W. 187
Silva Júnior, E.C. 15
Silva, V.A. 169
Singh, S. 93
Siqueira, J.O. 137
Song, J. 135
Souza, G.A. 15, 25, 185
Stanciu-Burileanu, M.M. 153
Su, Qing-rui 21
Sun, H. 47
Sun, X. 85
Sun, Z. 85
Sura-de Jong, M. 125, 129
Sztrik, A. 189
Tack, F.M.G. 79

Tan, G. 105
Tang, J. 147
Tang, Z.Y. 37
Tardini, A.B.B. 17
Tejo Prakash, N. 119, 123
Terry, N. 197
Thuillez, C. 61
Tian, H. 37, 139, 141
Tu, S. 33, 41

Van de Wiele, T. 79
Verma, P. 49
Vriens, B. 9

Wadt, L.H.O. 17, 133
Wafula Masinde, P. 101
Walker, S. 151
Wan, Y.N. 143
Wang, A. 113
Wang, D. 95
Wang, J. 11, 115, 125
Wang, Q. 161
Wang, R. 99
Wang, X.F. 161
Wang, X.X. 107
Wang, Z. 33
Wang, Z.H. 99, 163
Watts, M.J. 151
Wei, C.H. 29, 37, 139, 141
Wei, Dan 21
Welsink, T. 81
Wernli, H. 7
White, P.J. 111
Winkel, L.H.E. 3, 7, 9, 13
Wu, J. 103, 181
Wu, L. 187
Wu, Z. 103, 181, 203, 205

Xi, X.X. 5
Xiao, S. 157
Xiao, Y. 29
Xie, Y.H. 11
Xie, Z. 41
Xing, D. 41, 157
Xiong, S. 41
Xu, J. 157
Xu, Qiang 21

Yang, S.I. 193
Yang, Z.F. 5
Yao, C. 135
Yasin, M. 199
Yin, X. 105, 113, 121, 135, 147, 191
You, Y.H. 139
Young, S.D. 149, 151
Yu, T. 5
Yuan, L. 105, 113, 121, 135, 147
Yuan, S.L. 143
Yue, S. 85

Zanetti, M.A. 97
Zang, H. 105
Zhang, H.Y. 29
Zhang, Jiu-ming 21
Zhang, N. 39

Zhang, W. 47
Zhang, Y. 105
Zhao, W.L. 131
Zheng, B.-S. 89

Zhou, J. 41
Zhu, J.-M. 89
Zhu, M. 203, 205
Zuberi, A. 87